The Children
of Frankenstein

The Children of Frankenstein

A PRIMER ON MODERN TECHNOLOGY
AND HUMAN VALUES

Herbert J. Muller

INDIANA UNIVERSITY PRESS

Bloomington & London

SECOND PRINTING 1972

FIRST MIDLAND BOOK EDITION 1971
COPYRIGHT © 1970 BY INDIANA UNIVERSITY PRESS
All rights reserved
No part of this book may be reproduced or utilized in any
form or by any means, electronic or mechanical, including
photocopying and recording, or by any information storage
and retrieval system, without permission in writing from
the publisher. The Association of American University
Presses' Resolution on Permissions constitutes the only
exception to this prohibition.
Published in Canada by Fitzhenry & Whiteside Limited,
Don Mills, Ontario
LIBRARY OF CONGRESS CATALOG NUMBER: 76-103926
ISBN: 0-253-20144-6
Manufactured in the United States of America

to
Herman B Wells

————————————

Contents

Preface

When I attempted to write a History of Freedom, a reviewer of the first volume commented that my subject was "probably impossible," and long before I completed the third volume I was inclined to agree. When I was done, leaving all kinds of questions up in the air, I resolved never again to tackle so large a subject. Presently, however, I was asked to participate in a symposium on Science, Technology, and Human Values, one of the themes I had dealt with briefly in my history. After I apologized for rehearsing what seemed to me quite obvious ideas, on a depressingly familiar subject, I discovered that they did not seem so obvious to most of my audience, even though this was a university audience. Then I found myself becoming engrossed in another very large subject. And an immensely complicated one: for my concern was not merely the transformation of our economy by modern technology but primarily its social, cultural consequences, its effects upon people.

The immediate point of this informal preamble is that my subject is again a strictly impossible one—impossible, that is, to treat adequately. This book is a comprehensive survey but is by no means exhaustive or authoritative, since I am not a specialist or an authority on anything in particular. It amounts to a series of exploratory essays. As a primer it contains few if any original conclusions, for talk about "human values" is becoming almost as common as the neglect or abuse of them in practice, and as I read around I repeatedly found my judgments stated or refuted by others. I have employed no systematic methodology, attempted no quantified analysis, made no effort to set up a "model" of the kind now fashionable in the social and political sciences. I should confess that I would not know how to go about making a model, but anyway I have heard of no possible method for arriving at a confident answer to what is for me the critical question: Just what has modern technology done *to* as well as for people? For it has profoundly influenced all the major interests and activities of our society, affected people in innumerable ways—their environment, their work, their possessions, their desires, their daily habits, their recreation, their language, their modes of thinking, their notions of what is "natural"; and while it has affected them in some uniform ways, simply because it is so pervasive, it has also had many different effects, depending on their habitat, their social class, and their occupation, whether as housewives, factory or office workers, business executives, engineers, scientists, artists, politicians, professors, and all the other professionals.

Another reason for my informal preamble is that my book itself is informal, addressed to the general reader. Although it necessarily involves a good deal of exposition, I have used a minimum of the technical terms that have swamped the language because of technology. In particular I have tried always to keep an eye on concrete experience, including my own. One of course has to generalize, on this subject even to venture very broad generalizations, which the other fellow may then attack as "sweeping"; but it is too easy for all of us to lose sight of our own immediate experience. So with the standard generalizations about our standardized, regimented, mechanized, atomized mass society. They make obvious sense, they point to very real and serious consequences of our technology, they may well be the most pertinent description of our novel kind of society; but first I want to say that they make a procrustean bed, they do not correspond to much in my own direct observation and experience, they do too much injustice to too many ordinary people I have known. The stereotypes are more questionable because almost always they imply that life in past societies was much more "natural" or "human," therefore healthier; whereas actually we know little for certain about how the masses of men thought and felt about their lot in past societies, or how much they enjoyed their life, and I know only that psychologists and sociologists are unable to agree on a definite criterion of normality or mental health, any more than the wise and holy men of the past ever agreed on the good life for man. I urge all readers to keep checking my observations about our life against their different personal experience.

And so with the many obvious ideas that do not seem obvious enough to most Americans. Our fabulous technology has plainly given us a wealth of real goods, many of which we all enjoy, and which critics of our society may forget just because we all take them for granted. It has also brought many plain evils, even apart from the appalling possibilities of the destructive power at man's disposal, the absurdity that his "progress" now enables him to put an end to his civilization or perhaps to human life on earth—to do all by himself what it once took God to do with a flood. As I have written elsewhere, it would seem clear that the consequences of modern technology have been profoundly, thoroughly mixed; but nothing seems harder than keeping this clearly, steadily in mind. In my own effort to do so my guiding principle has remained a principle of essential ambiguity, with an eye to the problems created by all advances, the costs of all goods. Otherwise my primary effort has been to be sensible rather than arresting or novel, since never before has man had greater need of sanity; but I should repeat that what seems like obvious good sense to me is rather different from the common sense of most Americans, and add that the minority who agree with me may find much of what I have to say too familiar.

As for my qualifications for the task I have undertaken, I offer rather odd credentials as an erstwhile literary critic who got interested in the philosophy of science, took to writing history twenty years ago, as a nonprofessional historian felt free to wander all over the ages, then acquired the title (strictly unique, I believe) of Professor of English and Government, and so got involved with social and political scientists too. On our situation today I draw on all the mongrel intellectual company I have kept. I have also offered a longish historical introduction, for the sake of perspective on our situation, in which I perforce substantially repeat some of what I have written about the importance of technology in man's history from the outset. Such a perspective may be especially helpful in an age that used to be called "historical-minded" but is now described more rightly, I think, as "antihistorical," inasmuch as both the critics and the champions of our technological society too often appear to have little historical sense. In all this my possible advantages in breadth of interest entail possible inconsistency. Not to mention that I am myself (whoever *that* is—a pertinent question since I read that modern man is alienated "from himself") a creature of different moods, which readers should be warned have lately tended to be rather grim.

Another difficulty is the extraordinary pace of change. Many of my contemporary references—for example to the war in Vietnam—might need to be qualified by the time they appear in print. For this reason I have made little use of all the statistics I have ploughed through in my reading, and have made no effort to dig up the latest figures. In general, I have tried to focus on the deeper trends that may be expected to continue, the basic problems that will almost surely be with us for a long time to come. And because of the rapid change I have also devoted a section to the prospects of the next generation. Though in one sense the future is inherently more unpredictable than ever before, it is also with us more insistently than ever; for we are not only uncommonly busy making it but are consciously planning change, so much so that a report has been entitled "The Future as the Cause of the Present." We also have bred many specialists who are busy making systematic efforts to forecast the immediate future, again with an eye to possible control of it. Posterity is just around the corner.

Needless to add, I am indebted to a great many writers, too many to do justice to in a formal acknowledgment. I must at least mention Lewis Mumford, a pioneer in the subject to whom all students of it are indebted, and I more so than most because I am concerned about essentially the same human values that he has been throughout his long career. But immediately I am using Jacques Ellul's much-discussed *The Technological Society* as a springboard. It is an exasperating book because of Ellul's relentless logic, which in the way especially dear to French intellectuals he carries to the absurd extreme; his thesis seems to me grossly over-stated, over-simplified. Unlike Mumford, whose impassioned commitments are always explicit, he makes a pretense of complete objectivity, insisting throughout that he is merely diagnosing our state, not prescribing; whereas there is no mistaking his utter revulsion against our technological society. Often he mentions in passing some of the obvious goods that have come out of modern technology, but only to ignore them as he goes on demonstrating how we are hopelessly enslaved by it, always at the cost of our humanity. Yet by the same token his book is an uncommonly stimulating work, even apart from Ellul's many acute observations. It is the most incisive study I know of how technology has come to dominate our whole society. It is especially useful for those who still hope to make the best of this technology because it is the most complete statement of what they are up against, the worst they had better go through if they want their hopes to be not merely wishful. It forces most insistently the first and last question: *Can* man control his terrific technology? *Can* he direct the extraordinary power he has achieved to sensible, humane, civilized ends?

Although at the end Ellul appeals to man to cage the Frankenstein monster he has created, the whole drive of his book is to the conclusion that the monster is beyond control. I think it is technically possible for man to control it, humanly possible for him to employ it for saner, more civilized purposes than he does right now. One reason why I do not despair is again my personal experience: I have enjoyed life in our extraordinary age. But readers should again be warned that I am not optimistic about our prospects.

H. J. M.

July 1968

Acknowledgments

Portions of this book have already appeared in print. A condensed version of the chapter on Language was published in the Summer 1969 issue of the *Virginia Quarterly Review,* and many scattered passages were incorporated in an introductory talk, "Modern Technology and Human Values," reproduced in the fall 1969 issue of *Ingenor,* published by the College of Engineering of the University of Michigan. A few short passages are also taken with only slight changes from a paper given at a symposium at the California Institute of Technology, broadcast on an educational television program, and published in *Scientific Progress and Human Values,* edited by Edward and Elizabeth Hutchings (New York: American Elsevier, 1967).

I

Introduction

1

Definitions,
Premises, and Basic Issues

According to my dictionary, technology is "the science or study of the practical or industrial arts." In modern usage emphasis has been on industrial techniques, based on the machine. I am using the term in a broader sense to cover as well the distinctive practices that have been generated by the rise of industrialism. These include large-scale organization throughout our society, professionalism in all activities, and the ways of thinking and doing indicated by such typically modern terms as "system," "systematic" methods, and "methodology." Modern technology may be broadly defined as the elaborate development of standardized, efficient means to practical ends. A comparable definition is Kenneth Galbraith's, "the systematic application of scientific or other organized knowledge to practical tasks." Jacques Ellul prefers to call all this simply "technique." In any case it is well to keep in mind that the term is not so precise or "scientific" as it may sound.

Practical arts or skills have of course always been vitally important as a means of providing for the material necessities of life. At the dawn of man's history we find him chipping flints, inaugurating his career ever after as a tool-maker—a career in which he would have far more success than in his quest of wisdom and virtue. A historical survey may make clearer the "naturalness" of technology, the continuities through the long ages, and the unquestionable progress that man has made in this fundamental endeavor. Not until our century, however, did men begin to write the history of technology, and the obvious reason why historians have so belatedly recognized its importance is the novelty of the supreme importance of modern technology. This is not only vastly superior in efficiency to the technology of pre-industrial societies but essentially different in kind, and has had many unprecedented consequences.

The fantastic pace of change, due initially to the increasingly systematic "invention of invention" as the Industrial Revolution gathered momentum, has in our time been steadily accelerated by the many billions spent on "R & D," research and development. The most conspicuous result has been the immense power over nature, now including the conquest of outer space, that man has achieved by what are being called "megatechnics." But most important for my purposes is the range of influence of modern technology. It is a commonplace that our society is dominated by science and technology, so much so that some writers (such as Jacques Barzun) have attacked C. P. Snow's thesis of the "two cultures" by maintaining that we really have only one culture. I think we have at least two, as the very complaints about science and technology suggest; but there is no question of their dominion. Hence many writers are echoing Ellul's insistence that our whole way of life is "unnatural." The extraordinary progress of technology is regarded as a problem, and though most men still celebrate the progress, it has certainly created problems such as past societies never had to worry over. As some scientist remarked, it has taught us how to become gods before we have learned to be men.

My primary concern is the systematic neglect or abuse of what I consider essential human values. (It is significant that values are regularly called "human" these days even though the term is strictly redundant—only human beings can have conscious values.) To be sure, anything that people want or think is good may be called a human value. Like Spinoza, I assume that people do not desire something because it is good, but that it is considered good because people desire it. Although money is often called a "false" value, it is plainly a real one. As plainly, however, it is only a means to some end. So too is our marvelous technology. The matrix of our problems, especially in America, is the common assumption in effect that it is an end in itself—an assumption fortified by the immense energy that goes into it, the worship of efficiency as the sovereign ideal, the boasts about our material wealth and power, and the national goal of steady economic growth. True, we cannot absolutely separate means and ends, since the achievement of necessary means may be satisfying in and for itself. As the end for many Americans has been not so much a pile of money as the game of money-making, so the mastery of technology may be its own reward, affording the satisfactions of a problem solved, a truth found, an adventure consummated, a craftsman's aim achieved—all amounting to a kind of esthetic satisfaction, in keeping with the definition of technology as a study of the practical "arts." Yet I assume that we still can and must roughly distinguish means and ends, both for our personal purposes and for purposes of sizing up our society. The distinction is all the more necessary because of our fabulous wealth of efficient means. And I am committed to the simple, old-fashioned assumption that the proper end for man is the good life.

Once upon a time men believed, or at least were constantly told, that their main business on earth was the salvation of their immortal soul, through the service of God. In America today the popular word for the good life is "happiness." This seems to me a quite normal goal, but the question remains what people mean by the term. The American pursuit of happiness can often look like a compulsive, joyless effort to escape boredom, and in any case a people blessed with far more material advantages than any other society has ever enjoyed is not clearly the happiest people on earth. One plain reason is a paltry conception of the good life, or what I have called the highest standard of low living in all history. But this only forces the basic question. What, then, is the good life?

The whole history of thought and culture may be summed up as an endless disagreement over this question. I take it that men never can or will agree on just what is the good life, unless the human race reaches the state of a universal anthill that Ellul and others predict. Neither do I wish for anything like complete agreement, in view of the rich, diverse potentialities of human nature and the manifold personal differences in interest and capacity. Yet the history of thought and culture also suggests that we can arrive at a general agreement on certain basic goods. They begin with the elementary goods of physical well-being, through health and the satisfaction of hunger and other physiological needs. They include the basic goods of social life, beginning with family life and broadening into simple comradeliness, fellow-feeling, and love. They include as well cultural goods—the satisfactions of natural curiosity, of the esthetic sense, and of the related craftsman's or creative impulse. These may be suspect to the man on the street for the same reason they are often called "higher," that they involve pleasures of the mind, but the powers of mind are precisely what distinguish man from other animals. All men normally enjoy using their heads, as children conspicuously do, and all known societies have made some provision for the cultivation of these goods. To some extent they are always known to the man on the street too. No Americans are as purely materialistic, or as narrowly practical, as they are often reputed to be. Though always inclined to ask where does it get you, they do all kinds of things that get them nowhere—except into a pleasurable state of mind.

Since most of us are conscientious relativists these days, and I have myself always shied away from high talk about the eternal verities, I venture some more elementary ideas. I am now operating on the premise that there *are* permanent, absolute values, at least so long as we assume that life is worthwhile. We cannot absolutely prove that it is, inasmuch as many men have decided otherwise, even to killing themselves; but this is the assumption that the rest of us perforce operate on, and these values are what warrant that assumption. They may be given some scientific warrant too in the common assumptions about the basic needs and drives of man, and lately the attention to the universals of culture, long neglected by anthropologists because of their stress on the more conspicuous diversity of cultures. I prefer, however, to emphasize their basis in common experience. They are absolute goods in that they are good for their own sake, they do not require either religious or scientific sanction, and they do not have to be demonstrated because men conscious of them simply know they are good. If they are not good enough to make life worthwhile for all men, the plainest reason is the deprivation of some basic good, such as health, or the frustration of some basic need. And especially on these needs I have not sought the authority of psychologists and social scientists. Usually concentrating on the obvious physiological drives, such as hunger, sex, and self-preservation, they have paid much less attention to natural curiosity, the esthetic sense, creative activity, and the desire for self-realization through such normal interests, if only because these are harder to deal with by any respectable methodology. I am assuming that the "higher" goods are also sprung from vital needs, and that life is seriously impoverished when people, however well-fed and well-sexed, have limited capacities or opportunities for enjoying them.

In insisting on the reality and importance of what may also be called "spiritual" values, I am expressing not merely my personal opinion as an English professor but something like the judgment of the human race, apparent in all known societies. It remains apparent as we approach the more complicated question of "civilized" values, another of my major concerns. The rise of civilization, with its seemingly artificial life, may nevertheless be regarded as a natural outgrowth of the absolute goods, a more conscious, elaborate, refined realization of human potentialities, or an extension and enrichment of consciousness, the seat of all human values. Similarly its "high" culture was a more deliberate pursuit of truth, beauty, and goodness, sometimes holiness. The historic outcome was the growth of learning, the flowering of literature and the fine arts, in time the rise of philosophy, the higher religions, and pure science itself—the astonishing belief in the sovereign value of a disinterested pursuit of truth, an absolute value that scientists may overlook because many of them taboo value-judgments on principle. The fuller consciousness that men had grown to amplified as well such simple goods as fellow-feeling, which flowered into the no less astonishing ideal of universal brotherhood. In this fuller consciousness disagreements over the good life also became more marked, but first we should note the broad agreement on what were the great works of art and thought—an agreement the more significant because today we know far more about the history of civilization, and the values of all other civilizations, than men ever did or could in the great societies of the past.

With distinctively civilized values, however, we can no longer talk easily of permanence or absoluteness. They came very late in the history of man, and we have learned that civilization is a precarious adventure. Most of its high culture, moreover, was never made available to most people, in particular the illiterate peasant masses, who made do with their prehistoric kind of folk culture. At that anthropologists insist that we have no right to regard civilized societies as superior to the primitive ones they study, and some writers offer variations on the perennial theme of primitivism, the belief harking back to antiquity that the values of civilization are not worth its costs. To me they are worth it; but I would argue only that the great works of art and thought provide a richer experience than primitive or folk art (which I enjoy too), and that almost all people who have learned to appreciate them value this experience, as both a component of the good life and a possible guide to better conceptions of it.

More complex issues are raised by the distinctive values of our own civilization, above all the ideals of personal freedom, and the rights, opportunities, and incentives it has extended to the individual. In my efforts to write the history of freedom I was soon forced to question the congenial belief embedded in American tradition, that men have not only a natural right to freedom but a natural passion for it. On the historic record the great majority of men through the long ages manifested no such passion, they put up with what now looks like oppressive servitude, and they hardly dreamed of the rights and opportunities that most Americans take for granted as their birthright. Here I am arguing only that on the record man nevertheless has some potentialities for freedom, that once people have known and enjoyed more freedom they generally recognize it as a precious good, and that they have sound reason for this value judgment. Similarly with the ideals of individuality and self-realization that have come to be considered essential for the good life. They became conscious ideals with only a few peoples in the past, notably the ancient Greeks, and only in modern times have they been held up to the common people; there was no worry in past societies over the pressures to conformism that alarm contemporaries because the leaders of these societies wanted nothing more than conformity or unquestioning obedience in the common people. I assume only that these ideals have a biological basis in the fact that man is the most highly individualized of animals, as well as a historical basis in the wider range of choice and self-expression developed by civilization, and that on the record of Western civilization, and the basis of our own experience, there is again good reason for the common belief that they are precious ideals.

The excuse for all these elementary observations, at any rate, is that they may help as we begin to size up the social and cultural consequences of modern technology, the themes to be developed in later chapters. Most obviously it is by far the most efficient means yet devised by man to provide for the absolute goods of physical well-being, beginning with the bread man needs before he can realize that he cannot live on bread alone. Because these are "material" goods it has become all too easy to disparage them, and hard to realize that before our times untold millions of people starved to death, many more millions all through history had to get along without enough bread—as in the non-Western world they still do.[1] Almost all of us enjoy many other of our abundant material goods. Readers may forgive the repetition of the commonplace that it is quite possible to lead the good life in a comfortable home with plumbing, central heating, electrical appliances, even a garage. Very few critics of modern civilization actually spurn all these comforts and conveniences, and I know of none who are putting on hair shirts. Today the drive of the whole non-Western world to modernize confirms the judgment of the human race that the basic material goods are absolute goods. Those who think the human race is wrong—as no doubt it can be—might look harder at their own experience.

---

1. The many who tend to idealize the more wholesome life of the past might consult a UNESCO study of the typical life of a seventeenth-century Frenchman. He could expect to lose one of his parents by the time he was fifteen years old, two or three of his brothers and sisters before he was married, and if he reached the age of fifty he would have lived through two or three famines and three or four near famines, as well as two or three major epidemics. Americans should note, however, that there is still much downright hunger in their own most affluent society—a recent "discovery" that is all the more disgraceful because theirs has been called the first great nation in which more people die of eating too much than eating too little.

Almost as plain, but also often overlooked by critics, are the cultural goods that have flowed out of modern technology. It provided the wealth that made possible the effort of nations to educate their entire populace, as no pre-industrial society could do even if its rulers had dreamed of such a thing. It also provided the incentive because of the needs of industry; a predominantly agricultural society can get along with an illiterate peasantry, as all did in the past, but an industrial society has to have not merely laborers but many kinds of literate workers. Likewise modern technology has made much more widely available a wealth of cultural goods in paperbacks, musical albums, photographs and reproductions of paintings, the better offerings of the mass media, etc. To the extent of their interest and capacity, ordinary people may now share in the values of civilization as the vast majority never did in the past. And because of the ample provision for both material needs and the uses of the mind they enjoy more effective freedom, freedom not only from want but to want, with more opportunities for self-realization, a much wider range of choices and a greater power of choice.

What they commonly choose, however, brings up the depressing products of modern technology—the superabundance of trivial goods that have been made into needs, the flood of trash in the mass media, the incessant, high-powered advertising of a vulgar conception of the good life, and so on *ad nauseam*. I shall be obliged to return to such too familiar complaints about the uses of our wealth and power, and of literacy. At this point I return to my premise of absolute goods. Because men can agree on them they can also agree on some absolute evils. No one would deny, for example, that the pollution of our rivers and the air we breathe is a bad thing. No one would argue either that dropping nuclear bombs on people is good for them. A nuclear war would be bad for everybody, except maybe the "victors"—and people are beginning to realize tardily that in an all-out war there would be no victors, only some wretched survivors. Then the question arises: Why all the pollution? Why has the richest, most powerful nation in all history only begun to do something about it? Why must it go on making bigger and better nuclear bombs when it already has many more than enough to blast the whole earth? I am brought to the compulsions of modern technology, the reasons why it so often neglects or even outrages essential human values, and why it has also drastically limited human freedom.

From the beginning the Industrial Revolution introduced new compulsions. It would condemn millions of people to the routines of factory work, force them day in and day out to go through the same mechanical operations. In a real sense slaves to their machines when at work, they also had to live where machines were at home and so lived in dreary places like Gary, Indiana, once described as a city inhabited by four blast furnaces and a hundred thousand people. In various ways people have been subjected to the compulsions of the new technology in the interests of efficiency and economy, conceived in terms of money values, too often without regard to human costs. I shall have to elaborate the tiresome theme of how our technology has tended to standardize and regiment people, mechanize and dehumanize life, generate massive pressures against the individual it had helped to liberate.

More paradoxical, at once subtler and more tyrannical, are the compulsions resulting from the astoundingly rapid progress of technology since World War II. Together with a wealth of new wonders, from television to space ships, this has created the "affluent society," one so much wealthier than any in the past that it is different in kind. As has often been said, the primary function of Americans today is to be consumers. They *must* keep on consuming faithfully, arduously, to the end of their days. If most of them appear to be doing their duty gladly enough, it nevertheless illustrates the odd kind of tyranny of our economy; if they didn't keep on buying the latest model, and all kinds of superfluous goods or gadgets, our economy would collapse. To assist them we have built up an immense advertising industry, which spends many billions on strictly unproductive but now essential purposes. One of the forms of propaganda required by an advanced technological society, it is a reminder that the immense power man has achieved is increasingly a power over not only nature but people, through ways of manipulating people. These may be for their own good, but thereby they raise again the question of what is good for people, and what people are good for.

In general, the affluent society most insistently forces the issues of human values. The superabundance of goods it produces includes many trifling goods, silly when they become needs; and too many people, content to wallow in them, fail to ask the obvious question: What are they good for? So on a national scale with the goal of steady economic growth. This too is plainly a means to some end, but tends to obscure the end or become an end in itself. The national idol has become GNP—the gross national product. Now running to some $900 billion a year, this would be more impressive if it did not include the costs of producing nuclear weapons, polluting the air, advertising eyewash, fighting crime, bulldozing landscape, hospitalizing the victims of automobiles, supporting or compensating for all kinds of waste and blight. As Robert Kennedy observed (a few months before his assassination), the GNP takes into account "neither our wit nor our courage, neither our wisdom nor our learning, neither our compassion nor our duty to our country. . . . It measures everything, in short, except that which makes life worthwhile."

There remains the most fundamental paradox. The many billions now being spent on "research and development" represent not only the mightiest but the most conscious, deliberate effort in history to direct technological change. The "knowledge industry," centered in the multiversities, has become the biggest business in the nation. Specialists in forecasting can confidently predict a hundred important new technical innovations by the year 2000. All about us are the elaborate appearances of planning and control; never before has a society displayed such apparent foresight and will to take charge of its future. Yet the terrific drive of technology also seems mechanical, automatic, almost blind, because it seems irresistible. As is often pointed out, but to no effect, we apparently *must* do whatever is technically feasible, no matter what the human cost. We must spend many billions to put a man on the moon, even though many millions of Americans on earth live in wretched poverty, in ghettos that are ruining the American city. We must build big supersonic planes to cross the ocean in two or three fewer hours, even though sonic booms have been damaging both property and health (not to mention that President Johnson was at the same time trying to discourage Americans from traveling abroad). So one may recall the test pilot who radioed in, "I'm lost, but I'm making record time." Whence this terrific drive?

The obvious reason is of course the Cold War. With military defense goes national pride; we must keep ahead of those Russians. But what if the Cold War ever comes to an end? Some experts are aghast at the prospect of what would happen to the GNP. Although no forecaster myself, I should not expect so mighty a national effort to improve the quality of American life. But I should expect the basic drive of technology to continue. This is the drive to efficiency, or to what since Max Weber has been called "rationality"—a term that might endear it to intellectuals were it not that it refers only to technical means. Such rationality needs to be distinguished clearly from "reasonableness," which to me implies a concern for human ends as well. Accordingly it brings me back to the issues of human values, immediately the question whether we can control our terrific technology by civilized conceptions of the good life.

"Each of us, in his own life," concludes Jacques Ellul, "must seek ways of resisting and transcending technological determinants." Throughout his book, however, he has tried to demonstrate that technology has become autonomous, all-determining, in effect irresistible. In a typical passage he writes: "We are faced with a choice of 'all or nothing.' If we make use of technique, we must accept the specificity and autonomy of its ends, and the totality of its rules. Our own desires and aspirations can change nothing." If so, we are indeed doomed, for we are surely not going to give up the "use of technique," at least so long as our civilization survives.

My own thesis is less dramatic, I hope more reasonable. We do not have to choose "all or nothing," any more than we do now, or than Ellul did in choosing to publish his all-out attack on technique, with the indispensable help of the printing press. Technology is not yet wholly autonomous, nor do we have to accept the "totality of its rules," which at that are diverse, shifting, and sometimes inconsistent. It is not too clear on his grounds how we can still have "our own desires and aspirations," but anyway we have managed to retain some, and I assume they can change something. All the other activities that technology has deeply influenced—from government, public education, and high culture to recreation, social and domestic life, and sexual behavior—have retained some measure of autonomy, rooted in traditions much older than an industrial society. How much can be changed by the desires and aspirations they represent I would not venture to say; I am inclined to think not enough, since I agree that "technological determinants" are extremely powerful; and in any case I have no program to offer, beyond the hopes of education implicit in every treatise. But I think we still have some measure of choice in futures. What with all our knowledge and our awareness—maybe our too many treatises—we at least have less excuse for failing than did all the civilizations of the past.

2

The Historical Beginnings

In the beginning may or may not have been the Word. On earth a more likely beginning was a word unspoken because unconceived—Power. The first thing we know for certain about the life of man, once he had evolved from the ape, is that hundreds of thousands of years ago he began chipping flints: he was developing a technique to give him more power over the natural environment. Techniques are rooted in the animal world, where we find skilled hunters, burrowers, and nest-builders, and the marvelously efficient organization of insect societies; but with animals such behavior is instinctive. Man was learning all by himself, taking thought of the morrow, exercising a power of choice, being "creative." He was making use of his chief natural advantages over other animals, an adaptable hand and above all brain-power. Although we know nothing about how or when he evolved language, his most marvelous tool, it may well have had something to do with the need of transmitting to his offspring the lore he was acquiring. Not until very late in his history would he be able to conceive the Word.

His technological progress was at first extremely slow. For many centuries he went on chipping down or flaking off flints in uniform patterns, which we might now dignify as "standardized," but which indicated the inveterate conservatism of a creature who is by no means so naturally inventive as he would seem to modern man. He had an excuse in that there was no plain necessity to mother more inventions, for he had long managed to get along without tools, or with only such "natural" ones as the penis.

Nevertheless they were clearly useful, and once acquired they came to seem quite natural. Gradually he added to his stock of tools, fashioning more varied and efficient ones, never relinquishing the increasing power he was achieving over his environment, and so starting an essentially irreversible process. Since "power" now has sinister connotations, we might pause in natural piety to honor these pioneers in an enterprise we owe so much to. Ralph Linton once wrote a pleasant parable about a 100 per cent American fearful of insidious foreign influences, beginning with the unsuspecting patriot in bed swathed up to his ears in historically un-American materials, such as wool, and following him through his morning ritual with a detailed catalogue of the debt he owed for almost every item of his dress, furniture, food, and even his vaunted bathroom to countless different peoples all over the world, over thousands of years. All of us may forget that we are indebted to technology for virtually everything we own, including our minds.

As there is no need here of detailing the stages of the early technological progress of man, I shall merely point out a few developments of some pertinence to our situation today—the main point of this historical survey. Perhaps the most significant feat of our prehistoric ancestors was learning how to make fire. It was the first time man had got control of a natural force—an awesome feat commemorated in the fire-worship of ancient religions. Claude Lévi-Strauss, who believes that the discovery of fire is the central idea behind all myths, has suggested that another possibly significant turning point was man's new habit of cooking his food, which likewise made him a creature apart from other animals and marked his passage from nature to culture; I suppose this habit may have seemed "unnatural" to creatures long accustomed to eating their meat raw. In any case the use of fire remained the key to man's conquest of nature, especially when he developed metallurgy. Also portentous was the invention of embryonic machines, such as the spear-thrower and the bow. Before the rise of civilization men were using honest-to-God machines, notably the potter's wheel and the loom. They were still depending, however, on the deft hand of the skilled craftsman, who worked as a whole man, using his head too; so they had no reason to think they were creating a Frankenstein monster. As Lewis Mumford observed, the potter's wheel increased the freedom of the potter.

Another very ancient technique, possibly dating from the earliest tools, was magic. Gordon Childe suggested a likely formula of Neanderthal man: "To make a D-scraper, collect a flint nodule (1) at full moon, (2) after fasting all day, (3) address him politely with 'words of power,' (4) . . . strike him thus with a hammerstone, (5) smeared with the blood of a sacrificed mouse." We know for a certainty that magic remained confused with empirical knowledge down through the early civilizations, and for quite understandable reasons. It resembled practical technology as another means to power over nature. It was if anything even more standardized, requiring the utmost care in adhering to correct formulas for the sake of efficiency. Although the great bulk of magical practices have disappeared, since every group or society had its own magic, it continued to flourish everywhere on an ignorance of natural causes, and specifically on the fallacy of *post hoc propter hoc*: usually it "worked." My favorite example is the old Chinese custom of beating gongs whenever there was an eclipse, to scare away the big dog that was swallowing the sun or the moon; peasants sensibly went on beating gongs for thousands of years because it worked every time. Malinowski and others have seen in magic the rudiments of science. I would stress rather that it was something men had to outgrow before they could arrive at anything like the scientific spirit.

Yet it remains significant in our technological society. It survives in not only much popular superstition but the orthodox religious belief in the material efficacy of prayer, as if God constantly interfered in natural processes by granting prayers. For the man on the street it is mixed up, as it always was in the distant past, with the mysterious power man has achieved over nature. Knowing little about either science or the forces controlling his society, he is prone to the same prehistoric confusion of empiricism and magic. Propaganda, especially advertising, exploits his confusion and ignorance. He will swear by the latest placebo because of the same *post hoc* fallacy: usually feeling better after it, he knows that it worked.

Magic probably played some part as well in the early connection of technology with the esthetic sense, a permanent source of human values. As men elaborated tools they attended to not merely utility but pleasing shape—the rudiments of modern industrial design. One may imagine them feeling, if not saying, of their artifacts, "It's a beauty." With technological progress attention to design and ornament became more conspicuous, notably in pottery, where symbolic ornament might have some magical meaning. The connection with magic was most apparent in the greatest artistic achievement of prehistoric man—the superb cave-drawings. These drawings evidently served some magical purpose, inasmuch as they were not on exhibition but were made far back in the caves in dark recesses. Those who prefer to describe this purpose as "religious" should pay more tribute to technology, which would make possible all the great religious art of civilization, in architecture, painting, and sculpture.

For me an especially pertinent implication of the history of early technology is the importance of the creative individual. The individual has been played down by modern historians, who emphasize the "vast, impersonal forces" in history, and played down still more in the social sciences, where he is commonly regarded as merely a product of his society and culture, or otherwise a nuisance to be ironed out in statistics. He is of course a product of his culture, never more clearly than in primitive societies, where he is rigidly bound by custom. Yet "society" does not make inventions or initiate changes. To my mind the very slow changes in prehistoric times only emphasize the importance of some rare persons. I am unable to account for any invention, or even gradual improvements, except finally as a bright idea of some individual. Our immense debt to the technological progress made by man before the dawn of recorded history is a debt to many persons who will remain forever anonymous.

This debt became much more marked with the so-called neolithic revolution, resulting from the discovery of agriculture some ten thousand years ago.[1] Slow as this change must have

1. All dating of prehistoric times is not only approximate but tentative, subject to new discoveries—due, be it noted, to archaeological techniques. I am also pleased to note the speculation that the dis-

been at first, it was indeed revolutionary, and the clearest example so far of the fundamental importance of technology, its wide-ranging consequences even before our time. It meant a radically new kind of life for man as he changed from a food-hunter and food-gatherer into a food-producer, and settled down to the sedentary life of the village. Tame as this may have seemed to hunters, it stimulated far more mental activity than had their old way of life. Change grew much more rapid, if still slow by modern standards. In the village men learned to grow almost all the major food plants and to domesticate the major animals in use today. They invented all the tools needed for their new occupation, the baskets and pottery needed as containers. They took their first steps in transportation by inventing the wheel, for rivers the sail; these would lead to much wider communication. Other inventions included the brick, a means to more lasting, imposing habitations, especially for the gods.

Since Jacques Ellul, among many others, insists that the whole way of life in our technological society is "unnatural," let us note that life in the village could also be called quite unnatural for a creature who over millennia had been hunting and living in caves. Just because I shall echo the common complaints about modern life I would first emphasize the abiding question raised by the neolithic village: Just what is the natural life for man? For a creature who on the face of his record is highly adaptable, able to develop innumerable different cultures or ways of life? The wise men over the ages have never been able to agree on an answer to this question, nor can scientists today. In the village, at any rate, our neolithic ancestors built up what by the standards of cavemen was an artificial world, and however men felt about it they took to it, in time found it the most natural way of life. They gave more literal meaning to the saying of the poet, "Earth was not Earth before her sons appeared." In our urban, industrial society we may forget that before our time the great majority of men in all the civilizations were peasants, still busy at the tasks invented by neolithic men, who are commonly called "barbarians."

covery of agriculture may have been inspired by men's desire for grain in order to make not porridge or bread but beer—a "spiritual" good they discovered at about this time.

The barbarians also laid the foundation of society as we all know it, with some organization for communal living. Much of what people today call "human nature" was a second nature acquired in the village.

One example was the development of play in the sense defined by Huizinga in *Homo Ludens*: a voluntary, disinterested activity, indulged in for "fun" in "free time," always different from "real" or ordinary life, involving some sense of "only pretending" or "make-believe," yet typically performed with a seriousness that makes it more deeply "enchanting." In this sense, play is at the root of all the archetypal activities of human culture, goods that are no less absolute for being seemingly impractical or superfluous. It evolved into the many forms of ritual, which are mostly unrealistic ways of achieving natural goals; into all the forms of art, beginning with decorative art and later including drama, or plays on the stage; and eventually into the make-believe of philosophy and science too, or all play of mind beyond immediate utilitarian interests. Cavemen must have played to some unknowable extent, since they had ample "free time" and their remains indicate that they had some ceremonial rituals and esthetic sense, but in the village the more highly developed culture resulting from an advanced technology involved much more play. Neolithic artifacts also indicate its intimate connection with man's natural pleasure in symbols. We have recently come to realize that he is not only a tool-making but a symbol-making animal, who tends to go on creating superfluous symbols just for the fun or the hell of it.

Lewis Mumford, who to my mind takes a sentimental view of the neolithic village, has idealized it as "woman writ large," the Mother that provided the "loving nurture" of man for future development. It has indeed an idyllic aspect as a real community, close to the soil, stable, secure, integrated by kinship and moral ties rather than external organization, free from such disruptive forces as a money economy. It was nevertheless a very small, confined world, bound by convention and taboo (now condemned as "conformism"), meager in opportunities for the individual, richest in the wealth of magic it produced. More clearly portentous for the future development of man than the loving nurture of the Mother was a potentially disruptive discovery late in the neolithic period—how to work metals. Opening one of the major chapters in the history of technology, metallurgy too had important social consequences in its own day.

Although the village had had a few specialists, such as potters and magicians or medicine men, the new art called for full-time specialists, miners and the smiths who were to become ubiquitous. Metal made not only better tools but better weapons, preparing for the institution of large-scale, organized war that "barbarians" had been incapable of. Since the new tools and weapons were expensive, beyond the means of most men, they became in effect luxury items and sharpened class divisions in what had been virtually classless communities.[2] Presumably economic motives grew stronger. The village, usually remote from supplies of metal, was no longer self-sufficient for its material necessities. (It had imported chiefly such magical accessories as shells and amber.) Metallurgy opened wider horizons by stimulating trade and a spirit of adventure. In all such ways it prepared for another revolutionary transformation of man's life—the rise of civilization.

This was again a slow change by modern standards, so gradual that we cannot draw a clear line and date it; but suffice it that by about 3000 B.C. there were emerging societies so much larger, more elaborate, and more complex than the neolithic village that they call for a new name. The immediate means to this development was a brilliant technological feat, the organization of large-scale drainage and irrigation systems in the Tigris-Euphrates, Nile, and Indus river valleys, the seats of the earliest known civilizations. These systems made possible what is now called the "urban revolution" as villages grew into towns, then into cities. It deserves the name of revolution because life in the city was radically different from life in the much simpler, more homogeneous village. The city was made up of all kinds of different people—laborers, servants, artisans, merchants, officials, priests—

---

2. Grave furnishings found in the early layers of village settlements make clear this equality. The methodical techniques of archaeology have enabled modern man to recover a good deal of the vanished life of the remote past—a pious enterprise that never occurred to the ancients.

and it could no longer live by old custom alone; it required much more organization and technique. Thus the brilliant Sumerians, whose early history we know best, created large-scale government, a formal state with formal laws. Likewise they systematized large-scale business by standard weights and measures, a standard medium of exchange, the institution of banking and credit, and devices for time-keeping. For their business purposes they invented writing, another distinctive sign of what we call civilization. (Only the Incas in Peru lacked it.) We know the Sumerians best because we have unearthed many thousands of their clay tablets, which are indestructible—more "immortal" than the works of most poets who have staked on writings their claims to fame.

Writing brings up the complex of causes or conditions of civilization that are too often sharply separated into "material" and "spiritual," to the greater prestige or the too easy disparagement of technology. A technique invented directly in order to keep business records, and soon used as a means of exploiting or enslaving workers, writing would prove invaluable for the purposes of learning and literature, the transmission and preservation of the "higher" goods. Similarly with monumental architecture, still another of the distinctive signs of civilization. It was in its origins a most impressive technological feat, requiring both unprecedented skills and an unprecedented capacity for organization to build it; Lewis Mumford has described the organization that built the ziggurats and pyramids less flatteringly as the first "machine." But then the monuments continued to impress later generations by the mastery of materials for esthetic purposes, or as expressions of spiritual values. Such architecture also required the material means essential for civilization, what specialists call an "economic surplus" beyond the needs for subsistence; this surplus wealth, and the leisure that came with it, enabled some men to cultivate learning, art, and thought. We now encounter one of the first geniuses known by name—Khufu-onekh, the amazing architect of the Great Pyramid of Gizeh, who planned a structure requiring 2½ million blocks, each weighing 2½ tons, laid out in sides over 750 feet long with an error of only a few inches. Altogether, I see no point in arguing for the priority of material or spiritual factors in the rise of civilization. I am content to attribute it to powers of mind, the creative thought required both to devise the necessary technology and to elaborate the high culture.

More paradoxical was the mentality of the creators of civilization. While to us their remarkable feats look like man's own doing, their writings indicate that they thought differently. They attributed their powers to a high form of magic, now called the gods; it appeared that only the supernatural could make them feel at home in the natural world. Still lacking a clear conception of natural causes in their empirical knowledge, and of themselves as makers, they believed, or at least professed to believe, that they were utterly dependent on the gods. Their earthly rulers were either gods, like Pharaoh, or kings sent by the gods. So we may wonder again about what is the natural life for man. According to Dostoyevsky's Grand Inquisitor, it is not the life of freedom, which to the masses of men is an intolerable burden; they need first of all bread, then "miracle, mystery, and authority." Men in the early civilizations were not sophisticated enough to think in these terms, but on their record they implicitly agreed with the Grand Inquisitor: the gods and god-kings provided the necessary miracle, mystery, and authority. For them the primary condition of the good life was obedience, not freedom.

Given this mentality, we may better understand why the fruits of the new technology were little enjoyed by most people in the early civilizations. The economic surplus went chiefly to a few. Although the general standard of living rose, the great majority remained poor, and many felt worse because of their awareness of wealth around them; the poor as a class now entered history, to be with us ever after. Most were also less free than men had been in the neolithic village. They were reduced to varying degrees of servitude, often oppressed by an aristocracy or priesthood, always subject to conscription by their royal master—as many thousands of Egyptians were conscripted to build the Great Pyramid, to the greater glory of Pharaoh. In Egypt they were subject as well to Pharaoh's bureaucracy, which managed the economy and contained the seeds of a totalitarian state. Everywhere the peasants in the villages remained illiterate, a state quite different from that of preliterate peoples, when nobody could write; now they knew that some people could write and thereby enjoyed more power and prestige. Perhaps their chief blessing was that they were too ignorant ever to realize clearly how they had become the victims of a remarkable technological progress.

Otherwise the rise of civilization brought the abiding ambiguities. The city, always its center, remains the prime example. It was more clearly an artificial world than the village and created more problems by what might now be called its "dehumanizing" tendencies, beginning with the regimentation imposed by the division of labor. Its heterogeneous people, crowded together, were no longer self-sufficient but dependent on an external organization not of their own making. They could no longer know their world as intimately as the villager knew his. They could not live as simply either by unquestioned custom and unwritten law, but faced more moral problems and grew aware of common iniquity and injustice. The city soon began developing the ways that would make it proverbial for confusion, disorder, inhumanity, depravity; in it—not in any Garden—occurred the Fall of Man. Yet it went on growing and multiplying, attracting the ambitious—as it would all through history. Life in the city was obviously more varied, colorful, exciting, stimulating than life in the village. It offered more scope and opportunity for enterprise and talent. An artificial world, it was by the same token more a world of man's own in which people might realize more fully their distinctive human capacities for knowing, feeling, dreaming, aspiring, creating. Subject to more external compulsions, its people were none the less freer in some respects. That they faced more problems meant that they had more choices too, including moral choices, and it is only through making conscious choices that people can fully realize their humanity, become fully conscious of the good life.

Then, too, they have to pay the costs of self-consciousness. The pursuit of truth, beauty, and goodness inevitably entails more awareness of falsity, ugliness, and evil. In the early cities—still small, intimate, close to the countryside and the abiding simplicities—both the goods and the costs were still limited, but the basic ambiguities were already apparent. In time civilized men would rebel against civilization, or despair of their own. Today men enjoying both the material and the cultural goods that have come with the highest standard of living in history may sit down to their typewriter (with nylon ribbons) and write a book of alarm about the problems and prospects of our technological society.

It is unnecessary to review the technological progress of the early civilizations because there was little of social consequence. Their brilliant achievements in both technology and culture were largely confined to the early centuries of their history; thereafter they settled down to little but military or imperial adventure. As I see it, a primary reason for their relative stagnation was that their learning, art, and thought were enlisted directly in the service of the gods and god-king, devoted to glorifying and sanctifying the status quo rather than promoting the public welfare. Toward the end of the second millennium B.C., however, a discovery was made that introduced a new age—an efficient method for working iron. Since iron was much more abundant than the copper and tin that had made the Bronze Age, and tools and weapons made out of it were much cheaper, it led to a boom in trade and industry. It was a particular aid to little men and to new peoples on the rise; characteristically the old civilizations made little of it. They made less of another invention of major consequence that appeared early in the Iron Age, the Phoenician alphabet. It too was a democratizing influence, enabling ordinary men to become literate much more easily. And one people who made the most of both of these inventions were the Greeks, the first people to become fully conscious of man as a maker, and to declare and institute ideals of freedom.

The brilliant Greeks did not distinguish themselves by technological discoveries or inventions, for the most part making do with the basic skills inherited from the ancient East. They were quick to borrow and learn, however, and they had no such scorn of the "base mechanic arts" as Plato and Aristotle later expressed. Neither did they scorn trade; for this purpose they adopted the Lydian invention of coinage and improved on it by adding small change—another boon for little men. Their enterprising, far-flung commerce was the immediate means to their rise in the world. On the wealth and leisure it provided they developed their distinctive culture. It was this culture, of course, that made them so important in Western history—and still pertinent for those concerned with technology and human values.

Another people busy in exploiting the commercial possibilities of the Iron Age were the Phoenicians, who at Carthage eventually built up a power great enough to contend with the Romans in a prolonged, touch-and-go struggle for control of the Mediterranean world, and who might now teach Americans a commonplace moral. As losers in this struggle they have perhaps been unfairly remembered—history is always unkind to the defeated; yet there is no good reason to revere the Carthaginians. After achieving wealth and power, they made no contribution worth mentioning to ideals of the good life. By contrast the Greeks were no less enterprising merchants, if anything even better at business, but they treated it as only a means and grew much more concerned about ends. Although they too were overthrown by the Romans, they educated their conquerors and remained conscious Greeks when they were absorbed into the Empire. They have lived on to this day for their ideals of the life of reason, in freedom—not by "miracle, mystery, and authority"—and for the many great works they produced in the pursuit of truth, beauty, and goodness (though seldom holiness). If Americans are going to take seriously the talk about a Great Society, they might think a bit harder about the glory that was Greece.

Especially important for my purposes here was the birth of natural philosophy, signaled by the bold, astonishing, erroneous statement of Thales: "All things are made of water." Some historians who make much of the material conditions of life have explained the rise of philosophy as a response to the Iron Age, or stressed in particular the practical spirit of the Greeks, engrossed in technological interests—as from all reports Thales clearly was. I have maintained that while he and his fellow Milesian philosophers evidently were much interested in the business of this world, this was not their primary motive in speculating about what the world was made of. They emancipated mind from not only magic, or "common sense," but the practical spirit that had always informed magic; the answers they gave would have been of no practical use to them or the Greeks even if by a miracle they had come closer to the truth. Their importance was precisely that they started men on a disinterested pursuit of truth. And with this Thales started as well pure science. He was famous for having predicted accurately an eclipse of the sun, a "supernatural" event that had always awed or frightened men: he had hit upon the idea of natural causes, a fundamental idea of which there is hardly a glimmer in the records handed down from the ancient East. Greek medicine made this idea quite explicit. The way was open to a methodical advance in applied science.

In the Hellenistic era, dating from the conquests of Alexander the Great to the Roman conquest of the Greeks, their achievements were more ambiguous. In science they went far beyond the achievements of the classical Greeks, laying foundations on which the pioneers of modern science would build. In technology they explored the sources of power in water, steam, and air pressure, and came to the verge of harnessing such power. In philosophy their influential Epicurean and Stoic schools carried on the quest of wisdom and the good life started by Socrates, holding up ideals that the individual could achieve by his own unaided efforts, whatever his worldly state. These ideals reflected, however, the social and political failures of the Hellenistic Greeks, which facilitated their conquest by the Romans. Their failures were failures of mind too; by the end of the era philosophy was retreating to the prehistoric myth and magic in which thought had begun. And their brilliant achievements in science had only a slight influence on either thought or society—nothing like the impact that modern science would have.

The plainest reason was that the Greeks never developed scientific method or clearly distinguished science from philosophy; they had no means of testing or choosing between different theories. (Among the astonishing ones they chose to discard were Copernican, atomic, and evolutionary theories.) A related reason was that they never systematically applied their scientific knowledge. On the verge of harnessing natural sources of power, their technicians failed to take the last step; they were content to exercise their remarkable ingenuity in devising machinery to produce "miracles" in temples or to serve as elaborate toys for kings. Behind their failure to consider productive purposes lay the attitude of Archimedes, who founded the science of mechanics and was famous for his inventions. "Give me a fulcrum and I shall move the world," he boasted; but according to Plutarch he also declared that engineering and all useful arts were "ignoble and sordid"—harking back to the attitude of Plato and Aristotle toward the "base mechanic arts." This may seem high-minded, especially in our own age when practical utility is the overriding concern and the good life a possible by-product of it. The aloofness of the Greek thinkers is less admirable, however, when one considers that they enjoyed the leisure provided by the labor of slaves, whose competition kept the wages of Hellenistic workers miserably low. The institution of slavery had enabled Aristotle to declare that the life of leisure was the only life fit for a Greek, and to remark casually that the only slave a poor man had was his wife; as a consequence the life of reason, in freedom, was not for the common people. Hellenistic technicians resembled ours chiefly in their willingness to devote their knowledge and skill to devising engines of war. Modern technology differs in that it provides as well many more goods for the common people.

As ambiguous, finally, were the achievements of the Romans who took over from the Hellenistic Greeks. Like Americans, they were notoriously very practical men—much too practical to make any notable contribution to pure science. They distinguished themselves as builders and engineers, famous for their great roads, aqueducts, and baths. (Even celebrants of the American bathroom might admire the Roman baths.) They mastered as well the techniques for organizing and administering a great empire; their political organization served as a model for the Roman Catholic Church—the oldest living bureaucracy. Yet it was not primarily these technological achievements, or the wealth and power of the empire, that constituted the grandeur of Rome that would haunt the Western world. This owed more to the majesty of the *pax Romana*, of the Roman law with its universal principles of justice, and of the Roman ideal of a universal commonwealth, in which the emperor was only the "first citizen" and a servant of the people—together the ideals that the United Nations is now trying to realize. With such ideals the empire transmitted the Greek cultural heritage.

Then it too went down. Having made Rome fall in three

books now, I am pleased to feel no obligation to go once more into the many reasons for the decline and fall. Here I need only mention briefly several that bear on organization and techniques. Political institutions were not strong enough to maintain the ideal theory of the empire, or in time to prevent civil war. Business was not enterprising enough, nor government enlightened enough, to maintain a sound economy, much less economic growth; prosperity began to decline after the first century of the empire. The Roman masses, uneducated, were provided with bread and circuses but little opportunity or incentive to contribute any initiative; not for them either the good life. Technology was largely stagnant. Suetonius tells an illuminating story about an inventor who showed the Emperor Vespasian the model of a labor-saving machine that would enable him to build public works at great savings; the Emperor thanked him but had his model destroyed. Under the circumstances his motive was no doubt humane, his policy perhaps wise—the machine would have thrown men out of work; but it was hardly a far-sighted policy. It symbolizes the failure of the Romans to provide adequate material means for sustaining their universal commonwealth.

Altogether, ancient history teaches us nothing about technical means. It might teach us something about the need of keeping an eye steadily on humane, reasonable, civilized ends, and of adapting our unprecedented means to these ends.

3

Western
Civilization to 1800

In the Dark Ages from the fourth to the tenth centuries, writes Jacques Ellul, "there was a complete obliteration of technique, and in the Middle Ages that followed "the same nearly total absence of technique." He reflects the conventional view of these ages held until specialists began inquiring into the history of technology, but by now they have made plain that he is quite wrong. Actually Europeans displayed a genius for the practical arts from the beginning of their history. Lynn White, Jr. goes so far as to say that the chief glory of the Middle Ages was "the building for the first time in history of a complex civilization which rested not on the backs of sweating slaves or coolies but primarily on non-human power." In the light of our experience one may question the "glory" of this achievement, but one cannot fairly size up modern technology without realizing that it developed out of a long tradition and had deep cultural roots. In this view both its uniqueness and its naturalness to Westerners are more comprehensible. We may better understand in particular why our civilization was the only one in history to develop science or scientific method as we know it, and the only one to have an Industrial Revolution.

In the Dark Ages, when little but the rudiments of civilization survived the fall of Rome, these included technical skills. The few that were lost were compensated for by some improvements on the technology of the Roman Empire. Armourers created a pattern-welded steel that much impressed the far more civilized society of Islam; it anticipated a fateful development, the eventual preeminence of Europeans in military technology. A more useful innovation was the homely crank, a device unknown to the ancients. In the especially chaotic, disastrous tenth century the rise of a new civilization was facilitated by a simple invention only recently appreciated—the horse-collar. Brought in by barbarians from the steppes of Eurasia, together with the tandem harness and the horseshoe, this provided agriculture and industry with much more horse power; whereas the ancients had harnessed horses with ox-yokes, which choked them if they pulled more than three or four times their weight, the horse-collar enabled them to pull as much as fifteen times. Other innovations prepared for the reign of machines: water-driven mills, which had appeared late in the Roman Empire but were little used by the Romans, the invention of hammerforges and bellows for the mills, and the introduction of windmills. To me the possible "glory" of such enterprise is best exemplified by another invention uncelebrated until our day. Everybody who knows anything about the Middle Ages has heard of the Crusades, which were among the bloodiest, least glorious of medieval enterprises. How many have heard of an obscure Italian who in the thirteenth century invented spectacles? I am one of the many men who without spectacles would be unable to read.

A more conspicuous and remarkable enterprise, also long uncelebrated, was displayed in business by the medieval burghers, ancestors of the bourgeoisie. They were an anonymous "middle" class because there was no room for them in the medieval social structure, formally made up of nobility, clergy, and serfs. In laying the material foundations for a new civilization by trade and industry, they had to start almost from scratch, without government aid or any of the elementary aids of a settled society, such as good roads and regular communications. They responded to the challenge with far more energy, resourcefulness, and initiative than Roman businessmen had ever developed. Early in the Middle Ages they began mastering the techniques of business organization, through trading companies and banking houses, which led to capitalism on a larger scale than in other civilizations. They mastered as well techniques of social and political organization as they built up the little "burgs" where they had settled into thriving free towns, the main centers of ferment in Europe. Among other things they created their own law, a system of public taxation, and provisions for erecting public buildings, regulating business, assuring ample supplies of cheap food for workers, and taking care of the sick, the aged, and the poor to an extent the ancients never had; their social legislation might strike their conservative descendants in America as "socialistic." Unconscious revolutionaries, the bourgeois broke up the feudal economy as they simply went about their business, replacing it with a money economy; so they anticipated the leading role they would play in all the major historical developments in Western civilization, from the Renaissance and Protestant Reformation through the intellectual, political, and industrial revolutions that created the modern world. From the bourgeois—not the workers—came the great leaders of the revolt against their rule in modern times, Marx and Lenin.

A more unlikely source of technological development was Christianity. Its otherworldly tradition tended to degrade the natural world, just as its doctrine that man was specially created and alone had a soul severed his kinship with other animals, and in the Middle Ages the Roman Catholic Church was officially hostile to business. As the only considerable organization to come through the Dark Ages, it nevertheless became by far the largest and wealthiest organization of the time, and the biggest business. It early began building the glorious medieval cathedrals, which required great technical skill; technology would then contribute superb stained glass for their windows, oils for their frescoes and paintings, and organs to amplify their service with music. (Less seemly was the technology of torture developed by the Inquisition for its "spiritual" purposes—a wholesome reminder that spiritual values, too often coupled with a contempt for the body, can be more inhuman than the simple goods of material well-being.) But most important was the Rule for monastic life established by St. Benedict early in the Dark Ages. In time 40,000 monasteries were obeying his rules for both organized labor and regular prayers and devotions, their bells ringing seven times a day. Werner Sombart pointed out that they were unwittingly laying the foundations of capitalism by instituting the regular, punctual life—the bourgeois ideal, historically novel, of being "as regular as clockwork." One outcome was the medieval invention of the mechanical clock, an automatic machine dividing up time into uniform hours, minutes, and seconds. This was an abstract, mechanical time foreign to the natural rhythms of man's life, or to time in immediate experience. "Time-keeping passed into time-serving," commented Lewis Mumford, who argued that the ubiquitous clock, not the steam engine, was the key-machine of the modern industrial age.

The Renaissance, long known to schoolboys as "the revival of learning," has weathered considerable discrediting by virtue of its contributions to art and literature, or more broadly to ideals of the good life. Although still conceived within a Christian frame, these more closely resembled Greek ideals in the stress on the good life in this world for its own sake, not merely for the sake of salvation in a life to come. With the growing naturalism and humanism came a lusty, unabashed individualism. The highly self-conscious Renaissance man—as ardent for glory and fame as a Homeric hero, so proud of his art or his genius that he might even write his autobiography, as Benvenuto Cellini did but no Greek ever had—was a flamboyant symbol of the ideal of self-realization and self-expression, heralding a society that would provide the individual with more rights, opportunities, and incentives than any past society ever had.

In its worldliness, which foreshadowed the growth of an eminently secular society, the Renaissance nevertheless distinguished itself in technology too. The workshops of the great Florentine artists were the main centers for the study of geometry, optics, mechanics, and anatomy. Out of these shops came Leonardo da Vinci, the Daedalus or master technician of the age. An engineer and inventor as well as an all-around artist, he filled his notebooks with designs and schemes anticipating the wonders to come in modern times, though also suggesting better sense; he kept to himself such possibilities as a submarine because he thought it would be harmful. Another stimulus was the contemporary rage for magic, in particular alchemy. In their hopeless endeavors to transmute metals into gold, the alchemists not only learned something about chemistry but kindled a passion for control of the natural world. In the words of Lewis Mumford, "Magic was the bridge that united fantasy with technology: the dream of power with the engines of fulfillment." Meanwhile men were consolidating their considerable empirical knowledge of mining and metallurgy, notably in Agricola's classic *De re metallica*. And such knowledge could now be more widely diffused because of an all-important invention, the printing press. The major contribution of technology to the high culture of the Renaissance, and to literature ever after, this both multiplied and standardized books by creating uniform, mechanical, repeatable type. As Marshall McLuhan pointed out, the printing press represented the first assembly line and the first mass-production.

Low commercial motives inspired the most momentous adventures of the period, the discovery of America and of a passage to India by way of the Cape of Good Hope. Adam Smith was not simply provincial when he called these "the two greatest and most important events recorded in the history of mankind." In seeking a shorter trade route to the East, Columbus opened up a vast "New World," destined to become the greatest power on earth; in landing in India, the Portuguese signaled the eventual domination of the entire world by Western civilization. "What benefits or what misfortunes to mankind may hereafter result from these great events," Adam Smith added, "no human wisdom can foresee." We have a better idea of both the benefits and the misfortunes, but still lack the wisdom to foresee the final consequence. In any case, the key to the momentous events was technology. In the fifteenth century the Portuguese, under Henry the Navigator, had settled down to the task of mastering the arts of navigation needed for voyaging far from home and shore; they had much to learn from the Arabs. By the end of the century Europeans had the best ships in the world, and the best for fighting too because they had mastered the art of naval gunnery. They would dominate the ancient civilizations of the East not by a superior religion or culture, but by a superior technology.

An immediate consequence of the discoveries was a big boom in business at home. Historians now talk of an "economic revolution" in the sixteenth century. Money-power was organized as never before; business was rationalized by such inventions as double-entry bookkeeping and by an international bourse that facilitated speculation in "futures," or nonexistent goods; and a spirit of rational calculation pervaded a society that had once called it greed. A symbol of the age was the great House of Fugger. Capitalism, which had been primarily financial and commercial, now became more industrial; the Fuggers were not only international bankers and merchants but big manufacturers of cloth and owners of copper, silver, lead, and mercury mines. Old Jacob Fugger, who built up the House, could be called the first great apostle of the religion of business—wholly dedicated to money-making, refusing ever to retire to enjoy his immense wealth, instead keeping busy at making a "good profit" (over 50 per cent) to the end of his days. On his behalf I have noted that apostles rarely retire.

Still, the House of Fugger had nothing like the power over society that modern capitalism has had. Although it had some political influence through its large loans to kings, the kings went their own way; they bankrupted some big banking houses when they caught on to the idea of declaring their kingdoms bankrupt, an excuse for refusing to repay the loans. Neither did Jacob Fugger preach the gospel of free private enterprise. In general, the six-teenth century still looks very different from the nineteenth century, much closer to the medieval world. Relatively few men worked in large business organizations; the great majority re-mained peasants or artisans. There was no flood of inventions, no one major invention. There was no mass production for mass consumption, no advertising business, no dream of abundance. On the Continent the expansion of trade and industry led chiefly to the production of more quality goods for the wealthy, catering to the tastes for beauty and luxury developed by the Renaissance.

Hence the significance of Elizabethan England—a striking exception that did not strike historians until they began studying technology. Long celebrated for its poetry and drama, and for the adventures of its explorers and seadogs (with some slurring over of the sordid fact that they were often pirates, inspired by a greed for Spanish gold), the age was exceptional as well in its industrial growth, due in particular to a great increase in the production of coal and the working of cheap iron ores. This expanded industry went chiefly into the production of cheap commodities. England emerged as the leader of Europe in industry and technology, a position it would maintain for three centuries. Accordingly some historians now date the beginning of the Industrial Revolution in the Elizabethan Age instead of the last decades of the eighteenth century. I still prefer the conventional later date because it marked the beginning of a more rapid, cumulative development, based on the machine, which more unmistakably began transforming the whole society. (No historian could fail to notice this change as many had the industrial growth of Elizabethan England.) But at least the later development was clearly anticipated by the coal and iron technology of this age, and it is easier to understand why Britain would lead the way.

Historians have lately grown aware too of a quieter, subtler, potentially more significant change in the late sixteenth century. Men were developing a strange interest in knowing exact times, distances, rates, quantities—an interest that today seems wholly natural, since men have come to like everything "definite," but that in past societies had never carried beyond a few obvious practical needs, such as standard weights and measures. The new interest, perhaps traceable to the mechanical clock, had plainer connections with the habit of close reckoning spread by the growth of business, the constant dealing in numbers. One sign was that some men began collecting statistics, ushering in the immense industry of our time—and again an industry that Greeks and Romans had managed to get along without. A more curious sign was a churchman who now figured out the exact date of the Creation: 9 A.M. of October 23, 4004 B.C. But above all the new mentality helps to explain the most revolutionary development in Western civilization—the rise of modern science in the seventeenth century.

This had various sources, to be sure, beginning with the achievements of the Hellenistic Greeks. Another was the naturalism of Renaissance art, encouraging the technical studies carried on in the workshops of the painters. Copernicus, the first great pioneer, had been educated as a Renaissance humanist and thereby had acquired a reverence for the Greeks; after twenty years of silence he was emboldened to publish his theory of the universe, which defied both common sense and Christian dogma, because he knew they had proposed such a theory. It was not until the seventeenth century, however, that his bold theory made a considerable impression, for by then men were developing not only the scientific spirit but scientific methods, which the Greeks had failed to develop. They were asking more persistently what now seems the obviously sensible question, What are the facts? They were sticking stubbornly to the evidence in defiance of venerable authority, including that of Aristotle. They were performing experiments, such as Galileo's legendary experiment with falling bodies from the Tower of Pisa—a very simple one that apparently had never occurred to the ancients. And Galileo exemplified the new rage for exactitude.

He pointed out with pride that although philosophers had always been aware of the acceleration of falling bodies, he was the first to ask how much faster they fell and then to measure the exact rate of their acceleration. Other pioneers, especially in chemistry, set about weighing and measuring. King Charles roared with laughter when he heard that the members of his Royal Society were busy weighing air, but the joke was on him as a man of common sense—this method was an immediate key to the success of science. Isaac Newton gave these quantitative methods both more philosophical dignity and more scientific usefulness by erecting them into mathematical deductions, which could then be checked by further experiment. By such systematic techniques men could go on steadily advancing scientific knowledge, as they have to this day. Thereby they made a commonplace of the dream of some intoxicated pioneers—that even apart from its new conceptions of the universe, science was the greatest revolution in the history of thought.

The first of these intoxicates was an Elizabethan, Francis Bacon. Coming well before Newton, and being no scientist himself, he had a pretty crude idea of scientific method, which he saw as a simple process of induction; but he did see most clearly that science was not only a revolutionary way of thinking but a means of achieving extraordinary power over nature, and so of revolutionizing society. Together with many wondrous inventions, including automobiles and airplanes, he foresaw the organization of science and technology, the systematic research that would conquer nature. This dreamer also foresaw the ideal possibilities of science that are still very real to many of its practitioners: an international enterprise of a professional community working for the benefit of all mankind. Less happily, he anticipated with enthusiasm other possibilities that do not enthrall many of us. The Utopia he sketched in *The New Atlantis*, his last work, would be governed exclusively by scientists or technical experts.[1]

For a long time Europeans did little to realize Bacon's dreams. Although their inveterate interest in military technology led to some use of discoveries about motion in improving artillery, not until more than two centuries later did they begin systematically to apply science to technology. In the seventeenth century it was the other way around. The pioneers of science owed a great deal to the inventions of instrument-makers—among others the telescope, microscope, thermometer, barometer, and air pump; these extended their field of observation and experiment, enabled them to measure such things as heat and air pressure. The new mentality that culminated in the triumph of Isaac Newton was quite practical in a sense, but a stranger, more ambiguous sense than Bacon's for those concerned with human values.

In general the pioneers agreed with him that science was "useful" knowledge—this was the common refrain of the chorus that went up in the seventeenth century. They had none of the scorn of the later Greeks for the "base mechanic arts," to which they owed their indispensable instruments. Rather, the triumph of Newton led to an oddly contrary extreme: physics became the dominant science, sometimes called "celestial mechanics"—mechanics now ruled the heavens too. Newton's mechanistic theory of the universe inspired others with his own hope that "we could

---

1. In fairness to Bacon, who is now often abused because his dreams of power have been realized, he warned against possible abuses of his new philosophy: "I foresee that if ever men are roused by my admonitions to betake themselves seriously to experiment . . . then indeed through the premature hurry of the understanding to leap or fly to universals and principles of things, great danger may be apprehended from philosophies of this kind; against which evil we ought even now to prepare." Our technological society has not yet got around to so preparing.

derive the rest of the phenomena of nature by the same kind of reasoning from mechanical principles." Men were dazzled by his vision of a clockwork universe, God as the Clockmaker who had wound it up for all time. Their pleasure in its absolute regularity was understandable in view of their long experience of clocks, but even so it was rather odd. Why should men be pleased to live in a world conceived as a machine? It would seem to be a poor habitation for the human spirit, no less if this was called an immortal soul, for how this spirit could be free in a machine order was not at all clear.

Its new home would seem to be still less congenial because Newton had made his mathematical deductions by considering only the quantitative aspects of the natural world, what could be counted, weighed, measured. These John Locke would dignify as the "primary" qualities of matter, like size and mass. Color, sound, smell, taste, and feel he called "secondary" because they are dependent on the human senses. Galileo had said that they were "nothing more than mere names"—not really real because not inherent in objects themselves; and Newton too wrote that to speak of light as colored was to speak "grossly," in the manner of "vulgar people." Yet these "secondary" qualities were the plainest source of elemental human values, all that gives people pleasure in the natural world. Men appeared to be denying the evidence of their cherished senses. They would accept the incredible conclusion that the "real" world was only a whirl of matter, colorless, soundless, tasteless, valueless—a world utterly foreign to that of their concrete experience, and as alien to the human spirit. Such a world was suited only to the rationality and efficiency of technology, which could therefore more easily become inhuman.

Newton himself, however, was little aware of these inhuman implications of his great synthesis. Having arrived at his theory of gravitation—a principle of universal order governing all motion that indeed has a grand, wonderful aspect—he dropped the whole matter and turned to computing the generations since Adam. For he was not really concerned with immediate practical utility. Neither were most of the great pioneers of the scientific revolution, from Copernicus on; their discoveries about the workings of the universe were of no practical use to them, but could instead get them into trouble, as they did Galileo with the Church. Their primary motive was that of Thales in ancient Greece—they just wanted to know, to satisfy their curiosity. In time their kind of inquiry would give men a magical power precisely because it was not governed by the materialistic concerns that governed both magic and business. For the same reason it would take hard-headed businessmen a long time to appreciate the values of pure science or basic research. In our own day an executive of International Business Machines would announce that the most important thing that had happened in business in this generation was its discovery that science is important—this after three hundred years of progress unparalleled in the history of thought.

Once men began systematically to apply science to technology, and thereby to acquire immense power, the limitations and the abuses of science would become more conspicuous; and they will concern us sufficiently later on. At this stage I think emphasis belongs to the human values of science that led writers to welcome it, as Fontenelle, Dryden, and Pope did, and that fired the leading thinkers of the Age of Enlightenment. For science then did not at all seem like a separate culture, divorced from the humanities. Mostly still blind to the devastating implications of its devotion to mechanical principles, writers saw its plainer connections with the good life. They appreciated not only its contributions to knowledge but the imaginative or esthetic values of its theories, the beautiful simplicity it found beneath the confusing appearances, the visions of cosmic symmetry and harmony that had given Kepler a "sacred ecstasy" when he contemplated the Copernican theory. In particular they appreciated the ethical values of the scientific spirit, at once the disinterestedness and the dedication it called for, the intellectual honesty, the discipline of the respect for fact, the tacit injunction against wishful thinking.

Most obviously the scientific revolution was an emancipation of thought. It was an emancipation from not only ignorance and error but the Baconian "Idols," the sources of prejudice, confusion, and error. As the Copernican theory had triumphed over the opposition of the Church, the most venerable authority in Christendom, the men of the Enlightenment drew the evident moral: thought and speech must be kept free. Science did most to promote the ideal of intellectual freedom, which was championed so effectively by Voltaire. While it taught the possible values of doubt, or the refusal to accept any faith merely on traditional authority, it substituted a positive faith in its own method of truth-seeking, the authority of reliable knowledge and a reliable means to more knowledge. Hence it did most to promote the distinctively Western faith in progress—a faith such as past societies had never had.

As usual, this had some roots in the past. In the Middle Ages millennarian movements revived the hopes of messiahs going back to the Old Testament. In the Renaissance Thomas More and others offered Utopias that implied a belief in the possibility of a much better society, a society somewhere in the future—not the past, where men had typically located the Garden of Eden or the Golden Age. But not before Francis Bacon had writers proclaimed that man could steadily, indefinitely improve his state on earth by his own unaided efforts, for only with the rise of science did they possess a clear means to steady progress. As the novel faith in progress began spreading over the Western world in the Age of Enlightenment, it introduced a fundamental difference in man's attitude toward change. Through all the changes beginning with the neolithic revolution men had never really banked on change, never believed that it would naturally be for the better or would go on so indefinitely. Greek thinkers were no more confident that the future would be better than the past. Like Plato and Aristotle, they aspired to stability and order, not more freedom and growth; in their visions of a well-ordered state they sought to prevent change; and the last thing they wanted was revolution. The obvious impermanence in human affairs led them to conceive history as endless cycle, the rise and fall that is indeed the most familiar story in history, of kingdoms, empires, whole civilizations. Now the idea of progress in the air made men more disposed to welcome basic change.

Similarly the triumph of science inspired more faith in human nature, especially in man's powers of reason. Long described as a creature of Original Sin, naturally depraved, man was now more often described as a rational animal. It followed that he was fit to govern himself, fit even for freedom. Men were appealing to experience instead of venerable authority in political thought too, as Locke had. In all such ways the triumph of science had something to do with the American and French Revolutions. Thomas Jefferson pointed to the connection in his reverence for his "great Trinity"—Bacon, Newton, and Locke. No more than any other thinker of the time did he anticipate that in the name of Marx a "scientific" philosophy would inspire counterrevolutions in our century, with iron-clad guarantees of progress to a Utopian classless society, but meanwhile denials of freedom.

The French Revolution brings us closer to the Industrial Revolution, which was already under way in Britain. As not only a political but a social revolution, ending once and for all in France the old aristocratic order with its feudal privileges, it marked the ascendancy of the bourgeois, who had taken charge of it. And though France as a nation was slow to catch on to the possibilities of industrialism that Britain was exploiting, it produced one of the first thinkers to foresee clearly these possibilities, le Comte de Saint-Simon. In the excitement of the French Revolution some Frenchman had coined the term "Industrial Revolution," long before it became current in Britain, and Saint-Simon now popularized "industrialism." For him this signified a new kind of society, peaceful, orderly, and rational, in which wealth would be produced by machinery. He attracted many fervent followers by his call for a breed of "new men," builders and planners, to lead the way. In the new society, wrote "the last gentleman and the first socialist" of France, "the real noblemen would be industrial chiefs and the real priests would be scientists." It was an exciting vision, like that of Francis Bacon; but for us it makes a sobering introduction to a new chapter in man's history.

II

The Growth of Modern Technology

4

The Industrial Revolution

In 1835, when a "railway mania" was exciting Britain, the
editor of the journal *John Bull* was much alarmed by the
latest invention. "Railroads, if they succeed," he warned, "will give
an unnatural impetus to society, destroy all the relations which
exist between man and man, overthrow all mercantile regulations,
and create, at the peril of life, all sorts of confusion and distress."
Needless to say, railroads did succeed. A generation later Britishers
rejoiced in the thousands of miles of them that crisscrossed their
land. Although trains had been invented chiefly to transport
freight, they had soon attracted droves of passengers, and by the
end of the century would be carrying well over a billion riders a
year. Europeans on the Continent had been as pleased to imperil
life by building railways, and Americans were simply thrilled
when their first transcontinental railroad was completed in 1869.
Everywhere the locomotive, whistling through the countryside at
sixty miles an hour, was the most popular symbol of the new
powers man was acquiring in an industrial age. The editor, long
since forgotten, would have seemed a ludicrous old fogey.

Yet he was quite right. It is now hard to realize that before
the railway era the great majority of people the world over spent
their entire life in the region where they were born, never leaving
the village except perhaps to go to the nearest market town,
always remaining set in their traditional ways. The railroad
accordingly did represent an "unnatural impetus," and a profoundly disruptive one. It signaled the end of the old social and
political order as it broke down both geographical and social bar-

riers by carrying ever more millions of passengers, drawn from all classes. It became the popular symbol of a technological revolution that was in fact destroying the traditional relations between men and creating "all sorts of confusion and distress." Even its literal peril to life was dramatized upon the inauguration of the first railroad: a director was run over by the train, prophetically named the Rocket. Today we can better understand the disruption of a traditional society by industrialism, for this is the drama being enacted all over the non-Western world as it seeks to "modernize."

More to the point, the editor of *John Bull* still spoke for many respectable people of his day, and indeed for the vast majority of mankind throughout history. Men had always tended to resist fundamental change, any radical innovation. Though the Industrial Revolution had started at least fifty years before the editor wrote, it had proceeded slowly enough to obscure the deepest change taking place, also symbolized by the railroad. This was again the attitude toward change itself, now in daily life as well as government. It was the growing disposition to accept innovation, even to welcome it. Not only thinkers but ordinary people were calling change "progress." The once novel faith in progress was on its way to becoming common sense. But then we must add that that stodgy editor is still not really a stranger in our revolutionary world. Although change has long since become routine, most people welcome only superficial novelty—the latest models, gadgets, thrills. They still resent and resist any call for fundamental change in their ways of thinking; as Bertrand Russell once observed, most people would sooner die than think, and in fact do so. Especially in America they do not at all welcome radically new ideas; despite our boasts about our greatness, Americans seem more afraid of radicals or revolutionaries in our midst than are any other people in the world. And none more so than the business leaders who keep on revolutionizing the economy, accelerating the drive of our technology—the most influential radicals of our day.

This brings up a deeper paradox. The Industrial Revolution was the work of many inventive, enterprising, daring individuals. At first largely confined to a few districts in England, it was a natural, understandable development of what had started in the Elizabethan Age, but it was by no means inevitable, automatic, or predetermined by any known iron laws of history. Although broadly anticipated by Francis Bacon, its actual course was foretold by no thinker of the time, including Adam Smith, the most acute analyst in the days when machines were growing up. Yet by the same token the Industrial Revolution as a whole was quite unplanned. Its pioneers did not get together and say, Let us try for a change a society based on the machine. Like the inventor of the railroad, and the medieval burghers before them, they were unconscious revolutionaries who hardly foresaw, much less intended, the profound changes their innovations would bring throughout the whole society. The revolution illustrated the "vast, impersonal forces" of history that today we hear so much about. Slowly as it moved in the first fifty years, it came to seem impersonal and automatic as one invention called out another, and the impetus given it by the railroad made it well nigh as irresistible as irreversible. Basically the whole process was by no means so "rational" as historians of technology now make it appear. And all along society was even slower to realize the changes that were coming over it, to deal adequately with the problems they created. It illustrated the difficulties men always have in understanding and catching up with the history they have somehow made— difficulties that we now call cultural lag, but that in spite of our knowingness are as great as ever because of the terrific drive of our technology.

With these ironies in mind, let us review a few of the inventions that started revolutionizing life in a society that had known and valued machines since its beginning. James Watt's steam engine deserves its textbook fame as the key invention, no less because he only improved an engine long used to pump water out of mines. A pupil of a maker of "philosophical-instruments," Watt did not build on new scientific knowledge, but like the technicians of the past merely experimented in a practical spirit. Although he went broke on his early experiments with engines, he persevered, in 1769 patented one good for pumping, and then spent twenty years adding improvements that made it capable of driving all kinds of machinery, in factories as well as mines. Powerful, regular, and itself movable, it had a great advantage over water power in its independence of geography and season; henceforth industry could grow up in any region. Watt helped to sell his engine by establishing a standard measure called "horsepower," though without realizing that the animal would be replaced by the "iron horse"; as an old man he rejected the new-fangled idea of a steam carriage on railways. By that time the steam engine had nevertheless become the prime-mover of a new industrial age, now called the Age of Steam.

One reason why it took half a century to become the main source of industrial power was that the early engines had to be made by hand and could not be made to precise standards. The Age of Steam got really under way with the development of the machine tool industry, which signaled a fantastic development by turning out "machines that make machines," or breeding monsters like Frankenstein's; by 1850 all the basic ones were invented or perfected. The challenge of engines also led to the rise of a new profession, engineering, destined to take its place among the major professions.

Another early key to the Industrial Revolution that deserves its textbook reputation was the textile industry. A series of inventions known to most people (including me) only by name if at all —the water frame, the jenny, the mule—was climaxed by the power loom of Edmund Cartwright, a clergyman turned inventor to meet the increasing need of inventions. This made the industry the first major one to be revolutionized; machinery now did the work, making nonsense of the original meaning of "manufacture" —to make by hand. Likewise it accentuated the essential difference between modern machines and tools or neolithic machines. Whereas the latter required skilled craftsmen, most of the modern ones were automatic, requiring only operators to keep them going. Cotton manufacturers could accordingly begin producing for mass consumption, in foreign markets too. They did most to make Britain, already the greatest trading nation, the "workshop of the world." A grateful House of Commons awarded Cartwright in his old age 10,000 pounds for the benefits he had conferred on his country.

A title was enough for Richard Arkwright, the most enterprising of the early textile manufacturers, who had started out as a barber's apprentice. Described by Carlyle as "that bag-cheeked, pot-bellied, much enduring, much inventing barber," he had to endure the loss of his patent for the water frame, an invention he had stolen from other men; but he earned both a fortune and the title Sir Richard by developing the major social invention required by the new technology—the factory system. Unlike the spinning jenny, but like other of the new machines, his water frame was much too big and heavy to be installed in a worker's cottage, or to be powered by hand or foot; so he and his partners built cotton mills run by water power, in which they employed hundreds of workers. Factories had indeed long been known in Europe, in particular arms factories for its professional armies, but now many more were needed because they were the natural home of the machine. In his mills Arkwright mastered the new art of managing large establishments with machine processes, teaching workers (as he put it) "to conform to the regular celerity of the machine." He employed mostly children because they were not only cheap labor but quicker at learning this strange habit; men with skills were apt to make poor machine-tenders. Arkwright's title, still a rare honor for a businessman, perhaps did not entitle him to be considered one of the "real noblemen" envisaged by Saint-Simon; but in any case it foreshadowed the increasing prestige businessmen would win in the new society.

With his partners Arkwright heralded as well a new era of capitalism, primarily industrial. Such partnerships were common in the early days of the Industrial Revolution, inasmuch as few men had at once the capital to set up a mill and the technical and commercial skills needed to run it successfully. With the growth of industry small firms grew into joint-stock companies as other men were induced to invest their savings, and from mid-century on such companies became the standard form of business organization, producing the bulk of manufactured goods. The "organizational revolution" was under way, a major consequence of the new technology that would culminate in giant corporations, and with them big organizations all through the society. Capitalism was also equipped with a philosophy, such as it had lacked in the heyday of the Fuggers. In *The Wealth of Nations* Adam Smith had advanced his novel theory of laissez faire or free private enterprise, based on what is technically called the principle of the self-regulating market. The theory was that unregulated competition in self-interest best produced the goods people wanted, and that a free market would regulate itself, as if by an "unseen hand," because the incompetent would be weeded out, only those would succeed who knew how to produce the goods cheaply; granting a free rein to self-interest thus benefited the whole society. The new creed, which rationalized common business practice, was admirably suited to the interests of a rising business class that in England had to contend with an often hostile landed aristocracy in control of the government. In America, where business met much less opposition, the new theory became a gospel long before the end of the century. Here economic freedom would be hailed as the most fundamental freedom, the heart of the American Way of Life.

Early in the century America had contributed its pioneer in the factory system, Eli Whitney. Best known to schoolboys for his cotton gin, he also ran a small arms factory in which he introduced the manufacture of interchangeable parts, the rudiments of the assembly line.[1] Harking back to the printing press, his ingenious "American system" speeded up the mass production more characteristic of the new age. And among the new industries to adopt it was harvesting machinery, dating from Cyrus McCormick's invention of the mechanical reaper in 1834, which started the revolutionizing of agriculture too by the machine. Immediately it was another reminder that it took a lot of bold persons to create the "vast, impersonal forces" that made the Industrial Revolution. Twelve years after McCormick's invention, an observer in America—the celebrated land of opportunity and enterprise—marveled at a factory that was daring enough to manufacture a hundred reapers. "It was difficult indeed," he reported, "to find parties with sufficient boldness or pluck and energy to undertake the hazardous enterprise of building reapers, and quite as difficult to

---

1. Schoolboys may not be taught either that Whitney's cotton gin revived the dying institution of slavery by making it immensely profitable. Improved technology therefore helped to bring on the Civil War.

prevail upon farmers to take their chances of cutting their grain with them, or to look favorably upon such innovation."

Because of such attitudes the "unnatural impetus" of railroads takes on more importance, warranting another new name— the Railway Age. While in England they soon became a "mania" as a profitable source of investment, elsewhere they brought countries to what is now called the "take-off" stage in modernizing. They not only constituted a big new industry, a booming market for iron and coal, but stimulated the growth of all other industries. Whereas transport of goods had been slow and expensive, except for towns located on rivers or the sea, railroads could swiftly deliver all raw materials and finished products, giving manufacturers a much wider choice both in sites for their factories and in markets for their goods. And outside of Britain they gave still more impetus to industrialism by bringing in the aid of government. In America the government lavished huge grants of public lands on the transcontinental railroad builders, as well as generous loans for every mile of track they finished; such public enterprise enabled these entrepreneurs to make fortunes with little personal investment, at as little personal risk. On the Continent the railroad systems were generally planned when not built by the state.

Hence the revolution that had started in Britain now spread to all of Europe. Some countries took to industrialism much more slowly than others, remaining "backward," but everywhere it introduced the same tendencies—more use of mechanical power, more material goods, more standardization of products, and more business. By the end of the century a machine civilization was well on its way to becoming the first civilization in history in which most people would make their living through business instead of agriculture. Likewise it was becoming the first predominantly urban civilization, in which most people in the advanced industrial countries would live in or about the city instead of the village.

We may now pause over an early climax of the Industrial Revolution, the Great Exposition of London in 1851. Held in the Crystal Palace, a monumental building of iron and glass designed especially for the occasion (though not by an architect), it was still more dazzling as the first international exhibition of the wonders being produced by industry. The London *Economist* described the Palace as a temple erected to the honor of "the mightiest empire of the globe—the empire in which industry is most successfully cultivated, and in which its triumphs have been greatest." This was simple fact, no mere boast, but Victorians might be more edified by the deeper significance the *Economist* pointed out in the exhibits in this temple. They testified to the "honour" in which "humble industry" was now held, the "moral improvement" already apparent, the "devotion to peace," the promise of a still more peaceful future—all in all, the "irresistible assurances that a yet higher destiny awaits our successors even on earth."

So we are brought to the social consequences of the Industrial Revolution. Since it has become all too easy to resist the complacent assurances of the *Economist*, I should note first of all that its editor had some good reasons for his complacence. Men had obviously achieved much more of the power over nature that the human race had always cherished, and they could rightly expect to acquire ever more power, and with it more wealth. In his *Communist Manifesto* Karl Marx himself had already been pleased to grant that the bourgeois economy was by far the most productive in history. Unlike others it was producing an abundance of cheap goods; its main market was ordinary people, not the rich. Although economists are still debating to what extent, if any, real wages went up in the early decades of industrialism, by mid-century the trend was clearly upward; workers were on their way to the highest standard of living in history. C. P. Snow has reminded the literary world of a simple truth too often overlooked, that for all its evils the Industrial Revolution was the only hope of the poor—the great majority of mankind who had lived in poverty ever since the rise of civilization. With the prospects of increasing plenty it was not unreasonable to hope for increasing peace too. Marx could dream of his classless society, in which industrialism would provide an abundance for all.

In view of such real gains we may discount somewhat the admitted evils of early industrialism—the many women and children working twelve or fourteen hours a day in factories and mines, the many men slaving at starvation wages, the foul living conditions to which all were condemned in the industrial towns or city slums. Herbert Heaton, an economic historian, has observed that these were ancient forms of evil, taken for granted in the past, and that the revolutionary development was the realization that they were evils. It was the growing indignation over them, the belief that they were remediable, and then the positive resolve to do something about them. We know much more about them than we do about the lot of workers in the past because of a series of full, detailed reports by Royal Commissions and Committees of Inquiry; these were the main source not only of Marx's documentation of the evils of capitalism but of the indignation of many bourgeois, and of government enterprise in a series of laws to check the worst abuses.[2] And the spreading faith in progress that may seem ludicrous in the *Economist*'s hymn to the Great Exposition had much to do with the concerted efforts at reform, which in the past had been rare and sporadic. It was not merely a complacent belief in an automatic material progress, but an active will to improve life on earth. Among the many inspired by the faith in progress had been Robert Owen, an early industrialist who by his own success demonstrated that the current evils of the factory system were economically unnecessary, and who preached that "the new powers which men are about to acquire" made any poverty unnecessary.

Workers themselves were stirred by the new spirit. The violent uprisings that had broken out as early as the Middle Ages could be called the beginning of class war against the bourgeois, but the still unbaptized "proletariat" had then had no political program. In the 1830's the great Chartist movement swept over England, leading to enough riots by workers to look like a revolution, but also gathering more than a million signatures to a petition for a "People's Charter" of democratic government. If this owed most to the democratic ideals proclaimed by the American and French Revolutions, it also owed much to the Industrial Revolution. The industrial proletariat was the fastest growing class in England, it was concentrated in cities, and as Marx would

---

2. We might also note that the human costs of industrial development were as heavy in the Soviet, established in the name of Marx. The enforced collectivization of agriculture cost millions of lives.

say, it was being "armed" by the bourgeois. It was growing more literate and class-conscious, learning from the successful efforts of the bourgeois to wrest political rights from the ruling aristocracy. So it began its protracted struggle for the right to organize in labor unions, the right to vote, and the right to free public education. By the end of the century it had largely won these rights in the advanced industrial countries. With their rising standard of living, workers thereby had the means to more effective freedom than they had enjoyed in any previous society.

Yet they always had to struggle, in America especially against the almost solid opposition of the ruling business class. Although the most shocking abuses of the early years have been eliminated, there remain plenty of evils. We are still struggling with the basic problems created by the triumph of industrialism. I think we still need to dwell on the neglect of elementary human values that was so glaring at the outset.

To begin with, most of the workers long remained wretchedly poor. This was a repetition of an old story, beginning with the rise of civilization, but it was more demoralizing in a society that was acquiring far more wealth and power than any before it. Another common kind of victim of industrial progress was the hand-weaver, the many men left with useless skills, fighting a doomed battle against machines, in their fury now and then smashing them. Comfortable economists, always at home with abstractions, would point out that the new technology caused only "temporary dislocations," inasmuch as its victims would in time be absorbed by new industries, but the dislocations were always painful for human beings, who usually lacked the means of promptly moving into new jobs. Later on economists also grew aware of the "business cycle," the periodic panics of depressions that came with the new economy, throwing men out of work for no fault of their own; but they regarded this cycle coolly as a kind of natural law of business and proposed no means of preventing depressions, nor of taking care of their victims. And despite the accelerating industrial progress, and the gradual rise of real wages, there always remained millions of very poor people. Americans have only begun to realize belatedly how many millions of them there are in our most affluent society. An increasing abundance of material goods made more glaring an elementary failure of industrialism: all along it failed to provide a great many workers with the minimum necessities of a decent life—an adequate diet, adequate medical care, decent homes, pleasant surroundings.

Living conditions were most appalling in the new industrial towns, the heralds of the first predominantly urban civilization in history. In the same year when the editor of *John Bull* was alarmed by the prospect of railroads, Alexis de Tocqueville reported on the city of Manchester, the foremost center of industrialism: "From this foul drain the greatest stream of human industry flows out to fertilize the whole world. From this filthy sewer pure gold flows. Here humanity attains its most complete development and its most brutish, here civilization works its miracles and civilized man is turned almost into a savage." The smoke-filled towns were the drabbest, grimiest, and ugliest in all history. As they went about "making a new Heaven and a new Earth—both black," they made little or no provision for sanitation and recreation, parks and playgrounds, or any open space where people could gather and relax. The industrial slums that started growing up in the old cities too made drabness, filthiness, and ugliness seem still more like a natural, normal condition of industrial progress. Respectable people were not shocked by the foul surroundings in which men were learning to live, and could hope to live with any contentment only by virtue of deadened senses, with the help of gin.

One might excuse such failings by saying that it would naturally take men some time to learn how to live with machines and factories. Thus municipal government in Britain was quite unprepared for the new conditions, and lacked anyway the authority to deal with them; in time it would acquire the authority, make some provision for the elementary needs of a decent civic life. Yet the slums would remain, above all in wealthy America, and with them other root evils that would grow worse. Industry soon began polluting rivers and the atmosphere; not until American cities were smog-bound did people begin to realize, too late, how intolerable the pollution had become. Likewise people grew so inured to ugliness that they seemed quite unaware of it—as most Americans still seem unaware of the commercial blight on their landscape, once known as God's own country. Progressive industrial man has never shown the respect for the natural environment that his benighted ancestors did in creating and tending the landscape of Europe.

Hence the obvious question: Why all such neglect or even contempt of elementary human values? The immediate answer seems to me as obvious: it was due to the vaunted free private enterprise that created industrialism, for the sake of private profit. So in fairness I would again first acknowledge the very real enterprise of the pioneers in the Industrial Revolution, and the technical ingenuity and resourcefulness of their followers in building up industry, producing the goods that people did indeed want. No other class at the time was so enterprising. In America, where businessmen were given most freedom and opportunity, they rewarded the nation by conquering a continent. Having a particular need of machinery for this purpose, they responded by developing a genius for mechanizing, organizing, and standardizing their far-flung operations. In but a generation after the Civil War they made America the greatest industrial power on earth. Allan Nevins has called them "the heroes of our material growth"—the growth that Americans have always been proud of.

Yet the heroes also distinguished themselves by exploitation, plunder, and fraud, on a colossal scale unknown in Britain. They made this generation the most flagrant in the nation's history for routine corruption, in both government and business; they thoroughly earned another name for themselves, the "robber barons." With the help of the Republican Party, which ruled the country in their interests to the end of the century, they also made America the most backward of the industrial countries in social legislation to protect workers against the abuses of private enterprise and the hazards of industry—just as the interests of business had helped to make it the last "Christian" nation to abolish slavery. The economic freedom they prized, no less because the authors of the Constitution had neglected to mention it, was a freedom only for themselves, the men on top; they fought bitterly the enterprise of workers who sought to achieve more freedom for themselves by organizing in labor unions. They forced the basic issue that is still with us, in an era of much bigger business.

They had far greater power, both economic and political, than businessmen had ever had in the past; and how responsible were they in the exercise of this power? The answer in this period was plain: for the most part they were socially quite irresponsible. The magnate who said "The public be damned" was only putting crudely the freedom they defended most zealously—the right to be socially irresponsible.

Beneath the historically novel rule of business lay a profound change in mentality, a new conception of human nature. Adam Smith had anticipated it by his fiction of "economic man," born with an instinct to barter and acquire property. For him this was primarily a convenient concept for purposes of analysis, or I suppose something like the "models" that have become popular with social scientists; he did not exalt economic above other interests. Other men, however, took this fiction quite literally as an essential definition of human nature. Americans came to take for granted not only that man is born with an "acquisitive instinct" but that economic motives are the most powerful, or even that the profit motive is the only one that will stimulate men to do their best. Charles Peirce accordingly ventured the guess that the nineteenth century would be called the "Economical Century." In his day William Graham Sumner was saying plainly what businessmen still believe, that economic needs should be the "controlling consideration," and those who professed otherwise were only being sentimental. Karl Marx had anticipated Sumner, ironically, by arguing that these needs were in fact the controlling consideration. His novel theory of economic determinism made historians aware of the importance of the economic factors they had long neglected in their concentration on political and military history, but in an age increasingly dominated by business interests they might then forget their best excuse for this neglect, that in past societies economic interests had always been subordinated to noneconomic social ends, the business class had never been the ruling class.[3] Now economic means were being transformed into national ends.

---

3. Some apparent exceptions, notably Venice and the Florence of the Medici, were clearly exceptions to the rule in Western civilization as a whole. At that their merchant princes and bankers ruled as

In social life the result was a growing materialism that gave the nineteenth century still another name—the Age of Materialism. This was most conspicuous in America, where even before the country was transformed by industrialism de Tocqueville had been struck by the passion of Americans for material well-being. It was not simply an ignoble passion, let us add, since in a land of opportunity for common people they naturally wanted first of all to make a better living; but with success the materialism grew cruder. Americans became known the world over for their reverence for the almighty dollar, while they themselves grew accustomed to saying that a man was "worth" so many dollars. They looked grosser because of the conspicuous consumption for which the Gilded Age became notorious. Millionaires then gilded their social status by buying their daughters titled husbands from Europe, paying exorbitant prices for a commodity in plentiful supply. (In 1894 alone, 134 of them spent $175 million for assorted counts and lords.) But though Europeans continued to look down their nose on the vulgar Americans, their bourgeois were about as devoted to material wealth and power (not to mention the many nobles willing to accept American millions and marry beneath themselves). In all the industrial societies tradition was eroded by the nature of the economy, whose central institution was the market. The market set the worth of everything, in purely quantitative terms that could not measure the values of the spirit.

De Tocqueville also noted that the passion for material well-being had a slavish aspect, describing it as "the mother of tyranny." With factory workers, however, the story was rather different. They were "materialistic" of necessity; mostly too poor yet to realize they were animated by an acquisitive instinct, they had to be concerned constantly with the bare necessities of life. From the beginning of the Industrial Revolution, moreover, they had to pay the human costs of the new technology, in a kind of servitude that could be called strictly "unnatural" for men who

---

aristocrats, by virtue of more than success in business and with an eye to more than money values; among other things they were patrons of art. "Aristocrat" has never been the word for most business leaders in America.

were not technically convicts or slaves. In some respects they were worse off than the many slaves who were toiling in the cotton fields of America.

Factory work was quite different from the routine toil and drudgery that poor people had always been accustomed to. Instead of setting their own pace, workers had to learn the "new discipline," conformity to "the regular celerity of the machine." They were always under the eye of a supervisor, often a severe taskmaster who kept them working at high speed through the long day, and at best they were no longer free to pause, chat, or idle now and then. All became factory "hands"—a suitable name for machine-tenders who were no longer doing man-sized jobs but were turning out parts instead of finished products, and who served as interchangeable cogs in the system. As Karl Marx put it, the factory "transforms the worker into a cripple, a monster, by forcing him to develop some highly specialized dexterity at the cost of a world of productive impulses and faculties—much as in Argentina they slaughter a whole beast simply in order to get his hide or tallow." Work had given both peasants and artisans a definite social status and function, however humble, but most factory work was humanly meaningless except as a necessary means of making a living. Workers could not express themselves in their mechanical work—they could only deny themselves. If or when higher wages resigned them to machine-tending, it could only be at some sacrifice of their natural interests and capacities.

Although in time men learned to adapt themselves to the new discipline, to this day most factory workers—even the best paid—would never learn to be content with their mechanical work. Likewise there would remain other reasons for the "alienation" of workers that Marx was the first to describe. Under the new technology human labor became another commodity, subject to the laws of a market that ignored social values, or another resource to be exploited, like the natural environment. At best work was sharply separated from the life of the community, as from all other interests and activities. It provided no more occasion for festivals or communal rejoicing than for singing on the job; the factory was no place to nurture sentiment, fellowship, or primary human relationships. Work was separated from family life too, since fathers went away for the day and the family was no longer a productive unit. And the new technology brought other forms of regimentation that are worth noting even though they have long since grown so familiar as to seem "natural," for people still chafe at them.

Factory workers had to acquire habits of regularity and punctuality that at first were as strange as their barrack-like surroundings. Throughout the year they had to get up at fixed hours, often in darkness, just as through the working day they had to forgo their natural impulse to rest when they felt like it. Neither could they take vacations, nor ever enjoy the long spells of relative leisure peasants had known after harvesting their crops. In the words of Dr. Ure, the egregious high priest of the new textile industry, their employers had to discipline "the refractory tempers of work-people accustomed to irregular paroxysms of diligence." It was harder for many of them to learn the proper diligence because they also had to acquire a proper acquisitive instinct. Like most people the world over, those not working at starvation wages had been accustomed to working just enough to make what they considered a decent living. (Modern Afghans still reject offers of industrial work, in spite of higher pay, because it would make them eat more.) By now most Americans have out-grown such simple desires, but they appear to have occasional misgivings. Even businessmen often complain that their life is a "rat-race"; so apparently they do not enjoy their more regular paroxysms of diligence.

One early way of making all such regimentation more toler-able, or making better machines of men, was the new gospel of work. Throughout history men had normally regarded work as a painful necessity, or a punishment for Adam's sin in the Garden of Eden, but now it was exalted as a positive good. This was the source of the "moral improvement" the *Economist* hailed in the triumph of industry, the "honour" accorded to "humble industry." Work enjoyed for its own sake can indeed be among the deepest, richest satisfactions, the absolute goods of human life (as it is for me); but the gospel that was spreading was quite different. It was the Puritan ethic that made work a moral duty, diligence a cardinal virtue, and idleness a sin. The trouble remained that ordinary factory work could not be satisfying for its own sake. The natural impulse of workers was to shirk their duty as much as possible, ease their servitude to the machine, thus enabling respectable people to talk more easily of the irresponsible, shift-less poor. The popular acceptance of the gospel of work, especially in America, no doubt helped to reconcile workers to their fate, gradually to improve their material lot; but it also suggested a growing incapacity for the cultivation of other human values. Machine-tenders kept alive—more alive than the petty bourgeois—chiefly because most of them never really believed in the new gospel.

In this respect technology came to their aid, but on the whole without much quickening or enriching their humanity. Early in the nineteenth century a steam-driven printing press made pos-sible the mass production of books and periodicals, while an abundance of cotton rags stimulated the invention of a papermak-ing machine that considerably reduced the cost of paper. The result was the "communications revolution," immediately a swell-ing flood of cheap popular literature. With the spread of literacy some workers began educating themselves, aspiring to self-improvement by reading serious books. More took to trash, the staple product of the growing publishing industry. This new popular culture—the beginnings of mass culture—was quite un-like the folk-culture of the past. Rooted in the countryside, the latter had been of and by the common people, a genuine if coarse expression of their thought and feeling, whereas the new culture was a synthetic product, manufactured for the people. While it conformed to much of their natural sentiment, chiefly it served as a drug or an escape. It was pathetic because they badly needed an escape from the "real life" they knew, in a wretched environment, but it was depressing because the world of fantasy they escaped into was typically tawdry, a sorry substitute for what life can be.

The reading of the better educated middle class naturally included more serious literature. In Europe they provided the bulk of the audience for the greater writers of the century, most of whom were themselves bourgeois. Yet their tastes were otherwise more depressing than the crude tastes of the working class. Especially in Victorian England they took chiefly to a genteel, "refined" kind of trash: sentimental and romantic fiction with happy endings in which their conventional virtues were always rewarded, together with all kinds of writings that preached wholesome morals and gave them an easy uplift. The gush and the cant for which the Victorian Age became notorious had a more insidious effect on mentality than the coarse escapist literature of the workers, who at least knew this was fictitious, utterly different from the world they had to work and live in. The bourgeois were prone to confuse their sloppy fiction with "real life," mistake its stereotyped morals for the good life. And so they in particular bring up the issue of high culture, which traditionally had nurtured ideals of the good life.

Raymond Williams has pointed out a number of significant words that entered popular usage or took on their modern meaning in the early decades of the Industrial Revolution—words that were keys to the new kind of society emerging. Among them were "industry" itself in the sense of manufactures, "art" as fine or creative art, and "culture" as the state of intellectual development or the distinctive values of civilization.[4] With "industry" came "business," "capitalism," "commercialism," and "working class," all due to be used constantly thereafter. Literary men were especially concerned about "art" and "culture." Before the Industrial Revolution technology had most obviously served the purpose of culture and never been regarded as a serious threat to it; writers had worried only over the tastes of the growing middle-class reading public. But with the rise of business and the reign of commercial values, it grew apparent that in the new society the traditional values of civilization could no longer take care of themselves or be taken for granted. Although they would survive in lip service to something known vaguely as "the finer things of life," material goods fared much better.

---

4. "Culture" in the anthropological sense, embracing all learned belief and behavior, entered the language much later. I have therefore felt obliged to distinguish the older meaning from the technical one by speaking of "high" culture. This may sound supercilious—especially in America, where it connotes such typical Americanisms as "highbrow" and "egghead"—but I think it necessary to keep insisting on the commonplace that for man there are higher needs and goods, among them a need of a scale of values.

Early in the century the Romantic movement sounded a major theme of literature thereafter—a revulsion against industrialism. Writers attacked not the machine itself so much as its by-products: the rule of getting and spending, the utilitarian philosophy, the defilement of the landscape, and the horrors of the industrial towns and their factories, Blake's "dark Satanic mills." In France Balzac and Flaubert concentrated on a devastating analysis of the rising bourgeois and their values. The Victorians deplored the complacence over a purely material progress. "Our true Deity is mechanism," wrote Carlyle, and he summed up the progress: "We have more riches than any Nation ever had before; we have less good of them than any Nation ever had before." Ruskin added: "There is no Wealth but Life." He also attacked the ugliness of the products of industrialism, which even some men dazzled by the Great Exposition had noticed; the taste displayed in design and ornament was about the worst in the history of craftmanship. Matthew Arnold, the apostle of culture, saw the chief menace as the "faith in machinery" of a civilization grown "mechanical and external." Zola, Hauptmann, William Morris, and others spoke out on behalf of the workers, who had been degraded by the industrial progress. In general, the nineteenth century created a great literature, more varied and abundant than that of any century before it, in a way suited to the extraordinary energy and creativity it was exhibiting in technology; yet writers kept growing more critical of their society, often hostile to it, than writers had ever been before. William Morris wrote that his leading passion was "hatred of modern civilization."

At the same time, writers soon began simplifying and confusing the issues, in various ways that will concern me later. As Lewis Mumford lamented, the basically sound attack of the Romantics on industrialism was weakened by their reactionary cults of the past (especially the Middle Ages), of regionalism or the folk, and of primitivism; they had a limited understanding of the forces at work in their society, either the actual dynamism or the potential values of the new technology, which Saint-Simon and Robert Owen saw much more clearly. But in particular they confused the issues raised by another new word that entered the language early in the century—"individualism."

This could be considered a belated tribute to a civilization that from the Renaissance on had offered the individual exceptional opportunities and incentives, and that with the rise of democracy was granting him still more exceptional rights. In de Tocqueville's *Democracy in America*, however, it made its appearance as a bad word. To him it signified a calm, settled policy of every man for himself, which popular agreement made more insidious than the simple selfishness always known to civilized man. It grew popular as an economic individualism, later dignified as "rugged individualism." This was quite different from the ideal of individuality exalted by John Stuart Mill in *On Liberty*. In insisting that the only liberty worth the name was that of pursuing one's own good in one's own way, so long as one did not intrude on the liberty of others to do likewise, Mill defined the goal as essentially the ideal of culture or the good life proclaimed by Matthew Arnold: it was the fullest and most harmonious possible development of the individual's capacity for the pursuit of truth, beauty, and goodness, the powers that constitute the dignity and worth of human nature. Economic individualism was not only restricting the liberty of workers but tending to narrow, warp, and impoverish the individuality of successful businessmen by restricting their own interests, confining them to what Mill called "purely secondary objects," such as making money.

To this confusion the Romantics and their followers contributed by still another kind of individualism—the cult of the hero, the genius, above all the self-conscious artist. They tended to identify individuality simply with what distinguished a man from his fellows, often separated him from them, to the exclusion of what united them. The artist came to be regarded as by nature a nonconformist, a rebel—a type in fact uncommon in the history of art. Ralph Waldo Emerson's addiction to epigram lent support to this romantic idea when in preaching his gospel of self-reliance he wrote: "Whoso would be a man must be a non-conformist." Actually, of course, even thoughtful men conform to most of the conventions of their society, as Emerson himself did, and they can do so without serious sacrifice of their personal independence or self-reliance; all that matters is that they make their own choices with their eyes open. At any rate, the common hostility of writers to their society led many to feel that as artists they were naturally men set apart from it. In time there appeared the esthete, the high priest of a religion of art-for-art's-sake, who enjoyed a freedom from social responsibility resembling that of many businessmen. On lower levels the esthete became the Bohemian, subject to a compulsive kind of nonconformity.

Hence I would round out this chapter with the verdict of Walt Whitman, the most ardent champion of American democracy, and especially of the common people. Like them, and for their sake, he rejoiced in the "material success" of America and the certainty of increasing prosperity. Unlike many writers, he appreciated the grand poetry in the conquest of a continent with the aid of modern technology, all the "railroads, ships, machinery, etc." For the people's sake he held up the ideal of "perfect individualism"—not that which "isolates," but that which makes for "Solidarity," uniting men in comradeliness and love, a warm recognition of the personal dignity of all men. He defended them against the many "critical and querulous" writers who were forever harping on their limitations. In particular he could not abide the fastidiousness of Matthew Arnold:

> Arnold always gives you the notion that he hates to touch the dirt—the dirt is so dirty! But everything comes out of the dirt—everything: everything comes out of the people, the everyday people, the people as you find them and leave them: not university people, not F.F.V. people: people, people, just people!

Yet Whitman's ideal of culture was otherwise basically the same as Arnold's, and by this standard he pronounced an even harsher verdict on an America in process of becoming the greatest industrial power on earth. In *Democratic Vistas* he wrote: "I say that our New World democracy, however great a success in uplifting the masses out of their sloughs, in materialistic development, products, and in a certain highly-deceptive superficial popular intellectuality, is, so far, an almost complete failure in its social aspects, and in really grand religious, moral, literary, and esthetic results." The most apparent reason for the failure was the devotion to business, "this all-devouring modern word." In business the sole object was pecuniary gain by any means, however depraved. All the talk about freedom only emphasized that "we live in an atmosphere of hypocrisy throughout." And since what mattered most was "people, people, just people," he asked: What kind of people were there in this brave new America?

> Are there, indeed, *men* worthy the name? Are there perfect women, to match the generous material luxuriance? . . . Is there a pervading atmosphere of beautiful manners? Are there crops of fine youths, and majestic old persons? Are there arts worthy of freedom and a rich people? Is there a great moral and religious civilization—the only justification of a great material one?

Whitman wrote this during the Gilded Age, when "the heroes of our material growth" were running wild in an economic individualism quite foreign to his ideal of "perfect individualism"; so as an inveterate, sometimes absent-minded optimist he concluded that "the fruition of democracy, on aught like a grand scale, resides altogether in the future." A century later, when the leaders of a nation infinitely mightier and wealthier than he knew began talking about the "Great Society," all who are concerned about human values might agree that Whitman was at least asking the right kind of questions.

5

The "Neotechnic" Phase

Toward the end of the nineteenth century the Industrial Revolution entered a new phase, which in his pioneering *Technics and Civilization* Lewis Mumford called "neotechnic" to distinguish it from the long initial phase he called "paleotechnic" and regarded as on the whole a "disastrous interlude." Men now began mastering their new technology. They not only developed much more rapidly all kinds of practical possibilities but realized more of the ideal possibilities of the machine, feeling more at home with it. They exercised more social control over their technology, in the interests of long neglected human values. Mumford added, however, that barbarous paleotechnic ideals still largely dominated industry and politics, and in a review of his book written twenty-five years later he played down the successes of the neotechnic phase. In my somewhat broader treatment of modern technology I should stress still more both the persistence of old problems and the emergence of new ones. On all counts we may get a perspective on the further gains made in our own time, and also on the much worse problems confronting us.

It is unnecessary to give a play-by-play account of the technological changes down to World War I. Suffice it that industrial growth speeded up at a phenomenal rate. A host of inventions— the dynamo, gas engine, telephone, electric light, linotype, artificial silk, phonograph, moving picture, automobile, and airplane, to name but a few—made the modern world much more remote from eighteenth-century Europe than this had been from ancient Rome. Hundreds of new industries sprang up, making

practically everything by machine. Engineers and other technicians multiplied, supplementing factory hands. Giant corporations
initiated the reign of big business. Railroads linked all the
industrial cities with major ports, while fleets of steamers took
over from the sailing ships; rapid transport created a world market, opening up vast areas hitherto inaccessible and unproductive.
Rapid communication, as by wireless telegraphy, made it still
plainer that Westerners had laid the material foundations of One
World.

About the countless inventions the most important thing is
that invention now became an organized, systematic process, in
effect virtually automatic. It was symbolized by the fame of
Thomas Edison, who made invention his profession, hired a staff
to assist him in his laboratories, and in fifty years took out more
than a thousand patents. When he decided to invent an electric
light, he characteristically told the world that he expected to
finish the job in six weeks, and then found that no "flash of genius"
would do; he had to test 1,600 materials to find a suitable incandescent element. But that the job took him more than a year was a
slight matter—he did come through with his incandescent bulb, to
little surprise. Although men were still dazzled by the new wonders, they were already beginning to take them as a matter of
course, ceasing to wonder at all but the weirdest. Thus nobody
had anticipated the discovery of X-rays, but after it men predicted
most of the wonders to come. In books for youngsters I read as a
boy I was prepared for television.

As for all the new industries, a major one was steel. Much
more durable and adaptable than iron, steel ended its long reign
and gave a new name to this period—the Age of Steel. Potentially
handsomer too, it heralded a cleaner, brighter technology, featuring new alloys and lighter metals like aluminum. More important,
however, was the electric industry, signifying the conquest of a
new source of energy, inaugurating a speedier technology as well,
and eventually bringing the machine into the home. Replacing
steam as the prime-mover, electricity made the dynamo a symbol
of the new age, and then the powerhouse—a means of not only
generating immense power but transmitting it over wide areas.
Above all, the dynamo signaled the momentous alliance between

science and technology. The discovery on which it was based had been made by Michael Faraday in 1831, some fifty years before the electric industry got started; like the pioneers of the Industrial Revolution, inventors had not built directly on scientific discoveries. But from this time on they applied these discoveries ever more promptly. Scientists in turn were greatly aided by developments in technology, such as the vacuum techniques explored in mastering the electric bulb; they opened up the new field of electronics, leading immediately to the discovery of X-rays. While it remains necessary to distinguish pure science from technology, or applied science, in the neotechnic phase their alliance was permanently cemented.

Germany took the lead in both the electric and the chemical industries because it learned this lesson more quickly than Britain. While providing for much more scientific research in its universities, it systematically applied scientific knowledge in its industry. (In 1899 its chemical industry demonstrated the commercial value of research by coming out with aspirin, the first purely synthetic drug, which at once became popular all over the world.) American businessmen were in this respect as backward as the British, chiefly because of their unenlightened "practical" spirit. They ridiculed Andrew Carnegie for his extravagance when he ventured to import a "learned German," a chemist. Edison himself was scornful of pure science, lamenting that his mathematically-minded son seemed disposed to "go flying off into the clouds with that fellow Einstein." In these clouds physics was being revolutionized, for no practical purposes whatever; but out of theoretical nuclear physics would come America's atomic bomb— a gift of European scientists.

America was the leader, however, in the big new industry of food-processing, which had some genial implications. Dr. Ure had confessed to an apparent neglect of the sovereign principle of utility, that "not many automatic inventions have been applied" to the production of food, but Americans began remedying this with the invention of the mechanical reaper. While the small towns of Minneapolis and St. Paul grew up as the flour-mill center of the world, refrigerator cars helped Chicago to become its slaughter-house and contraptions like the corn-cutter and the pea-viner spurred the growth of the food-canning industry. Although it took food-canners a generation to make use of Pasteur's findings about bacteria, by the end of the century they were employing scientific methods of preserving food from contamination. The homely tin can took its place on the pantry shelf—a revolution ⌐ sorts for housewives. For all people it climaxed another significant change that today is hard to realize. The clergyman Thomas Malthus had won fame by his *Essay on Population*, which doomed many of the poor to starvation by its thesis that the food supply could not keep up with the increasing population, but in fact an improved technology had made it possible to feed a vastly greater number of people. In western Europe and America it had eliminated the age-old danger of famine, from which a million people had died in Ireland as late as 1848.

Somewhat more dubious was another development led by America, the organizational revolution that created big business. Among the early giants were John D. Rockefeller's Standard Oil Company and Andrew Carnegie's steel empire, which owned mines, railroads, and fleets of ore boats. When Carnegie sold it to J. P. Morgan, the financier merged it with other big companies to form U.S. Steel (1901), the first billion-dollar corporation. It inaugurated an era in which huge corporations would dominate all the major industries. Americans were of two minds about these giants—awed by their size and wealth, fearful of their great power. Conservatives who welcomed them would remain fearful of a related growth of organization, big government. An inescapable consequence of the increasingly complex society created by industrialism, it would be feared by Americans as "bureaucracy," the chief threat to their freedom, even though big organizations in business were just as bureaucratic, and rather less concerned about the freedom of the common people. Max Weber feared bureaucracy as an extension of "alienation," a further separation of workers from both the means of production and the products.

Confusion deepened in the twentieth century, when the United States took over the industrial leadership of the world. Most businessmen have yet to learn another lesson taught by Germany. They have always believed that the triumph of industrialism was due solely to free private enterprise, which alone could provide the necessary drive and efficiency: conveniently they forgot the lavish aid they got from government by high protective tariffs, grants of public land, subsidies, franchises, etc. Germany also encouraged its businessmen by protective tariffs, but under Bismarck it made itself the greatest industrial power on the Continent through state planning and control, including ownership and management of the railroads, telephones, and telegraph. Japan taught the same lesson by its extraordinary feat in the last decades of the nineteenth century. The first non-Western country to modernize, and still by far the most successful, it made itself a world power in a generation by an industrialization largely planned, financed, directed, and controlled by the government.

Everywhere, however, people were enjoying the services provided by government, beginning with municipal government. In 1842 New York had completed a system of reservoirs and aqueducts that made it the first large city of the day to provide an ample supply of pure water (thereby catching up with ancient Rome). In England the filthy industrial towns began to bend technology to such uses as drainage, sewage, and garbage disposal. Epidemics spreading out of their slums, killing upper-class people too, frightened government into a belated concern for the public health, helping to make this age of free private enterprise the age of "municipal socialism" as well. By the end of the century civil engineers were at last winning the struggle to make cities fairly sanitary.

European countries that had been slow to industrialize could at least profit from Britain's long experience of the abuses of unregulated private enterprise. Germany, where government was most efficient, now set an example in social legislation. The aristocratic Otto von Bismarck gave up his early policy of simply suppressing socialists, instead countering their menace by putting through much legislation on behalf of workers, such as a system of insurance to protect them against accident, sickness, and old age; conservative to the marrow, he nevertheless made Germany the first welfare state. Britain made slow but steady progress toward this end, culminating in a flood of welfare legislation in the years before the World War. Even backward America made some effort to catch up. It would not begin to match Bismarck's Germany until Franklin Roosevelt's New Deal, or ever catch up with the European democracies in social legislation, but in the early years of our century, despite business opposition, the Progressive movement succeeded in putting through many social reforms, including the regulation of child and women labor, accident compensation for workers, factory and tenement regulation, and a Pure Food and Drug Act. Industrialism in America was at last being forced to consider the public interest ignored by the self-regulating market.

The rapid growth of cities signified other forms of emancipation for the common people, first of all an unprecedented degree of social mobility. The city was attracting men from the countryside, where throughout history the son had usually followed the same occupation as the father, and it was creating many new occupations, more opportunities for the ambitious to move up the social scale. But industrialism was a particular boon to women and children, its early victims. Whereas about the only positions open to women had been as servants or governesses, the cities found increasing use for them in not only factories but department stores, telephone companies, schools, and hospitals. Then the invention of the typewriter opened up the business world to large numbers of them as stenographers and secretaries. In view of their growing independence it was no accident that the movement for the emancipation of women got under way in the industrial countries. With full legal equality this included the demand for suffrage (a movement that in Britain was founded in Manchester) —something hitherto unheard of in the whole history of "man." All along making up half of mankind, women had always been treated as inferiors, no less when out of chivalry they were called the better half.

As for children, indignation over their ordeals in factories and mines was as novel as tardy, for the children of the poor had always been worked hard, often been treated callously. (Even the utopian Charles Fourier delegated all the dirty work in his ideal society to small boys, on the assumption that they had a natural fondness for dirt.) It was only with the spread of free public education in the industrial countries that the idea began taking hold that youngsters were entitled to a childhood of school and play, freedom from adult responsibilities. Then the Society for the Prevention of Cruelty to Children was founded in England—a seemingly strange cause for a civilized nation that pointed to another anomaly. The Age of Materialism produced far more humanitarian legislation than had any previous age, or the most religious societies, in part just because of the materialism. It spread the idea that poverty and physical suffering were evils, not really good for the soul.

Lewis Mumford made most of the assimilation and perfection of the machine in the neotechnic phase, the development of an "exquisite" technology symbolized by the new motors and turbines. The early engineers had merely stuck some nondescript ornament on their crude machines and machine-products—posies, arabesque, Gothic tracery; now men began to develop the new art of industrial design. They grew aware of "the esthetic excellence of the machine forms, the delicate logic of materials," and the principle of form following function. "Expression through the machine," Mumford concluded, "implies the recognition of relatively new esthetic terms: precision, calculation, flawlessness, simplicity, economy"—above all the principle of economy, which aristocratic societies had been inclined to disdain. Builders in particular began realizing their new opportunities. Earlier in the century men had built the most beautiful sailing ship in history, the clipper, and they soon achieved a remarkable perfection in their big steamships, the splendid liners that William Morris called the Cathedrals of the Industrial Age. Engineers learned to do even more superb work in bridges. For centuries men had been building beautiful bridges, but never one so awe-inspiring as Brooklyn Bridge—a "symphony in steel." Likewise they made use of their new materials to develop a distinctive architecture suited to the Industrial Age. Steel, with the skills of the engineer, made possible the skyscraper, a thrilling symbol of the wealth and power man had achieved.

It was a temple to business, however, not at all a civic center like the Greek temples and the medieval cathedrals; and as a symbol of the popular worship of quantity, size, wealth, and power it brings up the seamy side of the industrial progress. Industry continued producing mainly wares undistinguished in design, showy when not shoddy. As it catered to the tastes of the Victorian Age and the Gilded Age, about the worst in the history of style, so it would continue to gratify the tastes of status-seekers who learned to ridicule their ancestors. Its immediate end remained private profit, not public good. John Ruskin reminded an audience of manufacturers that for the sake of profit they had not merely supplied but formed the market, and he bluntly pronounced a verdict that it still pertinent today:

> Every preference you have won by gaudiness must have been based on the purchaser's vanity; every demand you have created by novelty has fostered in the consumer a habit of discontent; and when you retire into inactive life, you may, as a subject of consolation for your declining years, reflect that precisely according to the extent of your past operations, your life has been successful in retarding the arts, tarnishing the virtues, and confusing the manners of your country.

Other social evils remained more conspicuous, though too rarely apparent to the apostles of private profit. With the rapid growth of the cities slums grew as fast; they were very profitable to landlords, who in New York made up to 40 per cent a year on them. Toward the end of the century Charles Booth, himself a manufacturer, published his monumental first-hand study *The Life and Labour of the People of London*, in which he exposed the shocking extent of poverty and misery in the world's greatest city, and with it exposed the belief popular in his own business class that they were due primarily to shiftlessness. Workers remained the chief victims of the periodic depressions, which were becoming worldwide because of the increasing interdependence of nations. Industrial strife grew bitter, often violent, especially in America. It dramatized the abiding paradox of modern technology: a rational, efficient organization of means that could perform wonders, yet failed to take care of some elementary human needs, found no room for most of the human values that make life worthwhile—unless or until it was discovered that they "paid."

So let us pause for a hard look at Andrew Carnegie, a symbolic Janus-figure in this transitional neotechnic phase. One of the "heroes of our material growth," he made a fortune by recognizing early that the future belonged to "King Steel," and by succeeding in reducing its cost to less than a penny a pound. A poor boy who made good—a hero of the favorite American legend—he was a popular preacher of the "gospel of wealth," and of the free private enterprise that had produced the wealth the country was so proud of; in exalting its individualism he spelled out "the law of competition" that decreed "the survival of the fittest." At the same time, he was far in advance of J. P. Morgan, the top financier, who years later would say: "I owe the public nothing." Carnegie anticipated the objection that self-made men relieve God of an awful responsibility by insisting on their social responsibility, a moral principle of trusteeship: the man of wealth was obligated to repay the society that had given him his opportunities by devoting his proved abilities to organized philanthropy on behalf of his poorer brethren. He proved his sincerity by a radical proposal of steep inheritance taxes on big fortunes left by men who had neglected their obligations of trusteeship. He himself gave away millions, in particular for public libraries all over the land; he anticipated the Rockefeller and Ford Foundations, which would give away hundreds of millions—organized philanthropy on an unparalleled scale. And having made a fortune out of steel, including countless tons of it for guns and battleships, he ended as a crusader for international pacificism, endowing the Hague Tribunal with an elegant building. He was even called "Saint Andrew."

Yet he exemplified the extreme inequality that was increasing under his gospel of rugged individualism: while hundreds of men were becoming millionaires, millions of poor boys never did make good. When he himself was making up to $20,000,000 a year, on which he paid no income tax whatever, his steel workers were getting about $500 a year for a twelve-hour day, seven-day week, with no social security whatever. He believed, as the leaders of the steel industry would into the 1920's, that efficiency and economy required such long hours; human needs must always be subordinated to economic needs. In other respects too Carnegie was something less than saintly in his dealings with his workers. Although pleased to call himself "the staunch friend of labor," and to declare that its right to form unions was as "sacred" as that of manufacturers to form associations, he went off grouse-hunting in Scotland during the bloody Homestead strike in which his hired gunmen killed ten of his workers and wounded scores of others; after the strikers were starved into submission, no unions were permitted in Carnegie plants. Even in insisting on his principle of trusteeship he warned against ordinary Christian charity, declaring that very few worthy people needed such help; or as Finley Peter Dunne's Mr. Dooley put it, he prided himself on never doing anybody any particular good—the worst thing you could do for a man was to do him good. And though a patron of culture, he had a pretty low conception of its values. He boasted that the gospel of wealth had greatly raised the level of not only material prosperity but culture in America, which was already superior to the culture of England; one proof was that President Ulysses Grant was paid $250,000 for his Memoirs, whereas John Milton got only 10 pounds for *Paradise Lost*. Mr. Dooley also commented that in endowing public libraries he gave only buildings with his name on them, not books, and offered no help to the people who wrote books.

Above all, Carnegie typified the business leaders of America in his dim understanding of not only the society they were leading but its economy. The big corporations in which he pioneered were hardly devoted to his "law of competition." They naturally sought rather to eliminate competition, as Rockefeller did most ruthlessly in building up Standard Oil; their ideal was monopoly, their common means were trusts or cartels, and another common aid was government favors won by bribing legislators. (It was remarked that Standard Oil did everything with the legislature of Pennsylvania except refine it.) Rockefeller had more explicitly repudiated another of Carnegie's slogans when he explained that the day of combination had come to stay, adding, "Individualism is gone, never to return." There was still plenty of room at the top for rugged individualists like himself, to be sure, but for ordinary Americans the term was indeed losing its old meaning. Whereas in the days of the Founding Fathers the great majority of them were self-employed, their own masters, now most of them were underlings in factory, office, or store, not at all self-sufficient. Carnegie had nothing helpful to say either about the worsening depressions in booming America, which for workers meant lay-offs or chronic unemployment—a new technical term that entered the language in this period. (Economists had talked of "over-population.") And though a humane man in some ways, he remained indifferent to the plight of his workers because of another novel feature of the economy, the massive impersonality exemplified by the corporation. He could go off grouse-hunting while they were shot and beaten into submission because he knew none of them, had no personal relations with them, felt no such responsibility for them as he would for an old family servant. Although corporations were legally regarded as "persons," endowed with all the rights of private property, they accordingly earned a reputation for being as "soulless" as their machinery.

Another strange aspect of the new technology was illustrated by the symbolic dynamo. The steam engine and other early machines were relatively comprehensible, or at least not utterly remote from the experience of the common man; he could see the pistons, hammers, pulleys, and wheels at work, the familiar motions of pushing, pulling, grinding, turning. By contrast the power of the dynamo was invisible, due to a mysterious something called a current of electricity. Men could no more see or really understand this current when they turned a switch that somehow made their electric light go on. They had little clearer idea, if any, about the workings of the private utility company that provided the electricity, or of the government body that supposedly regulated the company. Although much better informed than the illiterate peasants of the past, they knew much less about the world they lived in. They were completely dependent on an economy that kept breaking down periodically for economic reasons that were incomprehensible to them, and that if understood at all by business leaders were evidently beyond their power to prevent or control.

Less strange, but as troublesome, was another growth in this period, the metropolis. Whereas at the beginning of the nineteenth century only London housed a million people, followed by a few scattered cities of a hundred thousand or so, none in America, by the end of the century London had a population of 6½ million and big, fast-growing cities were commonplace, especially in America. Only industrialism could house, feed, and transport to work such dense masses of people. And only industrialism could create so much ugliness, squalor, and congestion. In the most urban of civilizations, the metropolis was a prime exhibit of the worst in modern technology. In America this would make a multi-billion dollar job of the "urban renewal" that cities at last got around to in our day, too tardily to undo most of the damage.

Here the private enterprise that created slums pointed to another anomaly, that little of the technical skills applied to industry was applied to planning the growth of cities. Although in Paris Baron Hausmann had shown what city planning could do, the growth of the modern city generally remained haphazard, for civic purposes often irrational and inefficient; except for monotonous gridiron streets and the segregation of slums, the big American city was a mere sprawl. The main reason was the sovereignty of business. Whereas the heart of old European cities was a civic center, the heart of the American city was simply a business district. The skyscraper that was beginning to dominate it was a center only of a congestion that had nothing to do with great civic occasions. Like the elevators that made possible tall buildings, the trolleys, subways, and elevated trains that brought crowds into the business districts were steadily increasing the congestion even before the advent of the automobile and the truck. Together they would make the city the noisiest in history.

Hence still another anomaly of industrialism. City people ostensibly had more leisure time as the work-week in this period was reduced from an average of seventy hours to fifty or less; yet life was growing less leisurely. People had to spend more time traveling to and from their job, in the discomforts of what came to be called the "rush hour." In the bustling cities the streets were too crowded, or real estate too expensive, to permit the waste of time and space on amenities like the sidewalk cafe. The leisurely noon-day meal was giving way to the quick lunch, the European and Latin-American siesta was unknown. The new habit of "making" or "saving" time was becoming a kind of compulsion, which might make people forget what they were saving it for. New York in particular became notorious for the pace and tension of its life, at the expense of the art of living. It gave birth to another characteristic new technical term—"neurasthenia."

So we are brought to the critical question, the mentality of the man on the modern street. Again I would first qualify the familiar generalizations about the city masses invited by those hurrying Americans—all anonymous, lost in the crowd, looking and behaving much alike. Actually they were a remarkably diverse lot, headed for many different destinations for different reasons; it took many kinds of people to make a modern metropolis, the most heterogeneous city in history. Obviously less respectful of tradition and authority than rural people, they were by the same token more flexible and open to change, more tolerant of unconventional belief and behavior, in this respect better suited—Thomas Jefferson notwithstanding—to the purposes of a free, open society. The very loss of community so often deplored in the modern city also meant that people might be freer to live their own lives without worrying about what the neighbors would think, as they always had to on Main Street.

Nevertheless the life of city people was in other respects plainly more regimented. They had to work regular hours the year round at standardized jobs in factory, shop, or office. With the appearance in Switzerland of the cheap watch, which soon spread to America and sold by the millions, mechanical time governed their whole day. Americans became the most time-conscious of peoples, and the least conscious of time as natural rhythm. The kind of tyranny of modern time was dramatized at the beginning of every workday by another invention, appropriately named the "alarm" clock. When it went off at a set hour, people were likely to grumble, but not to realize that throughout history men had somehow managed to wake up and get to work without it, and also to adjust their sleeping habits to the seasons and to their feelings. Those of us who consider regular hours and working habits a possible aid to the good life enjoy them most when they are freely chosen, but few workers in an industrial society had a real choice.

Such mechanical ways gave more literal meaning to "the masses" in the big cities. People were more densely massed in uniform houses or tenements on gridiron streets, and also in factories and office buildings, on transport to work, and in stadiums or on beaches for recreation. With the rise of the mass media they were growing more dependent on the machine for standardized mass entertainment too. The new industry of advertising was cultivating the art of mass appeal to sell standard brands, drilling people to want what everybody else supposedly wanted. In the mechanical routines that had replaced a rooted way of life they were more susceptible to waves of mass emotion. Ominous appeals were exploited by another new power in the land, the sensational jingo press. Before the World War newspapers selling by hundreds of thousands were demonstrating their power to stir up mass emotion verging on hysteria. In America they were largely responsible for the war with Spain over Cuba.

Mass production in the neotechnic phase likewise developed new techniques that had psychological consequences. In America Frederick W. Taylor introduced the principles of "scientific management." He demonstrated how costs could be greatly reduced by completely rationalizing and standardizing factory work, breaking every job down into its component operations and systematically timing each; these scientific methods enabled workers to turn out much more work with little more effort. Henry Ford then developed an assembly line on which some five thousand interchangeable parts were put together in a standardized continuous operation. The industrial wizard was thereby able to keep lowering the price of his famous Model T car until he had cut it in half, and so to create a mass market for it. In an article on "Mass Production" in the *Encyclopedia Britannica* he pointed out that its focus on "the principles of power, accuracy, economy, system, continuity, speed, and repetition" had made possible "the highest standard of quality ever attained in output of great quantities." As for its effects on workers, he remarked that these had been "variously appraised," but he emphasized the elimination of wasteful back-breaking toil: "The physical load is shifted off men and placed on machines. The recurrent mental load is shifted from men in production to men in designing."

Neither of these pioneers was bent on exploiting workers. Taylor called for a "mental revolution" in both management and labor, who should stop fighting over their shares of the profits and cooperate in scientific methods that would reward both; Ford pioneered as well in paying the highest wages in industry, having the wit to see that well-paid workers were necessary to sustain a mass market. Yet both succeeded in further dehumanizing factory work by thinking only of technological efficiency. Taylor granted as much in passing: "One of the very first requirements for a man who is fit to handle pig iron as a regular occupation is that he shall be so stupid and so phlegmatic that he more nearly resembles an ox than any other type." Chiefly he prided himself on having done away with the "old individual judgment" of workmen and foremen. Ford's assembly line completely eliminated the personal touch, the vestiges of craftsmanship that might give personal satisfaction. Later on workers on automobile assembly lines would remain among the best-paid in industry, and among the most discontented over the mechanical repetitiveness of their work, the absence of the least "mental load." As one of them put it, "You may improve after you've done a thing over and over again, but you never reach a point where you can stand back and say, 'Boy, I done that one good. That's one car that got built right.' If I could do my best I'd get some satisfaction out of working, but I can't do as good work as I know I can do."

Meanwhile millions of Americans rejoiced in their Model T Fords, little aware of how profoundly the automobile would transform American life and all the problems it would create. They were proud too of their big navy, second only to Britain's, as little aware of how grave were the dangers of the trend it represented— the growing nationalism and militarism that would soon bring on the catastrophe of World War I. For in the neotechnic phase technology was being applied with particular efficiency to the very old purpose that had always stimulated it, the inhuman—or all too human—purposes of war. "The army," wrote Lewis Mumford, "is in fact the ideal form toward which a purely mechanical system of industry must tend." Now Prussia became the first country to realize the ideal.

In the Franco-Prussian War of 1870 its armies, not only well-armed but most efficiently organized and disciplined, crushed with absurd ease the once greater power of France.[1] Bismarck had introduced compulsory military service for all able-bodied young men, making Prussia the first nation to impose such a duty in

---

1. Americans still bemused by the glamor of the Civil War might note that the Prussian army was a century ahead of the American armies that only a few years before had made our war the bloodiest of the century.

peacetime, and in beating down parliamentary opposition he made his famous pronouncement: "The great questions of the day will not be decided by speeches and the resolutions of majorities . . . but by blood and iron." After his smashing victory all the nations of the Continent got the point and began conscripting their young men. Britain and the United States, the only great powers that did not compel military service, could get by because they specialized in big navies, but they joined in another innovation of the neotechnic phase—the arms race. Any nation aspiring to be a real power felt obliged to build a military machine. Industrialism, which made the machines both mightier and more expensive, produced another great pioneer in Alfred Krupp, the self-styled "passionate gun-maker" and founder of a munitions empire.

The arms race accentuated an apparent paradox. Even apart from a growing number of pacifists, Westerners were beginning to believe that peace was the normal condition of a civilized society, war was natural only to backward little countries like the Balkans. Similarly they were making unprecedented efforts at international cooperation, in keeping with their growing economic interdependence, as in the Universal Telegraph and Postal Unions, uniform patent and copyright laws, World Fairs, the Red Cross, and the multiplication of international congresses of all kinds. Yet through public education, rapid communication, and the mass media technology had also helped to make nationalism the ruling religion of the Western world, far more vital than Christianity; so most people were proud of their big armies and navies. In Europe the military bureaucracy did best in promoting its own interests, and because of the arms race it had more influence than ever before in the shaping of foreign policy. Before the World War the German army, which was not under civilian control, had the most elaborate, carefully prepared plans—plans such as no generals could have drawn up before the advent of modern technology, since they called for armies of a million or so men, fully equipped, to swing into action almost overnight. And Germans felt aggrieved because they had not got their full share out of a related movement that began about 1870, the "new imperialism."

Imperialism itself was of course almost as old as recorded history. What began now, however, was a wild scramble for colonies that made imperial history on a scale and at a tempo never approached in the past. Within a generation the major European nations took over a third of the world, including virtually the entire continent of Africa. The million-odd square miles picked up by Germany was a small cut by the latest standards, far less than the colonial booty acquired by Britain. As for the reasons for the new imperialism, Marxists attribute it to industrial capitalism, in particular its need of raw materials, new markets, and new sources of investment. Other historians argue that more important was the surge of nationalism; except for some businessmen colonies were not really profitable possessions, but the greater powers had to have them as status symbols, proof of the national greatness. The military in all countries also welcomed them as naval bases. The upshot, at any rate, was indisputable. A superior technology now enabled the West to dominate the entire non-Western world except Japan, which had completely adopted this technology.[2] And having divided up the rest of the world, the European powers soon settled down to resolving their own differences in a war that modern technology made a World War, the most murderous in history to that time. The "progress" recorded in the rest of this chapter is a grim story.

Here I need not go into the knotty question of the immediate causes of the war and the measure of war guilt. By general consent of historians the major powers on both sides were to some extent guilty. The war began as a pure power struggle, emphasizing the lag in statesmanship behind the tremendous gains men had made in power. All the major combatants had joined in the arms race, had big armies or navies all set to go, and overrated their power. After Germany won sweeping initial victories by its superior military technique, the war settled down to a murderous stalemate on the Western front because all had underrated their power too

---

2. Thus General Kitchener had routed an army of 40,000 fierce dervishes in the Sudan, killing 10,000 of them at a cost of only 48 lives in his own smaller force. The British had artillery and Maxim machine-guns.

for a long war; none had prepared for such a war. The stalemate made plainer the novel aspects of the worst war to date, all due to modern technology.

It was the first all-out war, in which civilians were enlisted in the mobilization of the industrial resources of the major powers. An immense war effort, at home as well as at the front, enabled both sides to endure terrific losses in men and materials during the protracted stalemate. Germany was almost able to defeat most of the world virtually single-handed because of its superior organization of power, both economic and military. On the front, trenches and barbed wire eliminated the glamor of war in the past, when cavalrymen and soldiers in dashing uniforms had stormed hills carrying flags. (The French army had gone into the war wearing its traditional red trousers.) Warfare was mechanized by more than machine-guns and powerful artillery; submarines, airplanes, and tanks made their debut, together with poison gases. Hence the war was fought not only on land and sea but in the air and under water. The torpedoing of merchant ships and the bombing and shelling of cities signaled the end of the old distinction between soldiers and civilians. Although the citizenry on both sides had entered the war with patriotic fervor, inflamed by the mass press, governments soon began developing systematic propaganda in order to win over neutral nations, bolster the morale of their people, and weaken the morale of the enemy. By intensifying hatred of the enemy the propaganda helped to make impossible the more or less honorable peace with which nations had usually sought to end their wars in the pre-industrial past, while in Europe the immense effort and wholesale slaughter over four years left the victors exhausted too. World War I should have made it clear that there could no longer be real victors in an all-out modern war.

Needless to add, it failed to do so. The victors did make one serious effort to realize the ideal potentialities of modern technology by setting up a League of Nations, a world order such as all previous civilizations had lacked the material means of establishing, but its chances of success were fatally weakened when the United States backed out. Having emerged from the war as the greatest power in the world, it reverted to its traditional isolationism, which technology had made as futile as irresponsible. Otherwise the principal novelty on the political scene was the rise of totalitarianism, first in Russia, then in Mussolini's Italy; technology had also made it possible for dictators to exercise a more complete control over society, and over the minds of men, than absolute monarchs could ever hope to achieve in the past. The rest of Europe, like America, reverted to business as usual, or what President Harding called "normalcy." After ten years the Great Depression—the worst in history—demonstrated luridly that a capitalist economy was not yet under human control. It was worst of all in America, where businessmen had been given the freest rein and unstinted government aid under Harding, Coolidge, and Hoover, and where they had been most deluded by a big boom on the stock market, a fantastic inflation of paper values. "Normalcy" was ended by Roosevelt's New Deal, in which the government at last assumed some responsibility for the state of the economy and the plight of the many millions of unemployed, but the economy did not fully recover until America got into production for World War II.

Another consequence of the Great Depression, which threw millions of Germans out of jobs, was the rapid rise to power of Hitler, hitherto noisy but obscure. His triumph was facilitated and then consolidated by the most systematic, extensive, brilliant use of the art of propaganda developed during World War I, and now amplified by the invention of radio. Hitler profited as well from the brilliant feat of Germany in organizing its economy for that war; he proceeded to build a still mightier military machine. Its awesome power was quickly proved when he started World War II by invading Poland.

About this war the main point for my present purposes is that it was still more murderous, devastating, and barbarous than the first one, even apart from the technological feat of Hitler's systematic extermination of six million Jews. The machine came fully into its own, outmoding trench warfare too. Civilians made up almost half the casualties, primarily because of the mass bombing of cities. After some boasting about its "pin-point" bombing, America took the lead in this indiscriminate slaughter by massive fire-raids on German and Japanese cities; an Air Force general has told proudly how one such raid on Tokyo produced "more casualties than in any other military action in the history of the world."[3] The Germans pioneered in another awesome achievement by firing rockets at London from across the Channel, but America climaxed the triumph of technology by the nuclear bombs it dropped on Hiroshima and Nagasaki. These shook some men to their senses, even though the fire-raids on Japan had killed far more civilians; scientists were especially frightened by the appalling power they had given the military. When the war ended, however, the Cold War got under way, and soon America and the Soviets were engaged in a nuclear arms race—the last word in human folly, or conceivably in the history of the human race that began with ape-like creatures chipping flints.

---

3. His pride might be humbled, however, by the knowledge that the Mongol invaders of the Near East and Europe, under Genghiz Khan, probably slaughtered a larger proportion of the population than were killed in the World War. Their weapons were bows and arrows.

6

"Post-Industrial" Society

Having at last reached our own time, the generation since World War II, I feel obliged to begin by recording more history. Much of it will already seem like ancient history to young people under twenty-five, who now make up almost half the population of America. They must find it hard to realize what occurred to me only a few years ago, when I noted that in this one generation we have lived through more radical, sweeping, startling changes than past civilizations had experienced in a thousand years. These changes incidentally included a host of new technical marvels—radar, television, wonder drugs, transistors, computers, jets, sputniks, intercontinental missiles, etc.—most of which did not startle me when I first saw or heard of them; I suppose young people too soon learned to take them as a matter of course. We have all grown used to the momentous events on the world stage, such as the dawn of the Atomic Age, the emergence of the Soviet as a world power rivaling America, the Cold War, the conquest of outer space, and the revolt and rise of the whole non-Western world, the dozens of new nations in the United Nations. It is already unnecessary to use quotation marks with once sensational phrases like the population explosion and the knowledge explosion. In our society exploding appears to be a normal mode of expansion.

All these changes were due to technology. In alliance with science, it is now unmistakably the basic determinant of our history. In this chapter I am attempting only a sketchy survey of some major developments and the kinds of issues they have raised, to be considered in more detail in Part III. My immediate concern is the emergence of a new kind of society in America, variously known as the "affluent," "post-industrial," and "technocratic" society. It might also be described as a technological society come of age, except that within a generation it will most likely look immature.[1]

In the economy the most conspicuous changes were at first more of the same but to an explosive degree involving some differences in kind, all of which were obscured by the old slogans that passed for thought in conservative business circles. After the World War, during which America under government controls doubled its national output within a few years, there was much alarm over the "creeping socialism" that was menacing free private enterprise. Actually big business was making record profits, and corporations were growing into giants so rapidly that businessmen might have been edified by Adolf Berle's description of the American economy as "galloping capitalism." It was hard to consider General Motors a grievously harried organization when its annual profits were greater than the national income of all but a few countries in the world. The slogans about the sovereign virtue of competition also obscured not only the usual collusions to avoid competition but the plain truth that the giant corporations dominating the major industries were no longer competing in prices; they always raised their prices together. They paid no more heed to the alleged law of supply and demand. During one recession brought on by overproduction of automobiles, which had glutted the market, the Big Three alike raised their prices instead of lowering them to stimulate demand. By now the corporations are planning more systematically both supply and demand—one feature of what Kenneth Galbraith has called the "new industrial state."

---

1. As I am largely confining myself to America from now on, I wish to warn readers against parochial conceit. It is generally agreed that during this period America became the world leader in technology, at least in application, but I gather that basic innovations in the Soviet, Britain, Germany, and Japan may be no less important. American leadership may also be exaggerated because American specialists have been doing most of the work in the history of technology. In any case we have hardly distinguished ourselves by the uses we have made of our wealth and power.

Demand they plan by creating through advertising campaigns the market that supposedly determines their production. The giants can make whopping mistakes that would ruin ordinary corporations (as Ford did when it squandered a quarter of a billion dollars on its Edsel), but in general they can depend on their admen, specialists in psychological technology. An industry that had been growing steadily since the last century, advertising developed superior techniques after World War I. "The war taught us the power of propaganda," observed Roger Babson, the popular forecaster or prophet of business; so he could boast that business had learned how to sell anything to the American people. (The Nazis in turn acknowledged their debt to American advertising techniques.) Coming out of World War II as a $3 billion industry, advertising then started growing much faster than the rest of the economy as it seized on the new medium of television, ideal for reaching a mass market. By 1965 it was doing a $15 billion business and still growing.

While it helped immeasurably to maintain industrial growth in America by creating new needs, a market for all the latest models and gadgets, advertising has contributed as well to a more novel kind of growth, a "service economy." Comprising stores, repair shops, banks, insurance companies, real estate agencies, government bureaus, and all kinds of personal and professional services—an immensely wider range of services than in the past—this has outstripped industry and now accounts for more than half the gross national product. (GNP was formally christened in 1946, at the beginning of our generation.) For the first time in history, Daniel Bell points out, more than half the people working are not engaged in the production of food, clothing, housing, and all the other tangible goods. And most significant, he adds, has been the rapid increase in the ranks of professional and technical people. There are now some ten millions of them—more than there are farmers, once the bulk of the population. In particular our post-industrial society has been breeding scientists and engineers so fast that by 1975 we may have twice as many of them as we had in 1960. Many are employed in research laboratories, which have become standard in all industries, are devoted entirely to innovation, and have made technology a much more highly organized, systematic discipline than it was before the war.

In industry another fundamental change was the advent of automation. This at first appeared to be only an outgrowth of the assembly line, another difference in degree, but soon the word spread that it was another revolutionary development, which in time would enable a few million workers to do the factory work now done by many millions. From other quarters came much alarm over the prospects of the severe dislocation of the economy, all the millions of men thrown out of work and left with useless skills. The drive to "automate" seemed worse because it seemed as automatic as the uncouth word itself suggests, emphasizing man's bondage to technology. One enthusiast wrote characteristically that "if anything is to be learned from the past it is that we are powerless to resist change"—a lesson that may not hearten those concerned with human values. Now it appears that the alarm was excessive, and one may read mixed reports on how fast and far automation is going. But at least it has gone far enough to make clear that it is in keeping with a post-industrial society. It is more than just another set of machines, inasmuch as it involves self-regulating mechanisms and requires no human operators once the buttons have been pushed. As it reduces the factory force, long the mainstay of industrialism, so it does away with not only manual labor but all contact with raw materials. In the oil industry, among those that have gone farthest in automating, a labor union official writes: "A man may work for months on a pipeline, or in a refinery, or even in the production fields, and never see or touch oil."

Still more revolutionary is the electronic computer, whose development since it came into general use after the war has been so rapid that it amounts to another explosion. Soon there were thousands of computers in use; International Business Machines then announced a whole new "family" of advanced types, and now they are multiplying still faster, finding more and more uses, especially in "cybernation," the technical name for their marriage with automated machinery. One producer has said that the computer "has a more beneficial potential for the human race than any other invention in history," while an economist predicts more specifically that it will increase productivity so much that average family real incomes will double in each generation. But it too stirred some alarm, over an idea suggested in a *New Yorker* cartoon of a computer grinding out the message *Cogito ergo sum*: commentators stressed that it could think much better than man, at superhuman speed. The alarm grew upon the news that a computer taught to play checkers was beating its inventor. It appeared that man had now literally created a Frankenstein monster, a machine able to outwit him, make his own brainwork superfluous or obsolete. Norbert Wiener dwelt on another apparently frightening thing about these latest machines, that they were "very well able to make other machines in their own image." On both counts they seemed to be fulfilling Samuel Butler's ironic prophecy in *Erewhon*. Remarking the extraordinary progress that machines were making—so much faster than man—he asked: "May not man himself become a sort of parasite upon the machine? An affectionate machine-tickling aphid?" For such reasons the inhabitants of Erewhon ("nowhere" spelled backwards) destroyed all their machines.

Since there is not the least likelihood of our doing likewise, but still much foolish talk about the superior mentality of computers, it seems necessary to make some elementary observations. Wondrous as are the things computers have been taught to do, strictly they cannot think, but can only learn what men enable them to learn. They are far from infallible because they lack all judgment, and so can transform slight errors by their programmers into whopping errors. (Ordinary people who are now often informed that their orders have been delayed, or mistakes made, because the company has installed a new computer system may as taxpayers not be comforted by the news that at Cape Kennedy the omission of a hyphen by a programmer made it necessary to destroy a rocket costing $18,500,000.) Man remains the god that created them, only up to a point in his own image. They cannot go on to make up questions on their own, talk back independently, imagine, dream, aspire, create, or conspire to become Frankensteins. As one authority put it in the jargon now fashionable, they are "shallowly motivated." In particular I shall keep returning to a basic limitation just because computers are being used ever more extensively in business, government, and science. They can process only quantitative or factual information, not qualitative judgments or social values.[2] The real challenge is to the brains of the men who design and use the computers. These can be programmed to be just as trivial and foolish as man, or as somebody said, "Garbage in, garbage out." (Among the garbage already programmed are whodunits for television.) And the real danger is that man will abdicate his own responsibility, think only of the problems of economy and efficiency that his mechanical creature can handle so swiftly and surely, and evade the value judgments that it cannot make.

Otherwise automation or cybernation looks like a triumph over the machine to which men had been slaves. In the factory it does away with a great deal of repetitive, monotonous work, such as that on the assembly line. It is accordingly the latest phase in what one specialist has been pleased to call the "revolution in the machine-man relationship."

Under "scientific management," Peter Drucker noted, the worker was treated in effect as "a poorly designed, one-purpose machine tool; whereas actually repetition and uniformity are two qualities in which human beings are weakest." Before World War II Elton Mayo pioneered in studies of monotony and boredom, the "human problems" of industrialism, and after the war increasing attention was given to these long neglected problems in the name of "industrial sociology" or "human engineering." Henry Ford II himself recognized the oversight of his wizard father: "While we have gone a long way toward perfecting our mechanical operations, we have not successfully written into our equations whatever complex factors represent man, the human element." The term "human engineering" may set on edge the teeth of old-fashioned humanists; "engineering" gets the last word because the main objective appears to be the perfect adaptation of man to the machine. But at least it is something that these specialists are concerned with the "human element." They have published some

---

2. While they have been taught to translate language, for example, they are pretty crude at it because they cannot handle connotations or shades of meaning. When a computer was asked to translate "The spirit is willing but the flesh is weak" into Russian and then back into English, what came back was "The ghost is ready but the meat is raw."

helpful, if elementary, discoveries, for instance that the constant pace suited to machinery is not suited to workers, that workers cannot be treated efficiently in a standardized fashion, and even —at last—that the supposedly most powerful economic incentives are not enough to make them do their best. Although most executives still cling to old-fashioned ideas of efficient management, or have not caught on to the latest "revolution," at least some of the managers of a post-industrial society have begun to learn something about the nature of industrial workers. And with all this have come much cleaner, brighter, airier factories, which can be attractively located because they are not eyesores. The home of the machine is no longer a dungeon for its tenders.

Government in post-industrial society has undergone more striking changes, though for mixed reasons and to mixed effects that have been further confused by the slogans of business conservatives. As government started growing bigger far back in the industrial revolution, and expanded its activities when it began assuming some responsibility for the failures of the economy, so it expanded enormously to meet postwar needs. Together with state and local governments, the federal government now accounts for up to a fourth of all economic activity—more even than it does in "socialist" Sweden. It is up to its neck in business, which conservatives keep saying it should always stay out of.

The immediate reason was the atomic bomb. This led to the Atomic Energy Commission, which was soon managing a big business involving a combination of government and private enterprise. Likewise the Cold War led to an immense military establishment, such as America had never had in peacetime. While it has cost hundreds of billions to maintain, more billions went into the related space program, compelled by the felt need of keeping up with the Russians. Although the spokesmen of business did not protest against these programs, they continued to talk as if government were their natural enemy; but the big corporations working for it knew better. They were getting huge contracts that guaranteed their market and protected them against all risks, they were also reaping profits from research paid for by the government, and they worked hand in hand with government agencies, to their mutual benefit.

The upshot was the now notorious "military-industrial complex," another new power in the land. This oligarchy was due simply to the requirements of military technology—not to any theory of either capitalism or socialism. Even so President Eisenhower warned against it as a possible menace to democracy in his farewell address to the nation.[3] Although conservatives have been most fearful of bureaucratic power, they have otherwise expressed little alarm over this extreme example of it, the immense power exercised by a small inner group not directly responsible to the public, the possibly life-or-death decisions made often in secrecy without public debate. In view of the obsession with the Cold War most Americans do not complain either because their government has been directing our fabulous technology primarily to military purposes.

With some absent-minded exceptions, businessmen have been outgrowing their suspicion of another government enterprise, described by Daniel Moynihan as the "most powerful development" of this generation—a more wholesome effort to avert depressions and promote steady economic growth by new fiscal policies based on forecasts of economic advisers. Although growth was at first retarded by periodic "recessions," which were hard on unemployed workers even when they were given the still gentler name of "rolling readjustments," post-industrial society has so far been able to prevent the severe depressions that for a century were regular occurrences. Businessmen and Congressmen are much less horrified by "government planning," which they had regarded as something that only the Russians did, hence *ipso facto* an un-American activity bad for business. (They never objected, of course, to extensive planning by big corporations.) Similarly they are less frightened by deviations from archaic notions of "sound fiscal policy," the fetish of the balanced budget that hobbled the Eisenhower administration. Altogether, it appears that the economy is under far better control than it ever was in the

3. Malcolm Moos wrote the passage he recited, which is now considered by far the most important and enduring of his messages. Eisenhower had not previously indicated such misgivings, nor did he afterward to my knowledge. He tried to hold down the extravagant demands of the military because of his passion for economy, but his awe of big business precluded serious worries about its influence.

palmy days of free private enterprise; though laymen might wonder how it would fare without the big war business.

Ordinary Americans have mostly welcomed a quite different reason for the expansion of government—the growth of the Welfare State, dating from the New Deal. This has been much slower and spottier than the growth of the Warfare State. Congressmen who voted billions for military defense expenditures without batting an eye were still disposed to regard expenditures on public welfare services as reckless extravagance; so for years the nation spent on these services a smaller proportion of its growing national income than it had been spending, while a supposedly alarming increase in the federal debt was actually a decrease in relation to this income. Public demand for them grew, however, if only because an affluent society could plainly afford them. The necessity of keeping up with the Russians finally persuaded Congress of the need of federal aid to education, at first to turn out more scientists and engineers. More tardily the government began to respond to the crying public needs of decent housing and adequate medical care, which private enterprise was either unable or indisposed to provide except at a cost beyond the means of many Americans. At length the government launched a piddling but unprecedented "war" on poverty—the poverty that had persisted all through the triumphant progress of industrialism, and that in a land of unparalleled wealth and abundance had at last come to seem both unnecessary and unjust. President Johnson announced that the national goal was the Great Society.

But perhaps the most significant change in post-industrial society has come about through the vast sums that the government has poured into research and development, especially in the universities. (The last time I looked, the sums had reached $17 billion a year.) As I noted at the outset, the "knowledge industry" has become the biggest business in America. Daniel Bell predicts that the university will become the primary institution of the new society, more important even than the big corporation, and that all the major institutions will be intellectual, not business institutions. If businessmen may·long retain more prestige as well as wealth, a new professional elite of brainworkers will make the major decisions about policy.

So all of us in the universities might rejoice in our new

eminence. All, that is, except maybe those engaged in the old-fashioned business of liberal education, concentrated on human values. When Clark Kerr writes, in the kind of language now popular, that the "production, distribution, and consumption" of knowledge accounts for 29 per cent of the gross national product, and is growing at about twice the rate of the economy, teachers of the humanities may be reminded that their product represents an insignificant fraction of the GNP. The billions are going chiefly to science and technology. Bell emphasizes "the new centrality of *theoretical* knowledge, the primacy of theory over empiricism," or one might add over the common sense of businessmen; but what about *philosophical* knowledge? It appears that political scientists now distinguish between political theory and political philosophy, which in my innocence I had thought were virtually synonymous. Theory is what goes with the study of political behavior, the current vogue; it eschews value judgments and has outmoded philosophy. I remain more concerned with philosophy, the issues of the good state and its service of the good life.

In government the host of technicians and experts, who have given the name of "technocratic" to our society, raise a cluster of questions. By definition a technocrat is a technician who has been endowed with power, enough to influence when not to make important political decisions, and he is rarely elected to office. Some thinkers argue that technocracy dooms democracy, makes it already a mere appearance, or let us say an incidental nuisance to experts. Others might rejoice that the status of the new elite is based upon knowledge and skill, not success in pursuing economic or political self-interest; but what are their ruling values? Are their ends primarily the interests of people, or of system? Specifically, Secretary of Defense Robert McNamara developed an influential cost-effectiveness analysis for making key decisions, or what is called "value-engineering"; and does this technique take into account values other than economy? Can any such technique promote wisdom—an imprecise term, but a quality especially needed in a technological society with tremendous power at its disposal? At the moment all such questions are obscured by the Cold War. Daniel Bell has remarked the irony that whereas thinkers from Saint-Simon on who hailed industrialism assumed that the "new men" it would make leaders of would be hostile to the wasteful military spirit, it is primarily war that has elevated the technocrats. But if or when the obsession with the Cold War lifts, what then will come out of all the system?

Since I shall return to such questions, I shall say here only that I do not think the answers are simple, clear, or foreordained. As for the cultural values of the good life, government need not serve as their principal custodian or promoter; enough if it maintains the conditions that make it possible for people to cultivate these values. Immediately I am more concerned with the issues forced by postwar society regarded simply as an affluent society, in which most people have as never before the means to pursue the good life.

Among many other things, the affluent society has produced the historically novel type of the teenager, with whole industries catering to his tastes, and tending to make him not an endearing type. It also produced the education explosion that has sent millions of teenagers to college, most of them with the primary goal of becoming "well-adjusted" to such a society. With the coming of automation it holds out the prospect of increasing leisure in which to enjoy the abundance. This apparent boon is regarded, rather oddly, as a "problem"; but the problem seems less odd when one considers a favorite means of passing the time—staring at TV. A society in which the vast majority own a television set has made it the most popular and influential of the mass media, and so raises the issue of mass culture and the notions of the good life it inculcates. Meanwhile the big networks suggest some reservations about the prediction that the university will replace the corporation as the primary institution, and an intellectual class will make the major decisions affecting public policy. As the networks keep making their own decisions with little interference from government, so the big corporations have been holding their own. They too have been directing our fabulous technology, in the interests of private profit. Hence they force another basic question: How well is big business serving the public interest?

For the time being, all these issues may be illustrated by the mammoth automobile industry. Although invented in Europe, its product became distinctively American as the most popular status symbol, transformed American life more than has any other single product, and indeed has had a deeper, more lasting influence on people than had any historic conqueror. After Henry Ford achieved the mass production of the standardized Model T, a sturdy, economical car once beloved, he had to meet the competition of another wizard, Alfred Sloan, who built up General Motors by producing a range of fancier cars at higher prices, featuring annual new models. Since the war GM has dominated the industry, producing more than half the cars. In the early days of the industry Ford and other entrepreneurs had had a real passion for making automobiles, but Sloan had a more effective strategy. As he explained in *My Years with General Motors*: "The primary object of the corporation . . . was to make money, not just to make motor cars." He knew it could make more money by turning out cars with excess power and chrome, not just sturdy, well-designed, efficient cars. His kind of commercial rather than mechanical efficiency also included careful planning to assure a 20 per cent return on capital, with such success that GM actually averaged a higher return, almost double the national average, once hitting a record 50 per cent; though this did not lead to lower prices for consumers.

To these record profits government contributed by spending lavishly on services for the automobile—one kind of public service that conservatives whole-heartedly favored. Congress under frugal President Eisenhower approved a federal highway program costing many billions, while states spent more billions to speed the traffic on superhighways and freeways, and cities ate their heart out to provide more parking space. By now more than a hundred billion dollars a year goes into automobiles and their servicing—gas, tires, repairs, insurance and finance, maintenance of highways, etc. Americans own enough cars for every man, woman, and child in the country to ride in the front seat, since some ten million families have at least two of them.

The transformation of the country began when the automobile brought modern life to Main Street, the tractor to the farm; except in some backwoods districts, the villager and the farmer were no longer "hayseeds" or "yokels." The many shiny highways everywhere with all their accessories—gas stations, roadside stands,

lunchrooms, billboards, etc.—have changed the face of the countryside. The automobile stimulated the mass migration to the
suburbs, which have been spreading fast all over the land, changing cities into undefined "metropolitan areas." As it created such
other novelties as supermarkets, shopping centers, and drive-ins,
suburbanites grew absolutely dependent on it. It took fathers to
work, mothers to the store, youngsters to school. Teenagers used it
to make love in or to go on joy-rides. And from the beginning it
had brought joy to the hearts of Americans, an exhilarating sense
of freedom and power as they drove, and of possible social mobility
too.

Some by-products of their pleasure, however, illustrate a
neglect of human values that has become more conspicuous in the
affluent society. The most publicized is the slaughter on the highways, realizing the "peril to life" that the editor of *John Bull*
feared in the railroad: many more Americans have been killed by
automobiles than were killed in both world wars. The automobile
industry has contributed to the slaughter by its primary concern
with what made money, including excessive power rather than
safety. Highway engineers have been more responsible after their
fashion but in the interest only of efficiency and economy, bulldozing their way through landscape and cityscape without regard for
natural beauty or civic concerns. Business interests have shown
less respect for the landscape as they defiled it with garish billboards, designed to attract attention by not blending into it. The
automobile itself ends up in the junkyards that provide more
eyesores, speeding the transformation of the country into "God's
own junkyard."

The migration to the suburbs it promoted was more rational
—except that one reason for it is another basic irrationality of a
technological society, the blight of the American city. In the most
urban of civilizations, and the richest nation on earth, one might
reasonably expect to find beautiful cities, spacious, airy, comfortable; but in fact most of the big ones have notoriously been growing dirtier and shoddier, more congested and polluted, more
unfit to live in. The automobile has been chiefly responsible for
the congestion and pollution; ever more highways are built to
enable more drivers to get into the heart of the city. Hence the

people who have fled the city do not really escape it. They help to enact one of the most absurd daily spectacles in affluent America: millions of people driving to and from work, one to an automobile, bumper to bumper, often through smog, often tense and irritable—unable even to absorb the message of all the billboards and neon signs. Lewis Mumford summed up the whole story in a protest that this "Sacred Cow of the American Way of Life is overfed and bloated; that the milk she supplies is poisonous; that the pasturage this species requires wastes acres of land that could be used for more significant human purposes; and that the vast herd of sacred cows, allowed to roam everywhere, like their Hindu counterparts, are trampling down the vegetation, depleting the wild life, and turning both urban and rural areas into a single smudgy wasteland, whose fancy sociological name is Megalopolis."

Still, the Sacred Cow evidently remains a primary need of Americans in their pursuit of happiness. Granted that the industry spends hundreds of millions to create their needs by advertising, there is no question that most Americans like their cars big, powerful, and ostentatious. As certainly they want all the services that government lavishes on the automobile; as taxpayers they are more willing to pay for highway programs than for better schools. They seldom protest against all the bulldozing, the erosion of the heart of cities by parking areas. Once behind the wheel, they do not mind all the ugliness along the highways. Ultimately, in short, the American people must be blamed for the neglect of human values. They may complain about this or that, chiefly about the taxes they have to pay for having it so good; but apparently they would rather have life this way than support a national effort to make America fit for civilized living. Their primary duty remains consuming, buying instead of merely making a living to the end of their days, suitably marked by a costly funeral. Theirs not to question why—theirs only to go and buy.

So regarded, Americans scarcely look like a great people—the kind of people needed to create a Great Society, or to appreciate one. They support rather the forecast of Aldous Huxley, that technology is preparing them for life in something like Brave New World—a benevolent totalitarian state in which almost everybody will be kept happy and hollow. Automation will provide the machinery for running such a world. The technocratic elite is learning how to govern it. Big business is able and willing to cooperate by producing the needed superfluity of material goods. Admen will keep the people consuming and conforming, television will keep them entertained. Experts in motivational research—another new kind of technician—will condition them more perfectly to a state of mindless happiness, while wonder drugs will keep them tranquil, remove the uneasiness of such mind as they have left. In *Post-Historic Man* Roderick Seidenberg foresees a society in which all individuality and spontaneity will be ironed out and consciousness itself will finally disappear.

Again I do not think that the trend toward such a state is irresistible, for reasons I will go into at the end; there is much more to be said about the American people. Meanwhile the clearest reason is frightening, quite apart from the constant possibility of nuclear war. To the peoples of the non-Western world—the great majority of mankind—worries over the possibility of a Brave New World are simply academic. Stirred by the rising expectations of the fruits of modern technology, they will have to struggle indefinitely with the problems of scarcity, in India even with the threat of mass starvation. The affluence of America and western Europe is steadily widening the gulf between the few haves and the many more have-not nations, while the population explosion aggravates the difficulties and kindles other explosive possibilities. Americans may still dream of going their own way toward ever more affluence, ease, and comfort, their notion of happiness; but they can never hope for the security and stability of a Brave New World so long as the aroused non-Western world suffers from acute want. If consciousness is doomed to disappear, it can be trusted to remain painful enough for the foreseeable future. Today the problem in America is to alert it to the world-wide consequences of modern technology.

III

The Impact on Society and Culture

7

War

Although I knew from the outset that I had a very large project on my hands, I was staggered when I began to explore its ramifications, the effects of technology on other major interests. Early in the game I was grateful for my acquaintance with a specialist on its impact on public policy, who suggested some good books on the subject; but then I learned that he was completing for publication a bibliography that would run to more than a thousand pages. And government was only one of a dozen topics on my agenda. Needless to say, I cannot pretend to have pored over those thousand pages. Neither have I devoured the vast literature on all the other topics. So I begin by repeating that the chapters that follow are only an introduction to some of the major problems. At that I have largely confined myself to a consideration of the issues in America, on the good excuse that it is the most advanced technological society, but for the real reason that I could never hope to know well enough what is going on in all the other countries. And though I have read widely I did not come out with positive answers to the questions that most concerned me.

Let us ask, for instance, just how much difference has television made in American life? Almost all writers seem to agree that it has had considerable influence on people, most argue that its influence has been on the whole deplorable, some that it is no worse than popular entertainment over the ages, others that it has raised the level of popular culture; and all usually speak with assurance. Yet we cannot answer this critical question with any real certainty. While by attitude research, public opinion polls, and other such techniques we have acquired a mass of information about what Americans profess to think about all kinds of matters, it does not answer clearly the most important questions: how deeply their beliefs are felt, how firm their convictions are, how much these affect actual behavior. Still less can we speak with assurance about how much people have changed, since we have no such statistics about the attitudes of past generations. All manner of changes on the surface make plain enough what technology has done *for* people in everyday life, but we cannot be certain how much it has done *to* them, how deeply it has affected their mentality.

My main theme in this introduction, however, is the most fashionable term "crisis," the hallmark of our age. In past societies it is now relatively easy to make out crises in the dictionary sense of "turning points" leading to "decisive change," commonly change for the worse; we may say that at such points a society lost control of its destiny, as ancient Greece did after the disastrous Peloponnesian War that ended the golden age of Athens. In our own century we may look back on an endless series of turning points—World War I, the Russian Revolution, the Great Depression, World War II, the Cold War, or more specifically the missiles crisis in Cuba. We may say that never before in history was the state of crisis so comprehensive, universal, and perpetual as it has become in our time. To one who has lived with it all his years (as I have) it may seem the normal condition of modern man. If, as Whitehead said, it has always been the business of the future to be dangerous, the future has never before been on the job so steadily and conscientiously.

Now, this atmosphere creates a disposition to blow up all serious problems into crises. It appears that no problem is really respectable unless it is so named. Thus we may read of the crisis of capitalism, of Christianity, of higher education—of institutions whose condition does not seem to me clearly desperate. All the revolutionary change we have lived through has as yet been no more clearly decisive, in the sense of having sealed our fate, for we go on generating more crises. And our condition is novel in another important respect. In past societies men rarely had such an acute consciousness of crisis, any more than the highly self-conscious Athenians did in their Peloponnesian War; although sufficiently aware of defeat or disaster when it struck, they knew that times of troubles were an old story, and usually evidenced little sense of having passed a historic turning point. If or when they did begin to sense that they had lost control of their destiny, their spirit was very different from that of modern man, above all in America. For another popular term for all our problems is "challenge." We are forever being exhorted to rise to challenges; every day another book of alarm comes out, usually insisting that our situation is desperate, but almost always implying that something should and can be done about it—as even Jacques Ellul does in a prolonged demonstration that we are slaves to technology and can do nothing about it. In the declining centuries of other civilizations—Egypt, Rome, India, Islam, China—there were no such constant cries of alarm or insistent calls to action.

Yet we surely do face problems grave and urgent enough to be called crises—they will crop up in almost every chapter. They may look more dangerous after one has discounted the habitual worries of intellectuals. Thus another popular catchword is our Age of Anxiety, with the implication that never before have people suffered so from anxiety; whereas most Americans look complacent, surely more complacent in their affluence than were the common people in past societies. Most are in no mood to rise to "challenges." If anything, they could do with more anxiety, or at least more awareness of all the dangers the country faces.

In this spirit, at any rate, I begin with the most imperious of our crises or challenges—the constant danger of nuclear war.

In its service of organized warfare, technology has from the outset contributed to the essence of civilization as defined by Bertrand Russell, "the pursuit of objects not biologically necessary for survival." It enabled man to become the only animal that steadily, systematically exterminated fellow creatures of its own species. As we have seen, Western civilization has distinguished itself in the development of technology for this biologically unnecessary purpose. One historic turning point was its introduction of gunpowder, which the ingenious Chinese had invented but failed to use for weapons. Another was the mechanization of war in our century. By now technology has so profoundly transformed the ancient institution of war that we again face a literally extraordinary situation. One cannot repeat too often what everybody knows but too many still fail actually to realize, that man has achieved the power to blast the entire earth, destroy his civilization overnight, conceivably put an end to the human race. He has achieved the ultimate in biological absurdity, making preposterous his distinction from all other animals, whom he still calls beasts or brutes. And in our day America has led the way in this achievement. Since World War II a nation whose people still like to think of themselves as naturally peace-loving has spent about a trillion dollars on armaments, a larger proportion of its national income than any other free nation.

Americans have never really been peace-loving, of course. Always inclined to be restless, self-righteous, boastful, and belligerent, they have fought more wars since the beginning of their national history than has almost any other country in the same period. Until our time, however, they did not distinguish themselves in military technology. Although Bismarck's Prussia demonstrated their relative backwardness shortly after their bloody, prolonged Civil War, they were still backward when they entered World War I. James Conant has told of a leading chemist who rushed to the Secretary of War to offer the services of the American chemical association, and was thanked but informed that his services were unnecessary—the Army already had a chemist. Nor did the nation's leaders take to heart the lessons of the war, preoccupied as they were with getting back to "normalcy"; between the wars only a few millions were spent on military research. As World War II loomed up, the Army General Staff made a characteristic decision: instead of wasting time on research, it decided to be practical and concentrate on procuring weapons already developed.

It was President Roosevelt who made this war a turning point in American military history. Before Pearl Harbor he created the Office of Scientific Research and Development, with Vannevar Bush as director, which made so brilliant a record in mobilizing civilian scientists for the war effort, including the beginnings of systems analysis, that even the military were convinced of the value of scientific research. In 1946 Air Force General Henry Arnold summed up the lesson: "Today's weapons are tomorrow's museum pieces." But it was the development of the atomic bomb that sealed the lasting alliance of science and weaponry. When in 1939 some nuclear scientists—all European-born—had gone to Washington to suggest this project, they were coolly received by the military. Again President Roosevelt took the decisive step by authorizing the project, on the strength of a letter from Albert Einstein. He gambled $2 billion on it—far more than the government had spent on science and technology in the whole history of America.

With the beginning of the Cold War military technology grew steadily into the nation's chief business in spite of confused, conflicting purposes. The Russians assured its growth by their early success in exploding a nuclear bomb, destroying our monopoly. In 1947, when American authorities were telling President Truman and Congress that intercontinental rockets were impossible, Stalin assured another race by telling the Soviet that rockets could be "an effective strait-jacket for that noisy shopkeeper Harry Truman." The Russian sputnik then called out another expensive agency, the National Aeronautics and Space Administration. For such reasons President Eisenhower was frustrated in his passion for economy and a balanced budget; military expenditures continued to mount. It did not help either that the popular Air Force bitterly opposed ballistic missiles, for the manned bombers it wanted instead required continuous operations in big bases all over the world, and there was no halting the development of missiles too. When John F. Kennedy ran for president he exploited the alleged "missile gap." After his election the truth came out (at first denied by him) that there was no such gap, we were well ahead of the Russians, but we continued to build up a bigger missiles force anyway.

There is no point in reviewing all the specific technological developments. "The year 1951," wrote Herman Kahn, "is typical of the new era in which there is the introduction, full procurement, obsolescence, and phasing out of complete weapons systems"; and since then new systems have come and gone at a fantastic rate. They made some civilians realize that even America could not afford to produce anything or everything the military want, but even so new systems kept sending yesterday's weapons to the museum. The nuclear arsenal alone has grown immensely bigger than its scientific pioneers anticipated. When Robert Oppenheimer headed a task force to survey possible requirements for the long run, he ventured an "unholy figure"; but by the time of his death in 1967 our stockpile was twenty times greater than his wild guess. We now have the equivalent of fourteen TNT tons of explosives for every human being on earth. Since that statement was made a year or so ago it may well have become fifteen or more —who knows?

The decision that determined this outcome was made as early as 1950—President Truman's authorization of a new project, the H-bomb. (Later it resulted in the public humiliation of Oppenheimer, who was dismissed as a security risk for reasons including his expression of convictions that were "not necessarily related to the protection of the strongest offensive military interests of the country.") Truman's action illuminated what political scientists are fond of calling the "decision-making process"—a process out of which now come fateful decisions without public debate. Warner R. Schilling has reviewed it in detail in an article subtitled "How to Decide Without Choosing."[1]

The ugly spectre that set off this process appeared in 1949 when the Russians exploded a fission bomb. Although scientists had predicted their success, the government had made no plans for the contingency; only now did it face up to the disagreeable facts of modern military life, that for the first time a foreign power could readily strike at the heart of America, do to us what we had done to Japan. One possible answer was an effort to develop a vastly more destructive thermonuclear bomb. The Defense and State Departments favored going ahead with this project, though for different reasons. David Lilienthal, head of the Atomic Energy Commission, questioned it; he was shocked to learn at this time how much the military already depended on nuclear weapons. The scientists on the General Advisory Commit-

---

1. Published in the *Political Science Quarterly*, March 1961.

tee of the AEC, headed by Robert Oppenheimer, pointed out strong objections in their official report. They knew that the H-bomb would be useful only for destroying big cities, slaughtering civilians; its development would mean another turning point in history. The majority of the GAC went so far as to recommend that the United States unilaterally announce that it was not going to make this appalling weapon. Later one of them noted an unprecedented result of their report: for once it united the jealous armed services—in opposition to them.

Scientists were also concerned about technical problems of priorities. These included such questions as whether the nation needed H-bombs more than it needed an increase in conventional armaments, whether it would be worth the sacrifice in the tested A-bomb program, and whether it would not divert too much scientific talent. President Truman answered none of these strategic questions when he decided to go ahead with the project. He did not anticipate either such still unresolved issues as "city-busting" versus "counter-force," or the strategies of "first-strike" versus "second-strike." He rejected one alternative, that of exploring first the possibility of international control of nuclear weapons, which was opposed by both the Defense and State Departments. Schilling remarks a major necessity of the American political process, "the need to avert conflict by avoiding choice"; the President left unanswered the questions on which his subordinates could not have agreed. More important, I assume, was his agreement with the common feeling that if we did not develop the H-bomb the Russians would. I also assume the chances are that under Stalin they would have gone ahead with it. As Schilling concludes, both the Soviet and the United States would no doubt have preferred a world in which neither had the H-bomb, but neither wanted the other alone to have it; so both went ahead, and ended in a worse position than they started from. He sums up the whole affair concisely:

> The H-bomb decision is essentially a tragic story. The GAC was "right" in sensing that the development of the H-bomb would drive twentieth century man deeper into the box that he has been building for himself with his military

technology, doctrine, foreign policy, and cultural ethos. The GAC was also "right" in asserting that it was a time to stop, look, and think. But the GAC was not alone in seeing the dimensions of the box. It was every bit as apparent to most of the advocates of the Super program. The trouble was that no one had any good ideas of how to get out of the box. Nor are they apparent today.

Now, apart from the gratifications of national pride, civilians have enjoyed some benefits from technological advances made initially for military purposes. They owe to them their radar, jet planes, transistor radios, computers, and some advances in medicine. They can expect much more from the electronics industry, still more from the extensive use of nuclear energy for peaceful purposes. Even the apparent foolishness of spending many billions to put an American on the moon has had by-products that laymen may appreciate. Some share the excitement of specialists over the purely scientific possibilities opened up by the exploration of outer space—a program that would never have gone so fast and so far except for the compulsion to keep up with the Russians. Others are enjoying international TV programs relayed via satellites. If we manage to avoid the catastrophe of a thermonuclear war, the prospects are that civilians will enjoy increasing benefits from all the research and development stimulated by the Cold War.

Yet there always remains that *if*. At the moment our best hope is a novel one, the "balance of terror"; but the Cold War goes on. The obvious excuse for the huge American defense program, that the Soviet hardly acts like a peace-loving nation, only accentuates the collective insanity, for both sides already have many more than enough bombs to devastate the whole earth, both go on producing more weapons against which there is no defense, and the nuclear arms race gives no real security to either. Never very sensible, war can no longer be either a feasible or a legitimate way of settling international differences. And America too has hardly been acting like a peace-loving nation. Our assumption of the "responsibility" of defending the "free world" could look like imperialism as it led us into military adventures in Cuba, the

Dominican Republic, Laos, and Vietnam, and also led to the establishment of more than 3,000 military bases all over the world, manned by a million men. Most Americans know nothing about their Arms Control and Disarmament Agency for a good reason, that we devote to it only a tiny fraction of one per cent of the billions we spend on military hardware. Few Americans really believe in disarmament anyway.

Secretary of Defense McNamara made the most serious effort to hold the defense program to more rational purposes by proposing the principle of nuclear parity for the American missile force, but he was overruled by President Kennedy for political reasons; the principle was unpopular. The cry in Washington is still that mere parity will never do, we must always strive to maintain superiority over the Russians—even though they will then naturally try to match our own build-up, as McNamara pointed out, and in any case superiority in numbers has lost all meaning. (In his memoirs, *The Essence of Security*, McNamara later stressed the "mad momentum" of the arms race, the obvious futility of the whole process of action and reaction.) Most Americans believe that we must be first in nuclear weapons for the sake of our national prestige, which they do not think suffers because seventeen nations have lower rates of infant mortality than we.

A particular reason for the predominance of the "offensive military interests" that Robert Oppenheimer failed to respect sufficiently is the dangerous power now wielded by the military-industrial complex. About this President Eisenhower warned that "public policy could itself become the captive of a scientific-technological elite." "Elite" was a euphemism for the vested interests of the Pentagon and big business (a "system" that McNamara, a specialist in systems analysis, incidentally neglected to analyze in his memoirs). Ostensibly the controlling interest is always national security, but cruder considerations can influence the managers of so huge a business. The Pentagon is naturally much less interested in disarmament than in the procurement of more military hardware, at whatever cost. So are the big corporations who get contracts from it. A number of them have stakes of over a billion dollars in the defense program, running up to $10 billion for Lockheed Aircraft Corporation. They have created only a minor scandal by cementing their lucrative alliance with the military by hiring many retired officers; at last report Lockheed alone had more than two hundred such officers on its payroll, including twenty-seven admirals and generals. Labor unions too have a big stake in the arms business, since it employs millions of workers. Politicians and voters in communities having defense plants are no more disinterested.

For such reasons the military-industrial complex has long had a powerful lobby in Congress to push its costly programs. Senator Jackson of Washington, for instance, has been a conspicuous champion of its interests; his state gets more than half its income from government orders. (Cynics have referred to him as the Senator from Boeing.) In the House Mendel Rivers, chairman of the Armed Services Committee, has made a name for himself by his unflinching devotion to the interests of the Services, in return for which he has brought much defense business into his home district. The Congressional committees that review requests for appropriations from the Pentagon have habitually called in practically no witnesses except from the Pentagon itself. All these men, including the many hard-line anti-Communists among them, may of course be sincere patriots; Americans naturally have different notions about the national interest. The point remains that such notions are now largely shaped by the interests of the military-industrial complex. Until lately there has therefore been little effective protest against the notorious wastefulness of the defense program, the many billions squandered on obsolescent or inferior weapons.[2]

Another possible turning point was the pressure for a

---

2. Since this military-industrial complex has lately become a tiresome cliché, I might note that when I wrote a draft of this chapter, early in 1968, only a few critics were dwelling on its very dangerous power, which seemed to me fearfully obvious. A year later Congress finally woke up to the danger, the need of bringing the military under control, and now the newspapers are full of what I wrote when I felt like a voice crying in the wilderness. But I go on with what ought to be too familiar a story because as this goes to press the odds still favor the Pentagon in any showdown. As with many other gloomy statements about the contemporary situation, I would of course be pleased to be proved wrong by the event.

"Sentinel" missile-defense system. Secretary McNamara gave in by agreeing to a "thin" system costing $5 billion, even though he had first pointed out that the defense it provided would be quite inadequate; he knew that in the 'sixties we had spent $20 billion on anti-missile research and development, which had produced nothing but obsolescent systems opposed by the scientific advisers of both Presidents Kennedy and Johnson. The Pentagon offered no rebuttals of the scientific arguments against the latest system, but President Johnson gave in too, again for political reasons: if he didn't, Republican leaders were sure to charge that he was risking "national security." Ralph Lapp, a scientist who has been involved in the "weapons culture," spelled out the obvious moral: given such motives, we "might just as well junk all the elaborate systems of defense analysis that we possess." As it was, the arms lobby then began demanding a "thick" system that would cost $40 billion, despite evidence presented by McNamara that it could be defeated by new offensive missiles costing only a fraction as much. Later it was once more politics, not science, that apparently swayed President Nixon's decision to go ahead with a modified ABM system, rebaptized as "Safeguard" and equipped with a different set of arguments, even though by this time the Pentagon was admitting that its costs would be much higher than the stated estimates.

A related consequence of the huge defense program is the unprecedented influence of the military on foreign policy. The Chiefs of Staff sit in on the highest councils of state. No doubt they should be consulted, since America has staked its all on military power; their advice cannot be dismissed by simple scorn of the "military mind," remembering such generals as Marshall, Bradley, and Gavin, or President Eisenhower himself.[3] Nevertheless it is fair to say that diplomacy is not their forte. As might be expected, most of them have been hawks; they recommended the disastrous Bay of Pigs invasion and the war on North Vietnam. Robert McNamara has said that some men in high places have supported

---

3. In 1957 General Omar Bradley made a prophetic speech. "Missiles will bring anti-missiles, and anti-missiles will bring anti-anti-missiles," he predicted, until the "whole electronic house of cards" collapses. Unfortunately, there is still little sign of its collapse.

what he considered a disastrous concept, that we could find means of using thermonuclear missiles and thereby not only survive but win. The top military have also had more than their share of right-wingers, who are no more committed to the cause of democracy than to the cause of peace. The Department of Defense has promoted neither cause by pushing the sales of obsolete weapons to non-Western and Latin-American countries.

As novel a consequence of the defense program was the rise of the "idea industry," symbolized by the Rand Corporation (named after "R & D"). Experts in think tanks have provided the military with advice on matters ranging from simple technology to the highest policy. Herman Kahn, the most famous of them, set himself the task of carefully weighing all possible alternatives in foreign and military policy, and by scientific analysis seeking answers to such questions as the chances of war, the hazards of the arms race, and the consequences of a nuclear attack on American cities. "How many million American lives," he asked, "would an American President risk by standing firm in differing types of crises? By starting a nuclear war? By continuing a nuclear war with the hope of avoiding surrender?" His book *On Thermonuclear War* was hailed as one of the most important, original works of our time. It was also denounced for its apparent callousness, his very calm consideration of the indescribable horrors of nuclear war, or the ultimate in rationalized irrationality. He then defended himself as calmly in *Thinking About the Unthinkable* (1962), from which I am quoting. No book reveals more clearly or more fully the preposterous condition our extraordinary military technology has got us into.

Kahn's basic thesis seems to me quite reasonable. We *have* to think about the unthinkable because it is not at all impossible. We have to come to grips with the actual problems forced by our technology and current international relations, immediately with the fact that a number of powers possess thermonuclear bombs. In case of an emergency, military computer systems provide detailed information and instructions so swiftly that leaders have little time to think, and need to have on hand alternative plans worked out as carefully. It is not enough to insist that "there is no alternative to peace," for this might only give the Soviet and China an easier alternative—the Chinese keep proclaiming that the real alternative is world revolution. As for the "icy rationality" Kahn was denounced for, he asked: "Would you prefer a warm, human error? Do you feel better with a nice emotional mistake?" Americans in particular need systematic, hard-headed thinking about these matters because they tend to be short-sighted, excessively "practical," distrustful of theory. And Kahn is not at all brash in his claims for his "scientific" analysis, explicitly recognizing the difficulties of balancing risks, the constant possibility that our suppositions may prove wrong. Above all, he is by no means simply a hawk or a hard-line anti-Communist, but a man who still hopes for peace—and peace without what Americans think of as victory. He exposes the real horror of our situation, that a reasonable man may feel obliged to think coolly about frightful possibilities that men used to consider literally unthinkable. They include such possibilities as "doomsday machines," devices that could destroy all unprotected people, "perhaps eventually *all* people," and that could probably be perfected in less than ten years.

Then one may add that Kahn has grown so inured to the habit of "objective" analysis, a scrupulous avoidance of moral considerations, that he can seem unrealistic as well as inhuman in his war-game theory. In arguing for a civil defense program, the fall-out shelters that President Kennedy vainly tried to sell to the American people, he wrote that "the lives of as many as a hundred million people might be saved." Translated, this meant that as many as a hundred million might be killed. No doubt most Americans were simply heedless in ignoring Kennedy's program, but many refused to take these shelters seriously on grounds that seemed to me realistic.[4] What would life be like when we emerged from our shelters? In a country strewn with up to a hundred million rotting corpses? In the company of more millions of burned or mutilated people? With how many doctors and nurses still alive and able-bodied? How many hospitals not destroyed? Not to mention all the devastation, the contamination of food and water by fall-out,

---

4. Readers may remember a display by Civil Defense Head-quarters of a "de-luxe fall-out shelter" with soft carpeting, lounge chairs, television, Scrabble, etc.—in an affluent society just the thing for peace or war.

the problem of livestock, and so forth. Kahn answered that life would of course be harsher, but according to the "best estimates" it could go on tolerably. I was more inclined to think that there is no alternative to peace when he noted as coolly that the most effective opposition to the shelter program came from conservatives and military men who feared it would encourage a Maginot-mentality: "Brave men do not hide in holes." As a coward, I regretted that he devoted most of *Thinking About the Unthinkable* to problems of military strategy, and in his willingness to put up with high risks, or "acceptable casualties," failed to apply his method of exhaustive analysis to such possible alternatives as world government, disarmament, and strategies of peace. About research problems he wrote that he would assign the highest priority to "the arms race as a race" because of the "relative ease with which fruitful studies can be done." It appeared that disarmament was not a fruitful subject.

Leonard C. Lewin has since remedied such neglect in his *Report from Iron Mountain on the Possibility and Desirability of Peace*, supposedly a secret report by a group of assorted experts commissioned by the government to draft a program for dealing with the "contingency" of peace. The gist of their conclusion was that "lasting peace, while not theoretically impossible, is probably unattainable," but that "even if it could be achieved it would almost certainly not be in the best interests of a stable society to achieve it." When Herman Kahn published *On Thermonuclear War* one angry reviewer questioned whether there actually was such a person, suggesting that his book was a hoax in abominable taste. It is a tribute to Lewin's sustained deadpan imitation of the flatulent style of much expertise that some reviewers were not altogether sure his book was a satire; they thought just possibly it might be an authentic report. And for good reasons. The kind of thinking endorsed by the supposed authors of the report does go on in Washington, if not openly. At least their main arguments are the most logical way of making sense of some apparent lunacies of American policy. *Report from Iron Mountain* is another highly revealing study of our absurd situation today.

Its alleged authors pride themselves on being utterly objective. In the "Byzantine" spirit of Kahn, there is "no agonizing over cultural and religious values, no moral posturing." They avoid all reference to democratic ideals as they resolutely peer over "the brink of peace," judge only by the neutral standards of the stability sought by all societies. Like good scientists, they recognize the inevitable risks of their enterprise, inasmuch as a system of peace would constitute "a venture into the unknown," but since they have all history behind them they can find much evidence to support their conclusion that war is essential for social stability. The "miasma of unreality" enveloping idealistic thinkers springs from the fallacious assumption that war is a subordinate institution, merely serving the social system, whereas the highest priorities given military institutions indicate that "war itself is the basic social system." Only so can one understand the "superficial contradictions" of modern societies:

> The "unnecessary" size and power of the world war industry; the preeminence of the military establishment in every society, whether open or concealed; the exemption of military or paramilitary institutions from the accepted social and legal standards of behavior required elsewhere in the society; the successful operation of the armed forces and the armaments producers entirely outside the framework of each nation's economic ground rules: these and other ambiguities closely associated with the relationship of war to society are easily clarified, once the priority of warmaking potential as the principal structuring force in society is accepted.

The authors' arguments for the necessity of war include a demonstration that it has become much more efficient for eugenic purposes of holding down the population: whereas it used to kill off the healthiest specimens, the bombing of civilians makes its victims "more genetically representative." But most pertinent today is the necessity of war for a healthy economy, above all in the United States. World peace and general disarmament would create problems of "unparalleled and revolutionary magnitude," for which the nation is totally unprepared. The defense program has not only greatly increased the GNP but served as the great stabilizer of the economy, preventing the depressions that used to take place "during periods of grossly inadequate military spending." For this purpose the very wastefulness of the program is indispensable. "An economy as advanced and complex as our own requires the planned average annual destruction of not less than 10 per cent of gross national product if it is effectively to fulfill its stabilizing function."[5] Hence the authors dismiss as inadequate the social welfare programs that are commonly proposed as substitutes, for these could provide only temporary and partial relief. Before long they would be absorbed into the normal economy, their main objectives accomplished; only war can provide the indefinite, abnormal stimulus required for stability. They would be far too cheap, not wasteful enough to serve as an economic substitute for war. Perhaps most promising is the gross pollution of the environment, which is well enough advanced so that it might eventually replace nuclear war as the chief threat to the survival of the species; only the indications are that it will take at least a generation for it to become "sufficiently menacing, on a global scale, to offer a possible basis for a solution." Meanwhile our immense military establishment remains our best social-welfare institution.

Now, all this may sound preposterous—until one looks a little harder at our expanding Cold War economy. If by any chance this war did end soon, it would indeed create serious economic problems for which the country is quite unprepared. While the giant corporations engaged in the arms business are naturally not disposed to welcome anything like extensive disarmament, neither are many other businessmen; they fear it would mean a depression. In both world wars the United States recovered from a depression because of war production, and the huge defense program is the clearest reason why it has so far managed to avoid another one. On this score *Report from Iron Mountain* often echoes official language. Lewin cites, for example, a report of the Arms Control and Disarmament Agency pointing out that heavy defense

---

5. David Bazelon anticipated this argument in *The Paper Economy* (1963). War is "the most important illegal institution in our society" because it is indispensable to our paper economy as "artificial demand," and the only such demand that raises no political issues. "The essential aspect of this artificial demand is that the *product* does nobody any good—only the *process* is beneficial: *war, and only war, solves the problem of inventory.*"

expenditures in the public sector have "provided additional protection against depressions, since this sector is not responsive to contraction in the private sector and has provided a sort of buffer or balance wheel in the economy." Two economists from Utah, among the states awarded the most defense funds per capita, similarly concluded a survey: "The defense industry, as a whole, has proved itself to be one of the state's most stable industries. At least defense employment has not been sensitive to the 'business cycle,' which was true of those major industries of past years." (In quoting this, Ralph Lapp suggested that the land of the Mormons might erect a statue of Khrushchev.) Edward Teller remarked that if we could change the world into one without weapons it would be stabilizing, "but agreements we can expect with the Soviets would be destabilizing." The Department of Defense made another contribution to stability when it refused to permit West Germany to fulfill its purchase commitments in the U.S. by substituting nonmilitary goods for unwanted weapons. Only by using such language can one avert the suspicion that the Pentagon and its contractors might be pursuing their selfish interests.

As for social welfare programs, it would seem that a sane economy should find them adequate substitutes for war, since it would take many billions simply to make big American cities decent places to live in. They might also serve as the "moral equivalent of war" explored by William James; the challenge is no longer to harness Nature, as he proposed, but to use for the public good the great power over Nature we now have. Meanwhile there is no question that the military establishment has claimed and got the highest priorities, the welfare programs have run a poor last. If war is not strictly "the basic social system" in America today, it has underscored all the not so "superficial contradictions" listed in the report from Iron Mountain. Thus *Newsweek* reported, with no tongue in the cheek, that at the annual forecasting session of the National Industrial Conference Board in 1968 worries over the economic impact of a peace settlement in Vietnam gave way to buoyant forecasts: "Everyone now believes that even complete peace would not substantially alter defense spending for many months." We might be able to afford peace after all so long as the military continue to get the highest priorities.

When President Johnson followed their advice and committed the nation to an increasingly hot war in Vietnam, I was among the many Americans who thought the war was as stupid, immoral, and barbarous as it appeared to most of the world; but here I shall dwell only on some consequences of modern military technology, including more "contradictions." The most obvious, its inhumanity, is more frightening because it does not frighten most Americans; they are losing their capacity for horror. The growing unpopularity of the war was due to its costliness and apparent futility, not to such barbarous tactics as dropping bombs and napalm on villages that might be occupied by guerrillas or peasant sympathizers. Likewise the outright atrocities condoned by our military were less frightening than the impersonal routines of airmen simply following orders. We may assume that these airmen are no more depraved than soldiers used to be; they would not butcher peasants, women, and children face to face; but put them in bombers with orders and they can do so without scruple because they never see their victims close up. Apparently no one feels responsible for the barbarities; the men at the top who give the orders are not mere individuals but members of an organization, following approved procedures, doing their duty as they see it. Nor has Lyndon Johnson given any public sign of a guilty conscience—he too was only a patriot acting in line of duty. So let us imagine the man at the very top who might do the unthinkable and push the button setting off a thermonuclear war. He would first suffer some anguish (let us hope); he would have on his conscience (if he survived) the most monstrous deed in history, the extermination of hundreds of millions of human beings, putting Hitler in the forgotten shade; but almost certainly he would not feel like a monster, only like a patriot defending his country —just as Harry Truman felt when he plumped for H-bombs. As Gunther Anders wrote, the monstrousness of our technology has made possible "a new, truly infernal innocence."

Another sign of the dwindling capacity for moral imagination was the dropping of chemicals on peasant regions in South Vietnam, to "defoliate" them or spoil their water. Since the Pentagon's elaborate preparations for chemical and biological warfare have been hushed up, only a few of the new weapons are known, but again most Americans seem unperturbed by these appalling possibilities.[6] It is perhaps something of a blessing that people have lost the capacity for the kind of terror of man-made hell that many used to feel at the Christian hell; else they might be paralyzed by nervous tension and fear. That they are not is due to patriotic propaganda, apathy, feelings of impotence, or simply inability to realize the possible horrors awaiting us. But the schizophrenic lives we lead lessen the resistance to the compulsive drive of technology that gave us the H-bomb: Americans must explore the frightful possibilities of chemical and biological warfare because the Russians may be doing so. In all history man has never been more inhuman, and in this aspect more unfree.

Another consequence of our vastly superior military technology is ironical. For years our generals kept announcing that we were winning the Vietnam war, suggesting that they were either fatuous or dishonest, but their boasts of American victories were simply humiliating when we had half a million soldiers in Vietnam. Together with the South Vietnamese army, we greatly outnumbered the Vietcong, we had infinitely greater fire-power, and we alone had bombers; yet we were not clearly winning the war, not able even to keep the Vietcong out of Saigon. Although by the standards of our military we were fighting a "limited" war, refraining from mass fire-raids on Hanoi and Haiphong and from the use of nuclear weapons, we bombed Vietnam much more heavily than we had Germany in the World War. But to no avail. The mightiest nation on earth was at most holding its own in a

---

6. Although there was some scandal when an experiment with a gas went awry in Utah, killing thousands of sheep, this was soon hushed up. The Pentagon promptly lied, disclaiming any responsibility; and by the time the truth came out through a slip, people had lost interest in it. There was no uproar either over another horror cited by the Federation of American Scientists when it protested against the production of biological and chemical weapons of mass destruction as "pointless, dangerous, and provocative." Researchers at Fort Detrick have received medals for developing something called rice blast fungus, which can attack the crops of whole nations. The FAS asked: "Why buy what no contingency will ever require?" The generals who gave the medals remained silent, but one must suppose they have in mind a possible need or excuse for using such weapons. They have a further excuse in that the U.S. has never officially adhered to the Geneva Protocol of 1925 banning these weapons.

struggle with a poor little country. Our military had yet to learn that tactics effective in knocking out an advanced industrial country like Japan were ineffective against a land of peasants. General Curtis LeMay, who urged bombing it "back into the Stone Age," only proved that his own mentality was primitive even for military purposes, and that today more than ever war is too serious a business to be left to generals. Modern technology was stymied—unless it went all-out, as American leaders refrained from doing if only out of fear of Russian or Chinese reprisals. A trillion dollars spent on a defense program had defeated the purposes of war by building up such power that we did not dare to use it.[7]

Still, this restraint was something. If the "balance of terror" is a dismal kind of hope for supposedly civilized man, it is a real hope, possibly more effective than the traditional restraints, which failed to prevent war in the past. Herman Kahn observed that all the major powers, including the Soviets, have been very cautious about using their deadly power or even threatening to use it. He added that to his knowledge no nation had made a cobalt bomb, which he described as one that "might be vastly more deadly than an ordinary thermonuclear weapon at no greater cost." If so, men are at last learning to resist the seemingly irresistible drive of technology. Since no man has analyzed our situation more systematically than Kahn, we might well conclude with his survey of our prospects.

The danger remains that a single failure may be very destruc-

---

7. In trying to explain the bafflement of Washington in its efforts to understand and deal with the Vietnamese, James Reston recalled Paul Valéry's imaginary dialogue with a Chinese Buddhist, who explained why Westerners would never understand the mind of the East. The Chinese had seemed asleep and been despised, he said, because they set all things in order, and the fabric of their politics was woven of "the living, the dead, and nature." He concluded: "We simply preserve wisdom enough to grow beyond measure, beyond all human power, and to look on while you, in spite of your raging science, dissolve in the deep and fruitful waters of the land of Tsin. You who know so many things, do not know the most ancient and powerful, and you rage with desire for what is immediate and you destroy your fathers and your sons together." Although Red China today is rather different, one reason why it too baffles Washington may be that it preserves some of the traditional mentality.

tive, possibly catastrophic. He listed an alarming number of possible sources of failure: greater opportunities for blackmail, revenge, terrorism, and other mischief-making; more widespread capabilities for "local" Munichs, Pearl Harbors, and blitzkriegs; pressures to forestall a surprise attack by making one; tendencies to neglect conventional military capabilities; greater danger of inadvertent war through chain reactions; internal political problems such as irresponsibility, hysteria, and civil war; the gift of nuclear weapons to irresponsible rebels, terrorists, or other private groups; more complicated problems of control, especially as more nations want or get nuclear weapons; and so on. For such reasons Kahn thought that international politics cannot be handled indefinitely on the present basis. His elaborate analysis brought him to the conclusion that simple-minded idealists had leaped to, that sooner or later our situation requires some kind of world order. He even ventured the prediction that efforts to preserve our old nation-state system will probably be futile.

In a postscript to the paperback edition of *Thinking about the Unthinkable*, written two years later (1964), Kahn indicated that developments in the interim at least partially justified the growing feeling that the world was safer. There appeared to be less chance of either accidental or deliberate war; safeguards included the "hot line" between Moscow and Washington, and the test-ban treaty was another sign of improved relations between them. He thought it a good bet that in the "next five years" (these days a very long time to try to look ahead) there would be no crisis as tense as the Berlin crisis of 1958. Since he wrote, relations were strained by our war on North Vietnam, then improved by our efforts to negotiate peace, then strained again by the Russian invasion of Czechoslovakia. At the moment I write I do not know how Kahn sizes up our situation, nor do I feel impelled to find out. The main point is still that anything might happen tomorrow to make the prospects either better or worse. Anything except, perhaps, the discarding of all the nuclear weapons we have to live with. As I see it, there remain reasons for hope, but none at all for complacence. Kahn can be sure of a job for the rest of his life. The wonderful, fearful technology we have developed for military purposes will not be forgotten so long as we survive.

8

Science

Because it is now hard to realize, we need to keep in mind that from Newton through Darwin most scientists were amateurs. They were cultivated men whose favorite hobby was science, or perhaps professors who taught some other subject, such as philosophy. Although they were banded together in a few societies, like the Royal Society, and had a few journals to publish in, they had no laboratories to speak of, only some simple instruments. Just a few (not including Isaac Newton) devoted their whole life to science. According to the *Oxford Dictionary*, the word "scientist" was first used in English in 1840. Like so much in our world, science as we know it today is quite new—so new that we don't really know it.

Its development into a major profession began only in the last half of the nineteenth century, at about the time of its alliance with technology. As late as 1900, when the revolution in modern physics was under way, physics was a free creative activity, with only a few well-known practitioners and no great laboratories; physicists worked alone, with homemade apparatus, and were ignored by both government and the general public. Thereafter science began to grow and spread more rapidly, and since World War II at such a phenomenal rate that it amounts to another scientific revolution. The striking observation of Robert Oppenheimer, that 90 per cent of the scientists who ever lived are still alive, has become a commonplace. Derek Price adds that at the rate they are multiplying, in a hundred years there would be two scientists for every man, woman, and child. They have innumerable laboratories to work in, with such equipment as giant telescopes and cyclotrons. Their interests are furthered by all kinds of professional associations, research institutes, and foundations. The last I heard, some years ago, there were about 50,000 scientific journals in the world, publishing two million articles a year. In America thousands of scientists either work directly for the government or are subsidized by it in their research, to the tune of billions. A National Science Foundation is only another sign of a power and prestige such as the Royal Society never dreamed of. In short, science has become a major Establishment. It could be considered the most important establishment in America, as the ultimate source of the power of modern man.

Although from its beginnings modern science had an influence on society out of all proportion to the handful of pioneers engaged in it, this influence has become incalculable since it became highly organized and united with technology. It accordingly enters every chapter of this book. But here my immediate concern is a different question: How has science itself been affected by its phenomenal growth as an institution, all the power and prestige it has attained? And first of all, just what *is* science? What is the "scientific method" we hear so much about? Or "scientific truth"?

Now, these are strictly philosophical questions, so we cannot expect to get authoritative answers from scientists themselves. They seldom bother their heads over such questions anyhow, but they are all specialists, qualified to speak only about their own particular science; they know as little about almost all the others as the rest of us know. They help chiefly by making it clear that what we have to deal with immediately is not "science" but a large, loose collection of sciences. Usually these are separated into the natural and the social sciences, today with some overlap in the so-called behavioral sciences, but this classification raises further questions. They are not all scientific by the same standards; specialists in the long-established sciences of physics and chemistry may question whether the many new ones, such as psychology, deserve the name of science, while within psychology there are a number of different schools that are scarcely on speaking terms. Still more dubious is the status of marketing research, motivational research, and much other research that is paraded as scientific. As for "scientific method," scientists employ many different methods, and like the rest of us often think in ways that are not methodical. Most of what Paul Weiss said of his own science of biology may be applied to others: "It needs the observer, the gatherer of facts, the experimenter, the statistician, the theorist, the classifier, the technical expert, the interpreter, the critic, the teacher, the writer." And as for "scientific truth," Einstein himself had trouble giving it meaning, "so different is the meaning of the word 'truth' according to whether we are dealing with a fact of experience, a mathematical proposition, or a scientific theory." At that the common fate of scientific theory is to be proved untrue.

Since we are nevertheless bound to go on speaking of "science," as I have been all along, we should remember that it has never had a single, definite, precise meaning. Its extraordinary growth in this century has not removed but deepened the confusions about its meanings, beginning with the popular assumption that "scientific truth" is absolute, so-help-me-God truth. Inasmuch as no one can say what science "really" is, I content myself with repeating a few broad generalizations and rough distinctions I have made in my own writings about it.

If all systematic study can be called scientific, modern science has been distinguished by its persistence in asking "What are the facts?," in basing its theories on observed or recorded fact, and in seeking to verify its theories by methods that can be repeated and checked by others. As far as it goes, it has accordingly given us our most positive, reliable knowledge, a kind of truth upon which men can and do agree. It has gone furthest when dealing with subject matter, such as physics, that lends itself to mathematical deduction and verification by controlled experiment. Physics also makes clearest that science never goes far enough for our living purposes, never tells us the whole truth or nothing but the truth, for it abstracts the quantitative aspects of our experience of the external world, what can be counted, weighed, measured, or stated in mathematical formulas. Hence it is commonly said that science deals with facts or things, not with values, and commonly concluded by scientists themselves that human values are merely subjective, not "real." Nevertheless science has historically been distinguished by its devotion to the disinterested pursuit of truth and the value of truthfulness, or fidelity to fact.[1] In its efforts to be wholly objective it likewise exalts the value of intellectual honesty. When social scientists say that their business is simply to inquire into the social reality, what *is* rather than what *ought* to be, and that statements about values containing *ought* are unscientific because strictly unverifiable, they are likely to forget that they are saying this is how scientists ought to operate in order to prove the value of truth-seeking, confirm their debatable assumption that men always ought to seek the truth.

More important for my present purposes is the question of the relation of science to technology, or the essential difference between them. Today they are often identified. It is said that there is no essential difference in the spirit of the engineers and the scientists working, for example, in the space program; they are engaged in a common enterprise. Certainly the great majority of scientists employed or subsidized by industry and government are working on applied science, which is another name for technology. And modern science and technology have from the beginning been intimately related, much more so than was Greek science. As the pioneers in the seventeenth century were deeply indebted to the instrument-makers, so their descendants today are head over heels in debt; they owe to technology all their well-equipped laboratories, the means of verifying the latest theories. The most elaborate equipment of physicists in particular is blurring the distinction between physics and engineering. The very organization of science today is a mode of technology, in keeping with the growth of organization since the Industrial Revolution. So is its professionalization.

Yet I think it impossible to understand and appreciate science unless we distinguish it from technology. Men had technology from the beginning of their history, whereas science was a much later development and modern science another novel one. Many scientists are still engaged in basic research, in a spirit of pure

---

1. Hannah Arendt has noted that none of the higher religions except Zoroastrianism included lying among the mortal sins. There is no simple commandment: Thou shalt not lie.

curiosity. If they are confident that their findings will eventually prove useful, as most scientific knowledge does, they are not thinking primarily of practical utility; they simply want to know, for its own sake. They are animated by the spirit of "play" that Huizinga made out as the primeval root of all major cultural interests. For this reason Jacques Barzun—no ardent lover of science—declared that it was "in the strictest and best sense a glorious entertainment." For this reason too scientists are more concerned than most engineers over the human values of science that I have stressed. The engineer may also find his work satisfying for its own sake, not merely as a means to practical ends, but he is perforce always concerned with practical application and has long been a more willing servant of business, open to the charge that he devotes himself to finding better means of doing what shouldn't be done at all. For dedicated scientists pure science is an end in itself, like art. They regard it as a means only to the highest interests of humanity.

Hence many of them are not happy over the eminence of science today, the kind of corruptions that came with becoming a national Establishment. Most of the scientists busy in research no doubt rejoice in the wealth of opportunities, all the money now available, and many have appeared quite willing to work for the military on projects, such as chemical and biological warfare, that scarcely further the highest interests of humanity. But this is precisely what troubles many others, in particular most of the more eminent scientists. The trouble began during World War II with the most ambitious scientific project to that time—the development of the atomic bomb.

The nuclear physicists who carried through this project were unquestionably inspired by the patriotic or idealistic motives that made gentle Einstein recommend it to President Roosevelt: they feared that the Nazis might develop such a bomb. After the war many continued to work on nuclear weapons out of a sense of patriotic duty, or fear of the menace of the Soviet under Stalin. Yet most who spoke out were deeply disturbed. Although scientists had traditionally felt free to pursue their research wherever it led, on the placid assumption that truth was always good for mankind, they now had a much more acute social conscience than they had displayed in the past, or than most technicians still appear to have.[2] Leading scientists were among the first to shake off the national obsession with the Cold War, and to appeal for a treaty banning further testing of nuclear bombs. They expressed shock when President Eisenhower warned the country against possible domination by a "scientific-technological elite," even though his advisers quickly explained that he was not referring to science in general, for they had been accustomed to blaming only technology for the abuses of science; but I assume they suffered from a guilty conscience too. At any rate, few were such resolute hawks as Dr. Edward Teller. The *Bulletin of the Atomic Scientists*, with on the cover a clock whose hands are approaching midnight, has voiced their sense of social responsibility ever since 1945.

Another obvious reason why many scientists are troubled is the secrecy of much research, all the rigid controls imposed by security regulations, which on the whole promote insecurity. They had been accustomed to freedom not only in pursuing their research but in publishing their findings, communicating with fellow scientists in other lands. Before the war leading scientists had opposed federal support of science as a possible threat to their freedom, again a consideration that would not bother most technicians. Now they recognize that only such support has made possible the big research programs they welcome, but they are still uneasy over the costs to pure science; as Don Price observed, they regard the new relationship of science and government as a shotgun marriage. Similarly most of the scientists employed in industry complain of restrictions on their freedom. Few industrial research laboratories have the wit to give them their head, realize that for the long run it might be more profitable to let them follow their hunches in their own way, not hold them down to immediately practical research. Scientists who accept positions in

2. In his memoirs Max Born summed up the change: "Today, the belief in the possibility of a clear separation between objective knowledge and the pursuit of knowledge has been destroyed by science itself. In the operation of science and its ethics a change has taken place that makes it impossible to maintain the old ideal of the pursuit of knowledge for its own sake which my generation believed in."

industry have no clear right to complain, since by this time they should know that this is the price they will most likely have to pay; but in any case they do not pay it gladly.

Even the elaborate, costly equipment that scientists are now given to play with disturbs some of them. Norbert Wiener wrote that the new generation of research workers consisted primarily of technicians who are unable to do research without the help of machines, large teams, and immense sums of money. A physicist who visited a Russian laboratory reported: "They're almost as overequipped as the Americans." Hans Bethe noted a particular price paid by physicists for the big computing machines that cost millions: "Many scientists think only of how to put the problem on the computer. They no longer think about the problem itself." Others complain that they have lost flexibility and spontaneity, the excitement of simply running into the laboratory when they have a new idea for an experiment, and of devising their own apparatus. And the big federal projects, which amount to a kind of scientific pork barrel, have got scientists involved in unseemly politics. Among them are smart operators, good at getting their hands on the pork. An editorial in *Science* magazine commented that only scientists who mixed in the "right" circles and had the "right" connections were likely to be called in as consultants on the big projects. In general, the scientific community has suffered from much back-scratching and in-fighting.[3] Scientists have always had their share of vanity and personal ambition, have never been so wholly disinterested as the "mystique of science" has it, but their new eminence has made it plainer that they too are frail, fallible mortals.

Many fret over still another cost of their popularity. The military, Congress, and the public are alike prone to the illusion that scientists can do anything. The kind of miracles that the military want them to pass are likely to be distractions from what they consider fundamental research, but so are projects popular

3. A few years ago an accountant on the Atomic Energy Commission informed me that a bitter feud was going on between the high-energy and the low-energy physicists. Although I am not sure exactly what these terms mean, my guess would be that the high-energy faction came out on top.

with Congress and the public. A current example is the space program, designed primarily to put a man on the moon. Scientists know that this project, proudly described as "the largest and most complex scientific, engineering, and technological undertaking in the history of the free world," is chiefly a matter of engineering. They also know that it is silly to boast of man's "conquest of space," for he is conquering only a bit of it within his own solar system—an infinitesimal bit in relation to the stellar universe. At the same time, the program has exciting possibilities for the future of science. In outer space, scientists can hope to learn much more than they can from the most powerful telescopes on earth, for instance about such problems as the origin and evolution of our solar system, and the possibilities of life on other planets. But scientists were unable to set the priorities. The billions spent have been largely earmarked as moon money, even though they pointed out that machines with instruments would be more efficient for scientific measurements than a man on the moon would be. "There is plenty to do," said Lee DuBridge early in the project, "without trying to nail the American flag on the whole solar system by next week." This only made Congress and the Kennedy administration more convinced that the all-important thing was to put an American on the moon before the Russians landed there. Some scientists were not simply exhilarated when we did in fact plant our flag on the moon.

Within the scientific establishment, its mere size and organization have created the usual problems. Hans Bethe of Cornell University has lamented that he now had to spend half his time on administrative matters that had no direct connection with science. Since physics has become an industry, with a kind of mass production of its wares, all physicists engaged in big projects have to put up with much more routine, and enjoy as much less opportunity for initiative. In all fields money and organization have consolidated the role of scientific research as an institution rather than an adventure. Horace Miner, a student of the institution, advised that the first step was to get the necessary funds: "The placing of priority on financing does not deny that back of every research project there must be an idea, but ideas should not be allowed to retard researching." On the record they seldom do retard it. By now I assume there are more than 50,000 scientific journals, and more than the three hundred journals of abstracts of last report (though perhaps not yet one of abstracts from the abstracts); but a layman need not worry over his ignorance of the millions of articles they publish. Contributors to the *Bulletin of the Atomic Scientists* have deplored what one might expect: the great bulk of these articles are trivial, before long will be worthless. Too many of the articles serve only the purpose of enabling mediocre men to take their place in the establishment.

For laymen there remains the most unfortunate consequence of the extraordinary growth of science. It has long since become impossible to keep up with even the major advances in the sciences, the exciting work being done on various frontiers. Early in the nineteenth century the author of a book on professional education could still write that a "country gentleman" could learn in a few hours enough about mechanics to judge the merits of new machines, and that he could as easily pick up an adequate knowledge of astronomy, which "is indeed peculiarly suited to a country life." Today even a physicist, Hans Bethe remarked, can no longer keep up with the science of physics, let alone biophysics, chemistry, and astronomy. In the more alarming words of the biochemist Erwin Chargaff, "A unified and consistent vision of Nature has become impossible in our days, at any rate for working scientists." Under these circumstances no layman can pretend to such a vision, unless it be some old-fashioned one suited to country gentlemen, nor can he hope to know just where "science" now stands on the nature of the universe. He can be sure only that if he masters the latest theories, scientists will by then have moved on to new and probably stranger ground.[4]

---

4. Not long ago I heard Murray Gell-mann give a delightful talk on the latest theory about the elementary particles of matter and "anti-matter," now called "quarks"—a talk in which I enjoyed every word no less because I understood few of them. It was a "crazy" theory, he said, but it seemed to be working. As for the question whether quarks are "real," a friend was looking for them by grinding up oysters, on the assumption that "most things with curious chemical behavior in the ocean eventually are eaten by oysters." So far he had found none, but that was perhaps just as well: "Mathematical quarks are even easier to work with than real ones." Laymen may at least

Chargaff also suggested, however, that laymen need not be alarmed over this state of affairs, since it involved an element of mere fashion: "Each science protects itself from its neighbors by a cordon of slogans and catch-words; and fashion dictates whether this year we are featuring enzymes or proteins or nucleic acids and whether we wear the molecules long or short." Instead of attempting a survey of the frontiers in the natural sciences, I shall venture only some further comments on the mentality of scientists today. This concerns all of us because they have been multiplying at a rate that might be alarming if it continued indefinitely, and meanwhile they occupy such high positions in our technological society, represent so large a proportion of its brainpower.

Often they are charged with arrogance, a pride that makes them much too sure of themselves in spite of their professions of humility before the unknown and the unknowable. Thus one writes that "experimental science could beyond any doubt allow men to solve their principal difficulties," and that "biocracy," or "organization in accordance with the basic laws of life," was "our only chance of salvation." Often, too, they are charged with naivete. When they speak *ex cathedra* about public policy or the state of the world, their wisdom may sound like platitude, the vague generalities they inherited from the Age of Enlightenment. Traditionally they have equated science with progress, and they still tend to think that every advance in knowledge is necessarily good for mankind, just as engineers assume that every new product and service promote the public welfare. Loren Eiseley complained in particular of the too common neglect of civilized values by the apostles of progress:

> Even now in the enthusiasm for new discoveries, reported public interviews with scientists tend to run increasingly toward a future replete with more inventions, stores of energy, babies in bottles, deadlier weapons. Relatively few have spoken of values, ethics, art, religion—all those intangi-

---

get some idea of how the minds of physicists work these days, and of how far they have got from mere common sense. Lovers of literature may rejoice that the name "quarks" came from James Joyce's *Finnegans Wake*.

ble aspects of life which set the tone of a civilization and determine, in the end, whether it will be cruel or humane; whether, in other words, the modern world, so far as its interior spiritual life is concerned, will be stainless steel like its exterior, or display the rich fabric of genuine human experience.

The dangers of the simple faith in science, long apparent in nuclear weapons, have recently been dramatized in Eiseley's own science of biology, where the most revolutionary advances are being made. (The basic revolution in modern physics, started by Einstein's theory of relativity and Planck's quantum theory, was substantially completed early in the century.) The breaking of the genetic "code" has made conceivable the control of human heredity. Ordinary eugenics, even by such modern techniques as artificial insemination, is a very slow, inefficient way of improving the genetic stock, but the new knowledge of the chemistry of heredity may make it possible directly to alter particular genes and achieve an enormous gain in speed and economy.[5] Such powers may excite scientists because of their obvious beneficent possibilities, or simply as another triumphant demonstration of the power of science. This is nevertheless a pretty dangerous kind of power to give men, especially when one considers the possibility that it might be exercised by a Hitler and the kind of scientists who experimented on his victims in the concentration camps. Even so we can be confident that the researches will go on. The sociologist Philip Rieff tells of a physicist who was asked whether he would continue work on the experiment he was doing if he knew that its successful completion might destroy the world; the answer was Yes. I got the same answer when I asked an eminent geneticist about these researches, which he freely admitted had alarming possibilities. The passion to get at the truth at any cost

---

5. It is not widely known that thousands of Americans are already bred by artificial insemination, for the practice is not publicized; though not illegal, it is not clearly legal either. In any case it is not raising the IQ level in America. In selecting a donor of the genes, the common practice in hospitals has been to pick an interne of the same religion as the mother—as if there were such things as Catholic genes or Baptist genes.

is surely an honorable one (the geneticist was a very humane man); but it means that pure science has the same kind of irresistible drive as modern technology, and for the rest of us may present the same danger in becoming an end in itself.

Yet I am more impressed by the humanity of most of the leading scientists. Biologists have not been simply dazzled by the possibilities of controlling human evolution from now on, nor have they been trying to sell the public the idea of scientific breeding on a large scale; much more urgently they have been warning the public of the dangers of the population explosion, the imperious need of birth control. Similarly nuclear physicists have been most insistent on the need of controlling nuclear bombs. Prominent as scientists now are in government, they do not make the major decisions determining public policy—they merely advise; and one might wish that their advice had been heeded more often. If they often seem politically naive, they at least look somewhat wiser than the professional realists in the Pentagon, the State Department, and Congress. As for their views about our technological society as a whole, the leading scientists are seldom simple materialists or simple utopians. Many of them are as sober and civil as the late Robert Oppenheimer, or today I. I. Rabi, René Dubos, Harrison Brown, and Loren Eiseley, to name but a few. They seem to me rather different from the public image of the scientist.

In the popular mind this image is a curious compound of simple ignorance, a kind of Platonic Idea, and ambivalent feeling. To begin with, the scientist is a wonder-worker, usually credited with feats that strictly were more the work of engineers; he excites awe because most people have no understanding of pure science, but confuse it with technology. He is also familiar as the white-coated man with the test-tube in the ads, cool, detached, utterly absorbed in his work. In this guise, however, he begins to look a little inhuman. He is known to be rather queer because of his relative indifference to money and social status; he is at home only in the laboratory, not in the Rotary or any Boosters clubs. And there is an undercurrent of suspicion verging on animosity, in particular toward physicists. This is due in part to their un-American tendencies, but also to some awareness that the wonders they created have been horrors too, some feeling that they are responsible for the evils of our life. In the high schools fewer than five per cent of the students have been taking physics—a much smaller proportion than in the past, despite the national emphasis on the need of more scientists.

Such hostility has been most explicit in the literary world. Jacques Barzun's *Science: The Glorious Entertainment,* for example, is chiefly a diatribe against both science and technology as the root evils of our horrid civilization. Among his legitimate grievances is the kind of positivism that asserts that only science gives us knowledge or truth, and that in effect reduces literature to mere embellishment or entertainment. More debatable is another significant difference between the "two cultures" that Barzun reflects, even though he denies that we any longer have two in a society so completely dominated by science and technology. Literary people have tended to be pessimistic about the human condition today, as he is. Scientists are constantly sounding the alarm over the dangers threatening us, but they generally remain more optimistic.

Again their optimism can often seem naive because of their too easy equation of science with progress. Nevertheless it is a natural disposition in view of their remarkable accomplishments, and it seems to me on the whole a healthy one, especially because pessimism has become a badge of literary respectability. They know that science has enlarged and deepened our view of the world, done much to make it One World both materially and spiritually, and that we have the technical means of providing a better life for most people on earth. At Pugwash conferences American and Russian scientists were able to get along better than their governments and make a start toward peaceful coexistence; beneath their political differences they had not only a common faith in science but a common vision of both what might be and what ought to be.[6] At least scientists have been keeping on the necessary job of trying to do something about our problems.

---

6. Andrei Sakharov, a brilliant theoretical physicist who became the youngest member of the Soviet Academy of Science, has circulated an impressive pamphlet that was reported in the November 1968 issue of the *Bulletin of the Atomic Scientists.* In reviewing both the terrible dangers of modern technology and the wondrous scientific

So far, however, I have been dealing chiefly with the natural sciences. The social sciences, or more broadly what are now called in America the behavioral sciences, are a rather different matter. About them I would add some reservations, apart from their obvious difficulties in having to deal with subject matter that does not lend itself to mathematical deduction, controlled experiment, or the measure of precision, certainty, and warranted prediction achieved in the natural sciences.

The rise of these new sciences was another characteristic development of a technological age. Men had thought about society ever since Plato, of course, but only in the nineteenth century did they make it the subject of a systematic, empirical study, which Auguste Comte named sociology. If sooner or later they were bound to extend the range of scientific inquiry, an immediate reason for the new science was the Industrial Revolution, the rise of a new kind of society, which made them more aware of how little they really knew about their society. Comte concluded that the essence of this "industrial society" was not capitalism, as Marx would say, but industrialism itself. Among the contributions of Max Weber, a later pioneer, was his classic study of bureaucracy, which he linked with the rise of the machine and considered the deepest tendency of modern society. By the end of the century anthropology had joined economics and sociology as independent sciences, and students of government were calling themselves political scientists. Psychology, also established as a science instead of a branch of moral philosophy, would

possibilities, he emphasized that the primary need for avoiding the one and realizing the other was complete intellectual freedom, "fearless and open discussion," and the "deepest respect" for universal moral values. He was therefore highly critical of both the U.S.S.R. and the U.S., but also emphasized the need of their uniting in the cause of disarmament, world peace, and aid to the many poor nations threatened with hunger. For the latter purpose he suggested that both nations levy a development "tax" of 20 per cent of their national income for fifteen years. He too may sound naive when in an "optimistic variant" supplementing his review of the terrible possibilities he predicted that in the 'seventies they would impose such a tax, which is almost unthinkable in the present state of America; the wise money will no doubt continue to bet on the professional realists in both countries. But so much the worse for the rest of us, who might prefer Sakharov's kind of wisdom.

in our time be linked with them to make up the behavioral sciences.

A different line of development stemmed from Adolph Quételet, a mathematician who gathered and analyzed social statistics in the hope of deriving laws from them. Comte denounced his "Essay on Social Physics" as an abuse of sociology, but in America the future belonged more to Quételet. His methods would be incorporated in the reigning methodology, featuring fact-finding, statistics, and measurement, while Comte's efforts to build a grand comprehensive system still seem premature, even though a vast amount of data has been accumulated by now. Another significant difference has been accentuated by the emergence of the behavioral scientists. Whereas Max Weber was much worried by the bureaucratic drive to technical efficiency and rationality, the behavioralists are quite at home in a technological society. They have multiplied like rabbits since the war, so much so that one of them has proudly given still another name to our age—"the Age of Behavioral Science." Another sums up the "revolutionary change" they have brought about in one generation: the whole field of their science has become "technical and quantitative, segmentalized and particularized, specialized and institutionalized, 'modernized' and 'groupized'—in short, Americanized." Or as a more irreverent commentator put it, the *ize* have it.

Now, the importance of the social or behavioral sciences is so firmly established that it need not be demonstrated. Although I cannot do them anything like justice in this brief survey, I should emphasize that they have contributed not only a great deal of useful knowledge but stimulating thought, new perspectives, new insights. David Riesman, Daniel Bell, Edward Shils, Oscar Lewis, A. H. Maslow, Karl Menninger—these are only a few of the many men to whom I myself feel indebted, not to mention the great pioneers like Weber, Pavlov, and Freud. Critics of these sciences can forget their own indebtedness to them because much of the original thought that has come out of them has entered the commonplaces of contemporary thought. Yet just because they have acquired such prestige I think it is still necessary to dwell on their limitations, and in particular some dubious tendencies, such as the addiction to technical jargon, that reflect the tyranny of technology.

Most behavioral scientists have confined themselves to questions they can hope to answer with some assurance by their techniques of fact-finding and measurement, now with the help of computers. This is a quite sensible policy—but only if they keep aware of the implications of their own behavior, realize that they are therefore ignoring the most fundamental problems, which cannot be so segmentalized, particularized, etc. As it is, too many of them do not appear to realize how trivial much of their research is, and how often it produces only ponderous statements of the elementary. One big research institute, for example, looked into the matter of creativity and conformity and came out with such findings as that "group pressures inhibit originality" and "support for a deviate opinion, however small, reduces the amount of conformity." The best excuse for such banality is that relatively little research has been done on creativity, or on the kind of behavior represented by high culture. Statistical analysis is easier with ordinary or crowd behavior, laboratory techniques work best on conditioned or animal behavior. Man studied as a "behavioral" animal often looks more like a rat than a purposeful human being. "We have already learned quite a bit about the brain mechanisms for hunger, thirst, and aggression," wrote a former president of the American Psychological Association, who then revealed the usual priorities by adding that "we should in the near future start learning more about curiosity, affection, and—I daresay—something like what we now vaguely call altruism." Aggression is positive and precise, altruism tentative and vague. In general, behavioral scientists appear more optimistic than natural scientists about our condition, but on the brash assumption that in order to solve our problems we need only a lot more behavioral research, beginning with more measurements. Even a leading sociologist has said that there should be "no difficulty about collecting happiness ratings," which if taken regularly would provide "an authoritative measure of our times."[7]

---

7. I am refraining from the scholarly usage of identifying these references because I am concerned here with tendencies, not persons, and also because I have reason to think that these persons may have been writing in a careless moment. Even Homer nodded.

For as obvious reasons of convenience social scientists usually concentrate on contemporary society: their techniques cannot be applied readily to the past. In thus distinguishing their science from history, however, they have not only commonly betrayed a lack of historical sense but neglected some obviously important problems that can be sized up only in some knowledge of the past. As Edward Shils has pointed out (and everything I am saying here has been said by some sociologists), one such problem is tradition, which is all the more important because of the untraditional tendencies of a technological society. Revolution is naturally too large a subject to be handled by their methods, but until lately they have given little attention even to change, the distinctive essence of our revolutionary society; statistically controlled analysis is convenient only with slow or slight change. Another convenient but important study slighted when not ignored by social scientists is the reality of the individual, the importance especially of leaders and creative individuals, which may become clearer in a broader, longer view. An acknowledgment of something like this appeared in a "theoretical conclusion" ventured by one of them in their characteristic terminology, that "nonuniform events present problems for which the expertise that especially characterizes the bureaucratic organization is not useful or is actually disadvantageous, and that the characteristic of the primary group might provide a better organizational base for goal achievement under such conditions."

Laymen may be more troubled by the common tendency to not merely specialization but special*ism*, with the pedantry and arrogance it encourages. About this I merely repeat what I wrote in *Freedom in the Modern World*:

> Long ago Arthur Balfour protested against the "pernicious doctrine that superficial knowledge is worse than no knowledge at all," for such knowledge is all a man can hope to have and enjoy about most subjects. With this came the more dangerous notion that no opinion is worth much unless it is based on thorough research—only the expert is qualified to speak. It was most insidious in the social sciences because these deal with matters about which thoughtful men in the

past always assumed they were entitled to have opinions, and today still must have opinions, finally trust their own judgment. If most laymen are still too sure of their opinions, experts are as plainly liable to arrogance, and the pertinent questions remain: Are social scientists in general noted for their wisdom? Is their judgment on social and political problems eminently trustworthy? One reason for doubt, apart from their endless disagreement, is that in America they were carrying furthest the tendencies fostered by the advance of science and its alliance with technology. As specialists they grew devoted to technical "efficiency," in both their own labors and their social ideals, at some expense of humane interests. They talked most freely of the "functions" of men as servants of social purposes, "role-players," or anything but human beings trying to enjoy their lives. They illustrated the increasing functional rationality in the organization of modern life that troubled Max Weber: a rationality that could be unreasonable because it lost sight of broad human purposes, the ends of civilized life.

Weber himself, however, forced another troublesome issue, the taboo against value-judgments. In his efforts to make sociology a rigorous science, he insisted that its practitioners refrain from judgments about social good and evil because as scientists they could never verify them. Logical though this conclusion was, he was deeply troubled by it because he was in effect divorcing science and the humanities, leaving his own ideals high and dry; to him this was another crisis in Western culture. Behavioral scientists feel no such crisis, but for ambiguous or illogical reasons. Many who dutifully rule out value-judgments in theory rejoice in their realism and their illusion of complete objectivity, because of which they are sure their opinions about social problems are better than the opinions of laymen even though they cannot strictly verify them; they trust to the sovereign functional values of organization or technique. At the same time many pride themselves on being useful members of society especially because they hope to improve behavior, maybe raise "happiness ratings." It could be said that one reason for their prestige is that many are so well adjusted to American society and its ruling values.

Thus another recognized leader in the social sciences noted that another distinctively American development has been the remarkable growth of "attitude research," to which is devoted more than half the annual expenditures on all types of social research. He described attitude research as "the most efficient mode of self-observation yet devised"; so to him it meant "knowledge for betterment," a major contribution to "the process of self-regulation through self-observation that we in the modern world have come to call Social Science." To laymen it may have another meaning. The most generous sponsor of such research has been business, for purposes of selling goods. Specialists in motivational research (known irreverently as the depth boys) have been helping advertisers to determine the tastes, mold the desires, and in effect exploit the gullibility of consumers. Other behavioral scientists have made a similar contribution to the self-regulating political processes of American democracy. When Ronald Reagan was running for Governor of California, his managers hired a research team headed by two professors of psychology to "collect research facts, determine how Mr. Reagan feels toward a specific subject, and then figure out how best the issue may be presented politically to motivate voters to cast their ballots for the candidate." (So read the report in the newspapers, without comment, since they too are supposed to avoid value-judgments in their reporting.) One might add that if values are none of the business of scientists, why should they not sell their skills for commercial, political, or any other purposes? But however respectable this practice has become in America, where all kinds of commercialized dishonesty are respectable, it can look like a scandalous betrayal of the traditional scientific ideal of a disinterested pursuit of truth, in the interests of humanity. It is another example of the danger of identifying science with technology.

At any rate, some behavioral scientists are unhappy about the unsavory reputation their science has been getting as an instrument for manipulating people. Behaviorism itself was initially popularized in America by John B. Watson's boasts about how he could condition people. (He ended in the advertising business.) Recently the president of a psychological association expressed much alarm over the dangerous power we are acquiring over people, ways of influencing behavior and altering personality—much more power, he said, than laymen realize is at somebody's disposal. The ways now include biochemical manipulation of the brain, controls demonstrated in experiments on animals. Hence the boast of one behavioralist: "We will be able to modify man's emotions, desires, and thoughts." He added that it will be possible to produce a feeling of happiness without any real basis for it—a possibility that I should think might alarm psychologists and motivation experts because it might put them out of business. The rest of us may be more disturbed by the growing experimentation on human beings, which even in medical research raises ethical issues. Short of such extremes, some behavior that laymen have regarded as unnatural or inhuman is by scientific standards considered normal in our kind of society, the only kind that most of the experts know. Half a dozen psychiatrists who examined Adolf Eichmann pronounced him completely normal.

Still another apparent means of manipulating people is the new science called "social engineering." Although I know little about it, I share the misgivings of Aldous Huxley over the prophecy of an enthusiast: "The challenge of social engineering in our time is like the challenge of technical engineering fifty years ago. If the first half of the twentieth century was the era of the technical engineers, the second half may well be the era of the social engineers." In raising the question who will engineer these engineers, the possible controllers of Brave New World, Huxley remarked that the answer (at least by implication) is a bland denial that they need any supervision. "There seems to be a touching belief among certain Ph.D.'s in sociology that Ph.D.'s in sociology will never be corrupted by power. Like Sir Galahad's, their strength is as the strength of ten because their heart is pure—and their heart is pure because they are scientists and have taken six thousand hours of social studies."

As in the natural sciences, there remain the obvious beneficent possibilities of the knowledge and the power acquired through the behavioral sciences. In America we may discount the extravagances of their practitioners because Americans have always had a passion for quick and sure results. If our gains in "self-regulation" are questionable, the clear gains in "self-knowledge" may count for much more in the long run. As social and political scientists have lately been rediscovering the wisdom of Max Weber, so we may hope that the future will belong to the humbler, wiser ones among them. But in this hope I think we need to keep fully aware of both the inescapable limitations and the too common abuses of the social or behavioral sciences.

9

Government

History, it was said in the last century, is "past politics." In their practice most historians still agreed, concentrating on affairs of state. These are always very important affairs, of course, but they were becoming more important for reasons that neither historians nor statesmen were yet studying—above all, the growth of modern technology. There is no clearer example of its profound influence than the major developments in both government and political theory.

Democracy, it is true, was born independently of the Industrial Revolution. Its basic ideals were proclaimed in the American and French Revolutions, and in the United States were politically established before the industrial transformation of the country. As we have seen, however, the growth and spread of democracy in Europe owed much to the growth of industrialism. The rising business class, which naturally preferred representative government to the traditional political order, did not favor democracy, but it was creating an industrial proletariat that also wanted to be represented in government. And a particular source of agitation was socialism, a new political and economic theory that did come out of the Industrial Revolution. Sired by such thinkers as Saint-Simon and Robert Owen, it soon bred many diverse theories, of which by far the most influential would be the revolutionary socialism of Karl Marx, but they alike offered visions of a radically different kind of collectivized society. Marx notwithstanding, most socialists also agreed that their goals could be achieved by peaceful, democratic means. In England the workers' party that he helped to instigate did not call itself Socialist, preferring the good old English name of Labour. In Germany the Social Democrats, who before World War I built up the strongest socialist party in Europe, dutifully rehearsed Marx's revolutionary slogans about the class war, but this was essentially a harmless ritualistic performance. The leaders of the party and most of their followers were willing to seek their goals piecemeal by democratic processes.

Hence the Western world was unprepared for the revolutionary change that began during World War I—the rise of totalitarianism, a force that democracy has had to reckon with ever since. Karl Marx came into his own in the unlikely country of Russia; if he might not have recognized the alleged "dictatorship of the proletariat" set up by Lenin, it was at least set up in his name. The later totalitarian regimes of Mussolini and Hitler differed in some respects, and both from the Soviets in more important respects, but all three differed fundamentally from democracy. They were alike as states completely controlled by a single party, headed by a dictator, which outlawed all other parties, permitted no open public criticism, and exercised a more nearly "total" control over the economy, the press, education, and culture than government ever had in the past. For democrats the common name for them was "tyranny."

Tyranny is of course nothing new in history. (The Greeks gave us our word for it.) Totalitarianism, however, is strictly a revolutionary development, and again one that is directly due to modern technology. This built up the social machinery that alone made it possible: the large-scale organization in both government and industry, rapid transportation and instant communication throughout the state, the means of massive propaganda for controlling public opinion, and the immense power that a modern state can quickly bring to bear. Technology also provides the arguments that make Communism attractive to countries seeking to modernize: a government that brooks no opposition and completely controls the economy can more quickly mobilize and develop the resources of a country, just as it made backward Russia a great industrial power.

Even so I propose to center this chapter on the problems and prospects of democracy in a technological society. My main reason is no doubt a prejudice in favor of democratic government, but a sufficient excuse is that the great majority of mankind have come around to accepting, at least in theory, its once revolutionary ideals of human rights, which are incorporated in the Charter of the United Nations. One reason why Hitler failed is that his "national socialism" had no universal appeal except to would-be dictators; it had nothing to offer the common people of other lands. One reason why Communism has had more staying power is that it does have universal appeal in its promise of justice for the exploited and oppressed all over the world. The Soviets and their satellites pay at least lip service to democratic ideals by calling themselves "people's democracies." Communist parties in other European countries are accepting democratic processes, trying to win elections instead of calling for a revolution, much like the Socialist parties they once denounced for trusting to such peaceful means. Although dictatorship has been common enough in the new nations, few have turned to outright Communism. I would not say confidently that the future belongs to democracy, but I would say that no political issue today is more important than its prospects.

In America technology has most obviously antiquated the political system set up by an eighteenth-century Constitution for an agricultural society, on the assumption that both the powers and the responsibilities of the federal government should be sharply limited. By *ad hoc* processes, without plan, the government has grown vastly bigger than the Founding Fathers could possibly have anticipated, at a time when a handful of men were enough to run it. The President, who was supposed to do little but preside with dignity as Head of State, has acquired immense powers. He is aided by many specialists who do the planning for fiscal and economic policy, foreign trade, public health, and many other responsibilities that the government has assumed. Public administration has become a "science."[1] And the countless agencies

---

1. Old-fashioned thinkers may remark that universities having institutes of public administration offer no comparable training in statesmanship, which would seem more important even though it cannot be called a science. One might be dismayed by a simple question: How many of the nation's leaders in recent years deserve to be called statesmen? They stack up pretty poorly when compared with the Founding Fathers.

involved in it represent another institution that the Founding Fathers did not provide for, and that for many Americans has become an ogre—bureaucracy.

This is again an old institution that in the modern world technology has made new. Bureaucracy is as old as the large set of officials who served Pharaoh in Egypt, and who at the time represented a remarkable social invention. Today it is different in that it has become not only much bigger and more highly organized but more professional, rational, and technically efficient. While Americans always complain that it is wasteful and inefficient, bogged down in red tape, serious thinkers echo the worries of Max Weber, who in his pioneering study of it emphasized that it is all too efficient for its own limited purposes. "The fully developed bureaucratic mechanism," he wrote, "compares with other organizations exactly as does the machine with the non-mechanical modes of production." It is machine-like in its impersonality, operating according to calculable rules and "without regard for persons." It develops the more perfectly the more it is dehumanized, or "the more completely it succeeds in eliminating from official business love, hatred, and all purely personal, irrational, and emotional elements which escape calculation." Hence it can function smoothly for any kind of ruler, whether or not in the public interest. At the same time it always does its best to protect its own professional interests, as it has done by inventing among other things the "official secret." And Weber also emphasized that in an advanced industrial society bureaucracy has become absolutely indispensable.

In all this, however, he was describing something like "pure" bureaucracy, or the ideal it naturally aspires to. If the German bureaucracy of his day approached the ideal, his account of it needs considerable qualification in America. Here I would stress more the confusion and conflict of bureaucratic interests, and first of all the excessive fear of bureaucracy. The institution is not only clearly necessary for some humane purposes, such as administering public welfare programs, but itself embodies some human values. Its impersonality is not simply cold or inhuman, inasmuch as government has never been warm and loving. Administering rules "without regard to persons" means no favoritism, but at best an impartial justice. For similar reasons red tape has been called the mother of freedom: orderly procedures are a protection against arbitrary rule or inequity. Otherwise bureaucracy may still seem undemocratic as a privileged caste protected by the civil service system that Europe borrowed from ancient China; but few would like to go back to Andrew Jackson's spoils system, based on his belief that training and experience in government were unnecessary, ordinary intelligence was enough for American officials.[2]

Max Weber left open the most critical question for those concerned with democratic government, "whether the *power* of bureaucracy within the polity is universally increasing." European thinkers now say that since the World War it has been increasing. They have been worrying over the decline of parliaments, which have been losing power to the executive and the permanent bureaucracy, often serving as little more than rubber stamps for party leaders. Jacques Ellul is not alone in declaring that democracy is doomed by technocracy, the latest form of bureaucracy. He maintains that the state has become an "enormous technical organism," in which the key decisions are made by technicians not directly responsible to the public and politicians have no real say any more, no real choice except to approve what the technicians tell them has to be done. Or technocracy might be described in Hannah Arendt's phrase as "rule by Nobody."

In America, however, the issues are again more complicated, or hopelessly confused, by a political system that is hardly rational enough to deserve the name "system." To begin with, bureaucracy has unquestionably been proliferating since the war, obeying Parkinson's law. Some of the new agencies exercise very great power, above all those connected with military defense. With the President, whom they "brief" (possibly an all too precise term), they may make critical decisions that Congress only endorses. The

---

2. The historic record of China might reassure those who believe that bureaucracy is fatal to good government. Its big bureaucracy was headed by mandarins selected by competitive examinations based on the Confucian classics; and for all their deficiencies, or their failure to approach the American ideal of a "businesslike" administration, they kept China going, through the rise and fall of dynasties, for two thousand years—a political record unmatched by any other large nation in history except possibly ancient Egypt.

public has no direct say. Most voters are no longer capable anyway of judging the soundness of most decisions about policy, since these are based on a technical knowledge that even well-educated people may lack. In our kind of society we all have to depend on the experts in many matters. And we may worry more just because they are experts, for they are apt to be too sure of themselves, more arrogant than the aristocrats of the past. How dangerous they can be was illustrated by a complaint of the Director of the Los Alamos Laboratories Weapons Division: "The basis of advanced technology is innovation, and nothing is more stifling to innovation than seeing one's product not used or ruled out of consideration on flimsy premises involving public world opinion."

Yet American bureaucracy—as this complaint implies—is by no means an independent, monolithic power. Whereas most European countries had a permanent bureaucracy before they were full-fledged democracies, in America democracy came first and shaped the new institution. Top officials are political appointees. While Americans habitually grumble over their bureaucrats, they can always hope to get a new set in the next election. Congress also keeps a jealous eye on them. It has a voice in approving top appointments and determining the powers of the many agencies, influencing their policy. It harries all departments with requests for favors for local business interests, requests they cannot afford to ignore because they are dependent on Congress for their appropriations. Unlike European parliaments, it retains plenty of power.[3] Accordingly the chief complaint about American bureaucracy is that it falls far short of a "pure" bureaucracy in rationality. Congressmen and Presidents alike do not want expert administrators operating independently of them or of party interests. Professional bureaucrats in the higher ranks may exert consider-

---

3. An incidental example of how well Congress takes care of itself was the new House of Representatives Office Building, with offices for about two hundred Congressmen. Although it generally keeps its building money secret, word got out that the estimated cost of this one was close to $100 million—more than twice what the Empire State Building had cost, and more even than the Pentagon, the world's biggest office building. The Washington *Post* suggested an inscription over the door: "It takes a heap of money to make a home for the House."

able influence in their entrenched positions, but policy is determined by the men at the top or the President who appointed them, under that jealous eye of Congress. The upshot may be described variously as flexibility, responsiveness to public opinion, pragmatic politics, a "bumps and grinds" system, "organized irresponsibility," or a mess.

Hence I pause to review some important history that was being made even as I settled down to this study—history of a kind suggesting that democracy was very much alive, technocrats were not running the show. President Johnson announced his decision to limit the bombing of North Vietnam and make serious efforts to negotiate peace. While he had various apparent reasons, among them was certainly the growing protest against the war in Congress and the country at large, his concern over "divisiveness." Similarly with his more startling decision not to seek reelection. In this his mixed motives presumably included the interests of the Democratic Party, but again an obvious concern over a deeply divided people. In any case, it was the Chief Executive of the American people who made this important decision—not a clique of technocrats. He was heeding public opinion, in the knowledge that he could no longer hope for the consensus he cherished.

On the heels of his announcement came an explosion of the ugliest dissension at home, the Negro riots following the murder of Martin Luther King. The House of Representatives responded by passing the Civil Rights bill, under pressure from both the President and public opinion. This was only a start, and it remained unlikely that Congress would endorse the costly programs urged by the President's Commission on Civil Disorders, but the American Congress clearly had ample power and a real choice. If technocrats had narrowed its choice by the expenses of the war in Vietnam, they were not dictating its decision. Other administrators in the welfare agencies were naturally inclined to favor programs on behalf of the blacks. For bureaucracies too have diverse, often conflicting interests; so far from constituting a monolithic power, they are typically jealous rivals for money and power. For the rest technocracy had nothing to do with this problem—with either the deep-rooted prejudice of white racism or the moral indignation over the treatment of the blacks. The battle would be fought out on the grounds of democratic principle. One could not be at all confident that the conflict would be resolved equitably, not to say peacefully, but again for reasons having nothing to do with technocracy, much more with the shortcomings of American democracy. Besides a tradition of irresponsible politics, these included the irrational, inefficient structure of Congress: because Southerners were entrenched by seniority in chairmanships of powerful committees, a single Congressman from a rural district could defy the Chief Executive or the will of the majority.

All in all, granted the too common narrowness and inflexibility of the bureaucratic mind, I see little reason yet for alarm over the power of the bureaucrats in Washington (always excepting the Pentagon). If the increasing powers of the government are plainly dangerous, the issues have been clouded as usual by the slogans that relieve hardheaded men of the effort of thought. American tradition has it that the less government the better, the more the government does the less freedom the people have; so it follows that the people have been steadily losing their freedom ever since the New Deal. In fact the increasing operations of the government have had mixed effects. The agencies supposed to regulate business have usually done more to protect its interests than the public interest. The most drastic restriction on personal freedom was peacetime conscription, against which there was little protest except by "radicals." The public welfare programs were designed to promote the effective freedom of ordinary people. They have involved many new regulations, sometimes irksome; but readers might ask themselves whether they have actually been enjoying less and less personal freedom because of all those bureaucrats in Washington.

I see more reason for alarm in another technical development dramatized by the war in Vietnam, the official "management" of the news. A common practice in both the Eisenhower and Kennedy administrations, this led newsmen to uphold the public's "right to know"—a right that is not mentioned in the Constitution, or to my knowledge affirmed in any Supreme Court decision, but that is obviously fundamental in a democracy. From the beginning of the Cold War it was severely limited by security regulations, up to a point for sufficient reason: much scientific knowledge and military intelligence had to be kept secret so as not to help the enemy. Still, the enemy always knew much more about the operations of the Central Intelligence Agency and the Pentagon than the American people knew, and the right to secrecy was clearly abused by some top bureaucrats; the publicity bureau in the Pentagon even declared the right to hand out misleading news. President Johnson exercised this right so freely that his administration became notorious for the "credibility gap," which was a polite way of saying that often he was simply not honest with the American people. One example was the secret decision to step up the war in Vietnam, made in 1964 before the national elections. I am brought to the large subject of modern propaganda.

This obliges me to pause, to consider what the word "propaganda" has come to mean. It is another old word, going back to the Roman Catholic committee of cardinals to organize foreign missions—a kind of activity Christians had been engaged in ever since the days of the Apostles. It became widely used as a bad word only after World War I, when people grew aware that the warring nations had systematically employed propaganda that was much exaggerated when not false. Students of it then began playing down its sinister connotations by pointing out its ancient roots in common activities; one defined it simply as "any effort to change opinions or attitudes." This was a useless definition because thinkers, writers, teachers, and preachers had always been engaged in such an effort; there was nothing unreasonable or objectionable about it—except when the methods used were objectionable. I therefore take it that we cannot give "propaganda" a precise meaning, but nevertheless need to distinguish its modern forms, which technology has made essentially new. It has become an organized, systematic, often high-powered effort to persuade or indoctrinate people on a large scale.

Such propaganda is objectionable because its usual aim is not honestly to inform or educate and its methods are not aboveboard. Hence it may be called a way of manipulating people, who are not fully conscious of what is going on even though they have grown suspicious of it. Although Hitler proved that the "big lie" can be effective, it is seldom simply false because propagandists have learned that carefully selected "facts" help to give the necessary appearance of truth. Otherwise it shows little respect for people, sometimes a contempt for them, for essentially it exploits their gullibility, irrationality, or inability to distinguish honest thought from appeals to emotion. Certainly it is not calculated to promote intelligent behavior, develop powers of choice, or make people freer.

Because political propaganda was employed most blatantly by the Nazis and the Communists, we need to remember that it was introduced systematically by the democracies in World War I. (British, French, and American propaganda was much more effective than the German, while backward czarist Russia did relatively little with it.) A massive technological society not only has created its high-powered means but has a particular need of it. The democracies need it to sell all the goods they produce, to maintain their mass media, to get public opinion behind the government, or to keep the nation united. In government its purposes may be benevolent; the problems it creates are more difficult because the leaders employing it may be sincere or self-deceived, may use strictly dishonest methods out of high patriotic motives. At any rate, all government departments have their public relations officers, just as business has its advertisers and publicity men. The public is daily bombarded by official news that may or may not be true. In the newspapers people read more news emanating from a "White House spokesman" or a "ranking officer" in the Pentagon, convenient devices for enabling leaders to insinuate ideas while evading responsibility for them. The already great powers of the President are amplified by all the publicity he commands, and by the ease with which he can suppress, slant, or even falsify news.[4]

---

4. A flagrant example was the news justifying President Johnson's decision to send the U.S. Marines to the Dominican Republic, ostensibly in order to save American lives during a Communist revolution. When newsmen appeared on the scene, they were briefed by Ambassador Bennett before they had time to circulate freely. He told them all about the "Communist takeover" and the rebel atrocities that had littered the streets with a thousand or so mutilated bodies, violated six or eight foreign embassies—news that they dutifully relayed, and that was repeated by President Johnson. Responsible reporters soon found

Offhand, such propaganda makes more plausible the common talk about the "Establishment," or C. Wright Mills' thesis of a "power elite" commanding all the major economic, political, and military organizations. Actually, however, it points rather to the usual confusions, the different establishments that are alike trying to sway public opinion. Even the military-industrial complex is not clearly the controlling power, inasmuch as the President is able to modify or reject its policies and Congress has been growing more critical of it. The President himself is not simply an agent of big business or of any one class. In general, I make out no homogeneous, well-defined "ruling class" in America, but a number of elites whose interests may clash. While they may well have too much power for the rest of us to feel comfortable, it is no more a monolithic power than bureaucracy is.

Another complication is the entry into government of science, which is important even though it is far from being a power elite. This was foreshadowed early in American history, in ways that now seem suitably incongruous. Both George Washington and Thomas Jefferson were interested in introducing scientific improvements in agriculture, Washington even proposing the establishment of a National Board of Agriculture, but devotion to states' rights blocked any government action. John Quincy Adams had no more success when he tried to foster science officially. Still, the practical interests of Americans disposed them to welcome applied science. West Point was the first engineering school in the country. Out of the state agricultural colleges in the Middle West (known in the East as "cow colleges") grew experiment stations, extension programs, and a federal policy of doing more for agriculture than any European country did. Legislators in farm states forgot that such federal aid was "socialistic" and a possible infringement on states' rights. The colleges also supplied the Department of Agriculture with most of its higher officials,

---

out that the atrocity stories were untrue. Possibly the Ambassador and the President were misled by CIA agents, who have compiled an impressive record for untrustworthiness; but six weeks later the President was still telling a press conference that "some 1,500 innocent people were murdered and shot, and their heads cut off." One journalist commented that what this country needs is a "pure news law."

who helped to get it most of the federal funds for scientific research down to World War II. Today the Department remains a big, heavily supported agency even though agriculture has been steadily declining in relative importance.

In assuring the prestige of science in government, World War II also assured more confusion, the abiding problems of the "politics of science." By now scientists are all over the government; not only the President but the Secretaries of various departments have Science Advisers or Advisory Committees.[5] Given the billions of dollars spent on research and development, one might ask: How are they allocated? The answer, as might be expected, is that there is no definite policy, no one man or board in charge of the "decision-making process." The White House has never formulated a clear science policy, while in Congress some seventy different committees are interested in R & D and so are opposed to the idea of delegating authority to a single committee. Until lately little or no thought was given to the political and social consequences of this new development, the decisive influence it was having on the shape of our future.

Hence some scientists lament that they do not make policy or have any real power, but merely advise. They have to deal with Congressional committees made up chiefly of lawyers and business-men, "sharp cookies," as I. I. Rabi has said, but none of them with a scientific background. They report that many Congressmen simply ignore science and its possibilities, while others look to it only for spectacular results, like a cure for cancer. Congress alone has no Science Adviser. Other scientists, however, point out the plainest difficulty—that scientists as such are not qualified to make public policy, lacking both the knowledge and the experience necessary for such purposes, which involve the need of dealing rationally with irrational beliefs and behavior. Nor do scientists have any means of agreeing even on matters of purely scientific

---

5. Typically the State Department ("Foggy Bottom") has found the least use for them. Its Office of the Science Adviser, established in 1951, has partially recovered from the lethargy of the Eisenhower administration, when for some years it had no scientist nor science attachés abroad; but on the record it has had little or no voice in foreign policy.

priorities, or what some have called loosely a "science of survival." There is no such science, only a lot of different sciences that are pertinent. One has but to ask what proportion of the money spent on research and development should be devoted to biochemistry, nuclear physics, ecology, oceanography, meteorology, medical research, and so forth.

But at least all this is to say that there is no apparent danger of a scientific elite controlling the government. As it is, scientists can bring to government not only their professional competence but the values of the scientific spirit. Though I cannot know how much their advice has counted in the determination of public policy, it has evidently helped to offset to some extent bureaucratic tendencies to inertia, habitual appeals to tradition, and the typically American fondness for "over-kill." President Kennedy in particular gave a free hand to Jerome Wiesner, an intimate in the White House who became an influential adviser. The experience that many scientists have been getting in government is making them politically more sophisticated. They report that Congressmen in turn, especially those in the Joint Committee on Atomic Energy, are growing more sophisticated about science. Some universities are bestirring themselves, offering seminars on science and public policy. Although as yet very few students are getting professional training in both, we can hope for more men in public life equipped to deal with these problems.

Meanwhile American government at least makes better use of scientists than do most of the European democracies, with their long established bureaucracies dominated by traditionally trained administrators. Scientists can enter more easily a bureaucracy that is not a tight establishment, more easily advance to top administrative positions. And they have profited from the very looseness of our political system. Congressional committees often seek scientific advice because they are not bound by either party discipline or bureaucratic policy, but make their own policy. A President exercising great powers not clearly defined by the Constitution, with no traditional requirements of coordinating his policies or agencies, can readily make emergency use of scientists as Roosevelt did during the war. Lacking any ideology to speak of, the leaders of both political parties were pleased by the establishment of the Atomic Energy Commission, even though it was described as "an island of socialism in the midst of a free enterprise economy." The government could proceed to pour billions into research and development without consulting the voters, most of whom still have little idea of what is going on. Again the American way of politics is flexible and resourceful for the same reason that it is often irresponsible. All kinds of business can be done under it, for better or worse.

The politics of technology, or "development," are much more confused and complicated, not to say too technical for laymen to be much interested in them. Taxpayers should at least know that the bulk of the development has been done for the military, who except for occasionally frantic innovation, due to the latest achievement of the Soviets or to rivalry between the services, are generally about the most rigid, conservative of the bureaucracies in Washington. For years they awarded contractors the customary cost-plus contracts, ordinarily without competitive bidding, which put no premium on performance, too often resulted in poor performance, and rewarded it with higher profits. (The death of the three astronauts in their spaceship led to some scandalous revelations of sloppy work.) Now they have begun to substitute incentive contracts, but a Rand expert reports that buyers still lack sufficient incentive. Once corporations have sold their system, they will naturally not keep looking for a better one, and the rules of the game reward them for getting it under way with enough steam so that it is very hard to stop; more than one weapons system proceeded when it was already obsolete. Among other technical problems mentioned by the Rand expert are unexpected difficulties arising in the course of development. Taxpayers will not be pleased to read that "development is a business in which errors of 30 or 40 per cent can hardly be regarded as errors." The errors always mean a big increase in costs.

Outside the military departments government enterprise has been more daring, but still complicated by politics. It has been combatting with some success perhaps the most dangerous tendency of bureaucracy, which is not so much that it has been creeping toward socialism as that in a galloping age creeping is its natural pace. One historic landmark was TVA, which restored a region that had been impoverished by private business interests; although much more admired and imitated abroad than at home, it at least became popular enough to survive the attacks of power interests and the Eisenhower administration. During World War II the National Resources Planning Board made a start toward coordinated national planning by surveying resources and suggesting policy for technology, but it was killed by Congress; the word "planning" was still a red flag, and the Board had also run afoul of the Army Corps of Engineers, a bureaucracy powerful enough to defy the White House because it is a Congressional pet as a source of pork-barrel projects. After the war, however, massive government support of research was taken for granted, especially because much of it was conducted by private industry. Only this created another problem. Corporations might still get patent monopolies on discoveries made on public money; taxpayers who paid for the research might also have to pay monopoly prices as consumers. There was a battle in the Senate over the proposal to give A.T.&T. control over the communications space satellite developed by the government. The popular assumption remains that private industry is always much more efficient than government.

This brings us to the roots of the postwar problems. In the first place, government support alone made possible some technological triumphs; even a giant like A.T.&T. could not have launched and financed the space program. Similarly the airplane industry depends on this support for its projected supersonic planes. Other industries, like the merchant marine, depend on the government for their very existence. Yet the prevailing disposition is still to trust to the sacrosanct market economy to take care of all basic needs. As Robert Heilbroner observed, the market mechanism takes assiduous care of wealthy consumers, but not of poor ones; it does not provide adequately for such needs as low-cost housing, health insurance, hospitalization, and medical care. Hence the critical question: How much of the necessary innovation is to be left to private enterprise or the market, and how much to government?

In 1963 Edward T. Chase wrote that debate over this question was likely to grow increasingly hot and divisive, and had the makings of a "fundamental political crisis." At the moment it does not yet look to me like a "crisis"; although the country is still floundering, most Americans appear to be getting accustomed to more government enterprise. But particular problems, such as the black ghettos, do look like crises, and they force further questions in a country that still lacks a national planning board. Who is to do the necessary long-range planning? Who is to choose among the various possible goals, and the programs for achieving them? What bodies are to determine priorities, and how? How can Congress be equipped to play a more responsible role in making the necessary decisions? In short, will the American political "system" be inventive enough to deal with the increasingly complex and grave problems looming up?

The answer is necessarily uncertain.[6] The most that can be said is that a growing awareness of the problems in recent years has led to some measures. The government has set up the Office of Science and Technology; the Pentagon has asked the nonprofit Rand Corporation to study the social implications of technology; Congress has passed the Manpower Development and Training Act, to help workers left with useless skills by automation; a Panel on Civilian Technology, made up of industrialists, academics, and bureaucrats, has studied some sick or backward industries and issued a controversial report on urban transportation; and in general more things have been going on than most people know. Perhaps most important is the valuable experience scientists and engineers have had in dealing with large-scale, long-range problems. The space program is a striking demonstration of what coordination can achieve. If the government decided really to come to grips with such big problems as the city, pollution, and poverty, these proved methods might work more wonders.

But that too remains a very big *if*. Immediately it brings up

---

6. As I write, there is some to do over a nasty problem that I suppose is one among dozens. In the soft coal industry "black lung," a preventable disease caused by coal dust, has been disabling and killing thousands of miners. This has been known for many years to the industry, the United Mine Workers, and the government, but practically nothing has been done about it. Secretary of the Interior Udall then acknowledged that it "warrants most serious concern"; so one might wonder why nobody in power showed any concern—until an outsider, Ralph Nader, publicized the problem.

the remarkable disorganization of the government of a people supposed to have a genius for organization. While it is in theory attractive to delegate power and responsibility to the states and the cities, now clamoring for more federal tax money with less federal control, in practice most state government has long been unrepresentative, inefficient, and corrupt, and most municipal government is even more archaic and disorganized. Hence any major effort to attack the problems of the city or of poverty would at once stir up bureaucratic infighting between the many competing agencies in Washington, the states, and the cities, all further complicated by the greed of local politicians, the customary favoritism and graft, and the usual difficulties in wangling enough money and authority out of Congress and state legislatures. The whole problem looks worse because modern technology obliges us to consider, more than ever before, the needs of not only the present but coming generations. American government is not geared for such considerations. It puts a premium on administrative and technical skills, expertise or know-how rather than purpose, vision, imaginativeness, or creativity. Its "decision-making process" seldom produces bold policies, the kind likely to stir up controversy; generally it results in a settlement on an *ad hoc* policy involving the least disagreement and least risk. In Congress professional politicians are not naturally responsive to the needs of future voters—enough for them to win the next election, which is as far ahead as the political parties look. I see little reason for optimism over the prospects of ambitious domestic programs.

Voters are no more accustomed to taking a long view, least of all in the states and local communities, where one might expect them to be most concerned about the environment their children will grow up in. Typically they are much more concerned about taxes. And so a brief look, finally, at the effects of modern technology on the electorate, and on such democratic ideals as social justice, equality, and freedom.

Most obviously technology has exposed the serious limitations of voters by all the complex technical issues that are too much for their powers of understanding, often their capacity for active interest. Their limitations are more conspicuous because the

political parties—big organizations that were also never anticipated by the Founding Fathers—now spend many millions on
election campaigns. Organization and technique have become
all-important; together with the highest campaign expenditures
to date, they won Richard Nixon the nomination and election to
the Presidency even though he offered no definite program and
aroused little public enthusiasm. Publicity men help to merchan-
dise candidates, manufacture their "public image," and swell the
importance of "personality" on TV, which Adlai Stevenson complained was reducing campaigns to something like beauty contests,
and since his time has elected Hollywood actors to high office. At
the same time, technology has indirectly tended to alienate many
voters through the sheer massiveness of our society, in which they
feel impotent. On surveys many report fatalistic attitudes: voting
wouldn't do any good, it doesn't matter who you vote for, they're
all politicians, all a little crooked, and there's nothing I can do
about it. Under these circumstances they may be more responsive
to "charisma," as of the Kennedys—or of demagogues like George
Wallace. For all such reasons foreigners have remarked that the
election of an American President—the most powerful man in the
world—is too serious a business to be decided by Americans alone.

   In this view, however, it seems foolish to talk any longer about
"the revolt of the masses" or "the tyranny of the majority," at least
in political life. There is also something to be said on behalf of
the voters. In a massive technological society man can hardly be
the "political animal" he was in the little Greek *polis*; it is
unreasonable to expect all people to take an active interest in
politics, and rather surprising that so many do take such an
interest. Democracy does not have to be a Town Meeting in
which all participate; enough if it does not deny or discourage the
right to participate, as it did too long with the Negroes. Neither
is it essential that voters be able to form intelligent, knowledge-
able judgments on all serious issues—a test very few of us would be
able to pass. (Readers might also ask themselves: Do they know
just how to combat inflation?) What matters is that they be able
to size up well enough the men they elect to make the necessary
decisions. Then one may be depressed because their judgment of
men too is often notoriously poor, when not dictated simply by

their preference in parties. But so it has been all through American history, which has seen many mediocre or crooked men in high office; and the record of the electorate in this century seems to me on the whole somewhat better than Americans made in the forty years following the Civil War. Altogether, whether voters will measure up well enough to their increasingly difficult responsibilities in a technological society is another open question. All I am saying is that it still appears to be open.

Technology has had more ambiguous effects on the most distinctively democratic ideal of equality, as a principle of both representative government and social justice. The affluent society has accentuated the poverty of many millions of Americans, who enjoy nothing like equality of opportunity. It has also stimulated more effort to do something about the many poor who have been with us ever since the rise of civilization; the piddling "war" on poverty at least stirred up an immense amount of discussion, research, and report. The welfare state has made Americans somewhat "more equal" than the common people were in the old days, when they were theoretically held in higher repute and not called "mass-men." As for government, technology has most obviously multiplied and fortified the many pressure groups or lobbies that confuse or thwart the will of the majority. Nevertheless my impression is that this will—often nebulous, but often clearer too because of public opinion polls—is heeded rather more than it was in the days before the New Deal, when government was more obviously dominated by the interests of business. Politicians keep a close eye on the polls because voters still have the last say.

Ideals of equality are not incompatible, either, with the new professional elite, an aristocracy of technicians that most Americans fear is against the common man. Democracy of course requires some sort of elite as do all other kinds of government. The new elite is essentially more responsible than the professional politicians (once a new type too, emerging out of the party machines) who under weak or compliant Presidents often ran the country in the past. Despite the American tradition of anti-intellectualism, it is no more undemocratic because since the New Deal it has found more use for intellectuals than government formerly did. (Let us remember the plight of Henry Adams, eager to serve in government as the Adamses had for three generations, but after the Civil War discovering that it had no room or use for him.) The chief reason for worrying about this new elite is not that it is an aristocracy, but that it is a technocracy. It may put efficiency above all other values, or with the help of computers and systems analysis it may make policy on the basis simply of economic factors, the only factors that these techniques can now deal with. "In framing a government which is to be administered by men over men," wrote the authors of the *Federalist Papers*, "the great difficulty lies in this: You must first enable the government to control the governed; and in the next place oblige it to control itself." Today the problem is to oblige it to control the technology that has made it what it is.

Most complex are the issues of freedom in a technological society, apart from the economic freedom that is the chief concern of most businessmen who worry over big government. Writers have worried more over the "mass society" created by technology, the unlegislated constraints on personal freedom because of the pressures to conformity. In this view the welfare state may be regarded as a menace to freedom just because it has done so much more for people, and so has made them more compliant. Herbert Marcuse observed that the well-administered life gives no reason to insist on self-determination: "Democracy would appear to be the most efficient system of domination." Some political scientists say that the welfare state has made obsolete John Stuart Mill's classic *On Liberty*, for the individual who accepts so many personal benefits decreed by the government can no longer insist on an inviolate private realm.

In any case the plainest threats to individuality or personal freedom are still social, not political, and so are a subject for later chapters. Here I should add that technology has lately made it a live issue by a new threat to freedom, the growing invasion of privacy by both government and business. Personality tests have become routine with corporations. Computers with punch cards can readily make available all the information about people compiled by personnel men, credit bureaus, and government agencies. The telephone in one's home can easily be tapped, the living room bugged. Wide use is being made of all kinds of high-fidelity snooping equipment, ranging from tiny cameras, telephoto cameras, sniperscopes, and radio transmitters installed in buttons to lie detectors built into upholstered chairs on which a person can be tested without his knowledge. People may grow aware that there is indeed a private realm they wish to keep inviolate.

For the rest I add only some reservations about the common fears that freedom is being sacrificed for the sake of security. Security is by no means incompatible with freedom. A measure of it is plainly essential for effective freedom, just as only a free man can feel really secure. It is needed especially in an industrial society, in which many fewer people are self-employed or as relatively self-sufficient as Americans used to be. Conservatives who deplore the demands of workers for security forget that the big corporations enjoy an unparalleled degree of it—only the little fellows keep failing—and that no class wants security more than the young college grads headed for the corporations and the suburbs, the great majority of them good Republicans. And if these "organization men" make clearest the real danger that people will sacrifice personal freedom for the sake of security, there is little danger of their enjoying too much security for years to come. The very passion for security, comfort, and ease in America today is making it more certain that there will be plenty of psychological insecurity—as white people in the suburbs are beginning to realize. In a revolutionary age complacence assures neither freedom nor security.

10

Business

Since World War II dramatized the terrible destructiveness of modern technology, the spectacular growth of the economy has demonstrated that it has also given man astounding powers of recuperation. With the help of the Marshall Plan, western European countries were soon producing more goods than they had before the war, and they then enjoyed a boom so unexpected that economists called it a miracle. In America an increase of hundreds of billions in the gross national product makes the figures almost meaningless to the ordinary citizen: he can no more grasp the latest figure of $800 or $900 billion than he can the galaxies made up of billions of stars. I need not review the many new goods that Americans have grown used to in their air-conditioned homes, or "housing services." Supermarkets, superhighways, and supersonic planes help to define the trend: everything is getting super. Less well-known is the rapid growth of scientific management, far beyond the dreams of Frederick Taylor, through such techniques as operations research, systems analysis, cost effectiveness, linear programming, game and decision theories, and computerizing. But they call for more technical exposition than I care to attempt. Not pretending to have mastered them, I propose instead to look more closely at the role of the big corporations, the main source of these advances. Despite the periodic efforts to break them up or hold them down, they have been growing much bigger, dominating the economy more than ever, and all the signs are that they will continue to grow.

Now, even so big corporations have by no means doomed the

small firms and the individual businessman; there is still plenty of room for these, especially in the growing service economy. Neither are they engaged in a conspiracy to monopolize American industry. Their growth is at once cause and effect of the growth of a sophisticated technology, involving long-range planning of mass production, in which both the capital and the time needed for introducing new products are increasing. Even though they are forever singing their own praises, Americans might be much more awed by the wonders they perform in producing the goods the country wants. Nevertheless bigness creates problems that as usual are also not obvious enough to most Americans. These corporations represent by far the greatest concentration of economic power in history. A small number of them—150, say—control half the industrial assets of America. And in recent years more and more corporations have been growing by "the big business of buying businesses," collecting a stable of companies in unrelated industries.

However questionable C. Wright Mills' thesis of a "power elite," there is no doubt at all that these economic powers have great political power, especially in Congress. Still plainer is their immense social influence. While their policies directly affect the millions of people they employ or deal with as suppliers and retailers, indirectly they affect all Americans by their power in planning the economy, deciding what to produce and where, whether to expand or contract their operations, and how much. They may help a small city to grow by building a big plant in it, cripple another city by moving a big plant out of it. They have changed the face of whole regions, in the last generation especially of the Far West. Continuously they influence the habits and desires of Americans by their advertising and their domination of the mass media. It is chiefly they who determine the domestic uses of our technology.

To begin with the ABC's of the corporate economy, or the problems of bigness, the first question is simple: Who owns these corporations? The legal answer is as simple—the stockholders. The champions of big business like to call it "people's capitalism," emphasizing that more than ten million Americans now own stock; but almost all of the people's holdings are tiny. The giants, like

A.T.&T. with its 1,600,000 stockholders, have grown too big for anybody to "own" them in any meaningful sense; so it has been said that nobody owns them. The next question accordingly becomes more pertinent: Who controls the corporations? In theory the answer is again the stockholders, to whom the top executives report in an annual meeting, but this has long since become a threadbare fiction. Even if there were a hall big enough to hold them at the annual ritual, the great mass of stockholders are nobodies who know almost nothing about the corporation and have no real say in running it. In what Adolf Berle has called "the most violently private-property-minded country in the world," private property has in corporations lost its traditional meaning of the right of control. The boards of directors alone control them. The only threat to their sovereignty is a rare battle with well-heeled stockholders trying to muscle in, and these intruders are only other big business interests, not champions of all the little stockholders or rebels against the system. Robert R. Young gave away the whole business when he decided to get control of New York Central. Its chief executive fought him by hiring a public relations firm and a professional proxy solicitor, charging the expenses to Central, and later on Young and his backers charged it with the $300,000 they spent on winning the battle. All this was quite legal, even though the stockholders were not consulted about it, and whether or not they fared better under the new management.[1]

To go on with the ABC's, how then do these boards of directors get into their positions of authority? Except for the rare struggles with rival business interests, the answer is that they were appointed by their predecessors, and will in turn appoint their successors. Hence they have been described as an "automatic self-perpetuating oligarchy." In a country that prides itself on its two-party political system, big business is managed by one-party

---

1. A potential threat to the directors' control is a new financial power, the fast-growing pension trust funds that now total $80 billion and are in the hands chiefly of insurance companies and commercial banks. As yet, however, they have not exercised their power to choose the managements of the corporations in which they invest; and in any case they represent only another form of big business.

government. It is hard to conceive, indeed, how big corporations *could* be run democratically; one must imagine rival slates of candidates for the board of directors, each promising a fuller dinner pail to a mass of ignorant stockholders. At any rate, the fact remains that the government of corporations is a pure oligarchy, which the courts have agreed is not responsible to the stockholders. Considering that most of the industry of America is managed by a hundred or so corporations, whose directors often sit on more than one board, this awesome economic power is wielded by a few thousand men. With few exceptions, moreover, these men have not risen from the ranks of the workers or are not poor boys who made good, like Andrew Carnegie, Henry Ford, and other entrepreneurs of the past. The great majority of them, and of the executives they appoint, came from upper-class families, and are unknown to the public by name.

In a democracy we must therefore ask: What makes their power legitimate? In political rule throughout most of history the need of legitimacy was satisfied by the widespread belief that kings were ordained by the gods, or in secular terms by their role as father symbols, and in our time even dictators have usually taken pains to preserve the forms of election. According to Adolf Berle, the leaders of big business too have a deep felt need of legitimacy, wanting more than anything the accolade of "Well done, thou good and faithful servant." They are too sophisticated to claim they rule by divine right, as George Baer still could early in the century, and they know they cannot hope to serve as father symbols.[2] In their search of legitimacy they have therefore hired public relations men to create a good "public image." Although they often muffle the whole issue by invoking the sacrosanct rights of private property, they have also tried to assume the role of public leaders. But their chief claim to legitimacy remains the obvious one—performance, or efficiency. For the country they have been producing the goods on an unparalleled scale; for their

---

2. For readers who have forgotten their ancient history, Baer said: "The rights and interests of the laboring man will be protected and cared for . . . by the Christian men to whom God in His infinite wisdom has given the control of the property interests in this country." God had made him president of a railroad.

stockholders they have been making record profits. Europeans have learned to their dismay that big American corporations are far superior to their own in organization, management, and planning.

For social purposes, however, their performance raises some questions. They are less keen on innovation than the rhetoric of business has it. Innovation requires not only inventions but investment, plans for manufacturing, financing, and marketing that necessarily involve some uncertainties or risks; so it is likely to meet resistance from the bureaucracy, made up of different functionaries who are more aware of the uncertainties than are the bright-eyed inventors, and whose impulse is to play it safe. Often, moreover, the corporations achieve their conventional goal of "maximizing profits" by retarding the technological progress they boast of, resorting to a kind of featherbedding they complain of in labor unions. Years ago the National Resources Committee pointed to a common kind of practice: it reported that the Bell Telephone System had "suppressed 3,400 unused patents in order to forestall competition." A businessman stated the logic of this practice: "Why should a corporation spend its earnings and deprive its stockholders of dividends to develop something that will upset its own market or junk all its present equipment?"

Another consideration in corporate planning, notably in the great automobile industry, is rapid depreciation or planned obsolescence. In combatting the "fallacy" that carpets should last a lifetime, an executive in the carpet industry pointed out that "automobile builders have gotten themselves into the ideal situation, where they sell a product which will be next to worthless in five years and which will give trouble in the meantime," and he proceeded to give away this whole business: "Who would want a lifetime storage battery, a lifetime muffler and exhaust pipe, rot-resistant lumber, double-lived shoe soles, woolen clothing, nylon hose, clothesline? We could make all of these, but we don't." A simple-minded consumer who would like such things may wonder how many American products are as good as they could be. And since the primary concern of manufacturers is profit, not use, many of their products are of little more use than the latest eyewash or the fins on automobiles. As Paul Goodman suggested, the businessmen who protested that much of the made-work provided for the unemployed by the New Deal was "boondoggling," demoralizing because useless, might consider all the boondoggling sanctified by the profit motive.

For my purposes the most important question forced by the concentration of economic power in the big corporations remains the question of how responsibly they exercise it, in the public interest. Granted that their primary concern must naturally be to make profits, corporation executives are fond of talking about how well they are serving the interests of America. I think they are in fact distinctly more responsible than they were in the old buccaneering days, down to the Great Depression. Business leaders have begun to recognize their public responsibilities for helping to solve such national problems as the black ghettos in the cities. Generally they take a broader, more enlightened view of the national interest than do little businessmen; these mostly conform to the traditional type of businessman and politically are an unsophisticated, ultra-conservative class. Like Berle, many liberals accordingly now take a kindlier view of big business. Since it has come to accept both labor unions and the welfare state, they are pleased to think that "welfare capitalism" has at last solved the fundamental political problems of the industrial revolution. Nevertheless I also think that by and large big business is still far from being responsible enough, and that in view of its immense power and influence its shortcomings are what need to be stressed. Any day one may read of failures to come up to even elementary standards of public decency.

Thus the tobacco industry has done its best to discredit the plain evidence that cigarettes are a source of lung cancer. The drug industry shows little more regard for the public health while it spends $900 million a year on advertising, or about three times what is spent on medical education. When the Kefauver Senate committee began looking into the steep prices it charged, too often for drugs for which it made false claims, the president of a leading company agreed that "drugs should not be marketed until the makers can provide substantial evidence that the products yield advertised results"; spokesmen of the industry described his statement as a "significant new concept." Today, years later, a panel of scientists testing drugs announces that more than 2,000 of them, or two-thirds of those tested so far, bore "misleading and devious" labels, and over a hundred were "totally ineffective."[3] Business interests regularly oppose what they now call "consumerism" (suggestive of communism), efforts at legislation to protect the health and safety of consumers, or to give them their money's worth by truth-in-packaging and truth-in-lending. Their lobbyists in Congress help to see to it that the federal agencies supposed to regulate them are granted budgets quite inadequate for purposes of investigation and enforcement, and that laws setting standards lack teeth for enforcement or severe punishment for willful violation. Similarly the managerial elite use their influence in Congress on income tax legislation to facilitate their tax-dodging, maintain their status as an overprivileged class. Oilmen have been only the most conspicuous in wangling tax favors.

Again the mammoth automobile industry may serve as a prime exhibit. As the leader of the Big Three, General Motors sets the basic prices and styles; Ford and Chrysler find it too risky to buck its policies. Granted that a price war to the death between giants like GM and Ford is hardly to be desired, since so many workers, dealers, and suppliers too would suffer from it, the facts of economic life again belie the rhetoric of business. George Romney stated the facts simply: "When you get an inadequate number of companies in an industry, the customer ceases to be king"—the industry dictates what he wants. The auto industry is among the most backward ones technologically, especially in possible improvements of engines. Innovations are largely confined to styling changes and gadgets, which add at least $700 to the

---

3. I confess to bitter feelings about this industry since my doctor prescribed blood-pressure pills. The government bought from a foreign firm the generic drug the pills are made of at a cost equivalent to 5¢ a hundred; in the drugstore I pay $8.75 a hundred for them. Drug manufacturers justify their exorbitant prices by the money they spend on research, though actually it amounts to a very small fraction of their income, but on this pill about all they did was to give a drug a new technical name. I doubt that their research suffered when five leading companies, charged with a conspiracy to maintain artificially high prices on antibiotics, agreed to return to buyers $120 million of their illegal profits.

price of a car, and have the further advantage of increasing the cost of minor repairs. (They recall the saying of a big-shot's wife: "The way this business is run, we would have failed long ago if we didn't make so much money.") Safety devices were added only after required by the auto safety law, following which the industry also recalled more than six million unsafe cars. It resisted too the efforts of California to require auto exhaust controls for the sake of reducing pollution; the Justice Department then charged it with a conspiracy to restrain the development of such controls. All along it exploited its "ideal situation" by steadily raising the prices of its short-lived models on the excuse of higher wages, though labor costs amount to only a few hundred dollars per car, and GM continued to make profits far above the national average.

All such abuses of power and privilege are still obscured by the usual talk about the virtues of competition, the rights of free enterprise, the need of protecting the consumer's freedom of choice, etc. The spokesmen of big business cloak as well its power. Describing this as a "second government," David Bazelon wrote that it "governs mostly by denying its own existence . . . and by additional domination and perversion of the entire national culture, which it finances." In short, "It lives not by the sword, but by perfidy. If the traditional state exists by virtue of a domestic monopoly of armed force, then our second form of government exists by exercising a national monopoly of effective fabrication." "Perfidy" may be too strong a word, since I suspect that most business leaders really believe in their conventional image. But certainly there has been a good deal of conscious "fabrication" by their public relations men. In particular fabrication is of the essence of the most ubiquitous, incessant form of modern propaganda in America, through which business most directly influences the national culture—advertising.

On this tiresome subject, the first thing to be said is that the influence of advertising may not be so great as is commonly supposed. Business managers have worried whether it was worth all the money spent on it, and admen themselves sometimes confess to uncertainty; although they usually talk with breezy assurance, they work very hard and often look like harried, insecure men. (*Advertising Age* has reported that on the average they die about ten years earlier than the national average; so they might be considered anonymous martyrs to their calling, inasmuch as they get no public credit for even the most brilliantly successful of their ads or slogans.) In any case the influence of advertising is not simply insidious, even apart from its plain necessity in our kind of economy. It can be called humane because it helps to give at least an illusion of personal identity to anonymous people lost in the masses. Its language is usually more ingratiating than aggressive, while showing a particular respect for women. And there are of course many honest as well as ingenious, attractive ads.

Yet unquestionably there are also many dishonest ones. "How many of us here," said a DuPont advertising executive in a talk to his fellows, "can say he's never been a party to deceptive advertising?" They have to be deceptive because their business is to sell any product good or bad, or basically no different from competing products. Above all, there is no question either that advertising has considerable influence on people; the Madison Avenue men have no trouble demonstrating that their campaigns can jump sales. And since they are uncommonly fond of talking and writing about their profession and their proved methods, they themselves make plain that for social, cultural purposes they are basically irresponsible.

In the naive palmy days of free private enterprise, when Calvin Coolidge said that the business of America was business, he once edified a rapt audience of advertising men:

> Advertising ministers to the spiritual side of trade. It is a great power which has been entrusted in your keeping, which charges you with the high responsibility of inspiring and ennobling the commercial world. It is all part of the greater work of regeneration and redemption of mankind.

Today I suppose some public relations men might still venture such verbiage, but I doubt that admen would so describe their trade. More typical of our sophisticated, hard-boiled age was a talk by a vice-president of Benton & Bowles, who maintained only that advertising "reflects our society more accurately than anything else does." He went on: "Esthetes and apologists can rail at its vulgarity, its brashness, its aggressiveness, its insistence, its lack of cultural values, its crass commercialism, its loudness and its single-mindedness—but let them rail"; these are the qualities "that have built the nation," and they are "qualities of virility." Dr. Ernest Dichter, head of the Institute for Motivational Research, has even argued that by advertising techniques America could win the Cold War: all we had to do was sell the world a "good image" of America (though perhaps one that left out some of our qualities of virility).

In *The Strategy of Desire*, Dichter stated baldly the premises of advertising. They are that people are irrational and selfish, and that to persuade them one must never appeal to their reason, but capitalize on their irrationality and appeal to their selfish desires or whims. As a "scientific" advertiser, Dichter thought the "only debatable question" was how best to do so. The most successful advertisers have given a number of different answers, reported at length in Martin Mayer's *Madison Avenue U.S.A.* Ted Bates & Company specialized in the "Unique Selling Proposition," for instance that Colgate toothpaste "cleans your breath while it cleans your teeth"; the president of the agency explained that every toothpaste does this—*but nobody had ever put a breath claim on a toothpaste before.*" David Ogilvy made most of snob appeal ("The Man in the Hathaway Shirt"). Norman B. Norman tried to get at the "deepest levels" of customers "by expressing the *real* reasons why people buy these products," and for this purpose he found Freud helpful on sex drives (as did Dr. Dichter); one result was the famous campaign for Lucky Strikes, "So Round, So Firm, So Fully Packed." McCann-Erickson went in rather for emphasis on "power drives." Many others have long exploited the pathetic hopes and fears of people by advertising easy ways to become more successful, popular, glamorous, virile, or feminine. All these methods are linked by a uniform assumption, a low estimate of the mentality of the American people. On the face of the advertising record it is evidently a sound estimate, or another "accurate reflection" of our society, but in any case it refutes the common claims that business is elevating our society. Advertising recalls rather the satire of Mandeville:

All trades and Places know some Cheat
No Calling was without Deceit
Thus every part was full of Vice
Yet the whole Mass a Paradise.

Martin Mayer defended advertising by arguing that besides "its purely informing function" it *"adds a new value to the existing values of the product."* (His italics.) People taking the worthless pill really feel better, just as the teenage girl using the same lipstick under a new brand name feels she is a beauty. Like-wise "a product which satisfies the sublimated sexual drive gives its consumers a high order of tertiary value." In simpler words, Americans really are dopes. As for the common objection that advertising creates "false" values, Mayer answered that *"in an economic context* it is unimportant whether a use value enjoyed by a consumer is true or false." (Italics mine.) At most he granted that critics of advertising had some reason for their disappointment in American culture: "Perhaps advertising *ought* to do something for the culture, but it won't; says it can't; says it shouldn't be asked; in its most defensive moments says the majority is *right* to like garbage, simply because it buys so much garbage." Briefly, the business of America remains business, not culture.

My excuse for laboring the obvious is another pertinent question about the directors of our industrial technology: What manner of men are they? They are obviously not a cultural elite, nor by and large much concerned about intellectual or cultural values, the problem of the good life. A poll of top corporation executives taken by *Fortune* magazine some years ago revealed that few of them read serious books, except books on management. Today executives are making some effort to improve their "image" in this respect. Troubled because the best college students are turning away from business, some are responding to appeals that corporations contribute more to higher education and culture. Yet on the whole power and culture remain more sharply separated in America than in Europe. The creed of business leaders is still that the primary social purposes are economic, like their own. In particular the adman who praised the "virility" of America reflected a common attitude in his scorn of "esthetes."

Esthetic interests were clearly more important in pre-industrial societies.

Here a possible exception might be the new art of industrial design. Design has unmistakably improved, notably through the efforts of Container Corporation (sponsor of annual conferences at Aspen). Nevertheless most American styling is meretricious, merely fancy, arty, or a mode of conspicuous waste, dear to manufacturers because dearer for consumers. To those who respect machine forms it can often be as offensive as the automobiles with fins or shark-mouth radiators. For the controlling consideration in the introduction of new models is not the question whether they are either more durable or more beautiful—it is simply what will sell best; and anyway the new model must give way to next year's showy model.[4] The best design is usually found in products like the airplane, such as the superb Boeing 707, in which machine functionalism is not flawed by the need of annual new models; and with airplanes it is only the exterior that is handsome— decorators take over the interiors, whose chief function is to crowd in too many narrow seats to permit comfort.

In spite of all such complaints, however, few American intellectuals now call for socialism. The example of the Soviets has discouraged inclinations to believe that public ownership is the answer to the problems raised by the big corporations. As it is, their power is limited by other independent powers. One is government, on which the economic elite may have too much influence for the public good, but which it does not entirely control; government has been imposing more social responsibility on some industries, as by the auto safety law. Another power is the big labor unions. And even public opinion—systematically bamboozled though people are by advertisers, public relations men, and politicians—counts for enough to keep them all busy,

---

4. About the highest compliment to American industrial designers that I have heard was paid by a Dutchman who had just won an international prize for his vacuum cleaner. He told me modestly that designers had an easy time in Europe: all they had to think of was clean, honest design. In America they had to contend with the requirements of fanciness and planned obsolescence, and he thought that under these handicaps American designers often did a remarkably good job.

and often worried. Although it is not organized, some private associations defend the public interest. Ralph Nader showed that one man could make a dent even in General Motors.

One may wonder whether the terms "socialism" and "capitalism" are useful any more. Outside the Communist countries, socialism in political practice has come to mean only a somewhat greater measure of public ownership, depending on the country and the socialist. Capitalism has of course become quite different from the system described and doomed by Karl Marx. In America both its ardent champions and its Communist opponents only obscure the actual economic system, the various combinations of private and government enterprise, representing at best an effort at a democratically controlled economy. For all except extreme Rightists and Leftists, the actual problem is neither to preserve nor to overthrow "capitalism," but first to size up the new relations between business and government since the World War. Having already indicated the major development, that big corporations doing so much profitable business for the government can no longer regard it as simply a menace, I should add chiefly that in the newer industries, such as aircraft, electronics, and atomic energy, many corporations have been dependent on the government for the great bulk of their business. (General Dynamics—a giant that was managing to fail—was rescued by the Pentagon, which awarded it the contract for the ill-fated F-111 plane.) In addition the government spends an estimated $6 billion a year on direct subsidies to private business. All such arrangements have blurred the customary distinction between the private and the public sectors of the economy.

Yet it is still necessary to make this distinction. Private enterprise, which takes care of the bulk of the nation's business, remains unable to take care of many public needs, or often to take advantage of the latest technological possibilities. It appears, for example, that the moribund East Coast shipping industry could be revived by fast hydrofoil ships, capable of outperforming trucks and therefore relieving the traffic problem too; but the ships remain on drawing boards for lack of capital, and the trucking industry would no doubt oppose the needed government aid, even though it has long depended on all the new highways built by the government. Hence the critical question is again what public needs should be left to private enterprise. Apart from conservatives bent on reducing government expenditures (though not on subsidies to business), some serious students of business believe that the big corporations will in fact move into the public sector, take over tasks now performed by the government. "New fields, new societal responsibilities, and new cultural environments," writes Michel Crozier, "will provide the basic challenges for American business in the next two or three decades"; and he thinks its most sophisticated elements are now ready to accept such challenges, take business out of "its ghetto." He adds, however, that in the public sector it will need more than its economic rationality, for it will have to face the basic problem of human wants, which it has subordinated to economic considerations. Or in simpler terms business leaders will have to question the "philosophy" running through public conferences I have read about, that their primary concern must still be profits: corporations are not philanthropic organizations. It is not at all clear that there is enough profit for them in such enterprises as cleaning up the black ghettos—unless the government heavily subsidizes them, as it does in slum-investment programs guaranteeing returns of 12 to 14 per cent.

The A.F.L. building trades unions are no more likely to sacrifice their interests in meeting the problems of the ghettos. Their interest has been simply to maintain their high wages and their status, and it has made them among the most recalcitrant unions in admitting Negroes to their ranks. They illustrate the anomalies of the big labor unions, which help to offset the power of the corporations, but raise another cluster of problems. In the popular mind the unions and their leaders still have a reputation for being radical, inclined to "socialism." Actually they do not want socialism, still less Communism; they resist any government interference with their right to bargain and to strike, and are as opposed as the corporations to compulsory arbitration. And though any opposition to business interests still smacks of radicalism in conservative America, the unions themselves have been growing politically more conservative.

This is not at all surprising when one remembers that the big labor union is another bureaucracy, with the inescapable hierarchy. The men at the top are paid high salaries and go off to conventions with fat expense accounts, like businessmen. The unions are run somewhat more democratically than the corporations, with elections and more often a real contest for leadership, but for the most part they too have one-party government. They have further excuses for it in that any serious conflict might split the union, and that there is nowhere in the union for a defeated leader to go; having adjusted his mode of life to a high income, he can hardly return to a job in a factory. At any rate, one-party rule creates the same problem of how to keep the executives responsible. Some labor leaders who have taken to the ethics of smart business and politics have profited by flagrant corruption. Like most stockholders, the union members are willing to put up with shady dealings so long as their leaders get them good pay. Most of them take little interest in the management of the union bureaucracy, especially because collective bargaining has become a complex technical business, conducted by skilled lawyers. The most conscientious, thoughtful labor leaders, such as Walter Reuther, worry over the problem of how to keep workers actively interested, give them a fighting cause, now that unions are no longer fighting for their lives and workers are relatively well off.

Although there is always much more public or Congressional worry over the power of the big labor unions than of the big corporations, most of these corporations have learned to live with them. They prefer to bargain annually with a disciplined, industry-wide union rather than incessantly with a lot of separate little unions in scattered plants; after all, they do not really "lose" when they give in to union demands—the public will foot the bill in higher prices. Some years ago Clark Kerr, then a specialist in labor affairs, thought that the greater danger was not industrial conflict but industrial peace, the growing collusion between labor and management at public expense, as in a big steel strike that was carefully stage-managed after a settlement had been privately agreed on. More lately neither the corporations nor the unions acceded to President Johnson's request that they keep raises in prices and wages minimal in order to combat inflation. Today, however, there still appears to be enough real conflict between them, if only because the unions have shifted their emphasis from simple demands for higher wages to such demands as a guaranteed annual wage and some voice in decisions about corporate policy.

In a democracy it might seem reasonable to have labor leaders serve on boards of directors, with maybe a public representative too, as a possible way of keeping them all more responsible; but so far the big corporations have insisted on keeping private and inviolable the realm of management. The chief concessions they have made have been paternalistic ones that give "soulless" corporations the appearances of a soul. Otherwise most still adhere to old-fashioned notions of scientific management, seeking to minimize the "human element." The chief industrial engineer of a corporation described proudly a program that would tell every worker "exactly how to do his job down to the last detail." This he said would serve a psychological purpose too: "Not only will the operators not have to think any more, but they will be much happier on the job and more loyal to the company when we are able to show them that we know so much more about the job than they do." In view of such abysmal ignorance of human nature, I assume there will continue to be plenty of misunderstanding and conflict between labor and management.

Another abiding source of trouble is that workers are still generally regarded as "labor" and treated as means, not ends, the way human beings ideally should be. Most businessmen would be nonplussed by Le Play's remark that the most important thing that comes out of a mine is not the mineral—it is the miner. Few appear to realize how strange or possibly inhuman are the ways in which they have learned to think in our kind of economy. Thus there was much concern among businessmen when unemployment, which had reached nearly 7 per cent of the labor force at the end of the Eisenhower administration, fell to below 4 per cent. This still meant more than three million jobless people, but businessmen worried instead over the "labor shortage"; they could no longer readily find in their neighborhood just the kind of workers they wanted. Charles E. Silberman wrote a reassuring piece in *Fortune* magazine; "Business Can Live with the 'Labor Shortage.'" Although the problem was real, he granted, the shortage was not absolute. Businessmen had been spoiled by the large pool of unemployed, which among other things had got them into the habit of accepting only applicants who had graduated from high school. While they now grumbled over scraping the bottom of the barrel, they hadn't really reached the bottom; most of the three million looking for work had had full-time jobs, proving that they were not "unemployable." Silberman noted incidentally that the inconvenience of the labor shortage was magnified by an oversight: a government task force that studied the plight of employers in Milwaukee reported they "were not making adequate use of Negroes."

This whole problem has a comforting aspect, however, in view of the most revolutionary technological development in industry—automation. Evidently the alarm over the unemployment this would cause was premature. So too with cybernation. When an announcement came out that a new computer could do the work of seventy-five of the biggest ones then in use, a unionist remarked, "When computers start creating unemployment among computers, it's really time to start worrying." But they have not in fact created widespread unemployment among brain-workers. The prospects are that an increasingly complex society, with an ever more sophisticated technology, will need more professional people.

Still, cybernation is only in its infancy. I assume it will surely keep growing fast, as the computer industry itself has been.[5] Already one hears that computers can perform many thousands of tasks, from collecting eggs in electronic hen-houses to designing in an hour a chemical plant "that would take a platoon of

---

5. Led by International Business Machines, this industry illustrates the growth of enlightenment in big business that was celebrated by Thomas G. Spates in *Man and Management: The Spiritual Content of Business Administration*. Although he makes plain that he is using "spiritual" in no "heavenly sense," it is still perhaps too exalted a word for what comes down to a concern with "human relations"; but anyhow he reviews a number of well-known corporations that early distinguished themselves by enlightened or humane labor policies. IBM, which is among them, has given Harvard $5 million for a ten-year study "to seek to identify and analyze the primary and second-order impacts and effects of technological change on the economy, business, government, society and individuals." This public service is especially welcome coming from the computer industry.

engineers a year," and all kinds of new things are on the drawing boards. In *The New Utopians* Robert Boguslaw has written of the enthusiasm of a recent breed, men bearing such titles as system engineer, operations researcher, computer programmer, data processing specialist, and system designer. Never before has the word "system" been so highly esteemed—but only in technology; unlike missile systems, philosophical systems are disreputable. Boguslaw notes that the new utopians are quite different from the traditional ones, whose concerns were humanistic, for they are "concerned with non-people and with people-substitutes," systems that reduce the scope of responsibility of human beings.

In the big corporations computers have been substituting in particular for the middle managers. Since top management can find out instantly from computers what is happening all over the corporation, it no longer needs all the intermediates who used to make decisions in their plant and send up information. Assured of accurate information and relieved of the need of analyzing it, the top men can also concentrate more on their strategic work of long-range planning and policy making. They have more time and scope for imaginativeness and creativity, the "vision" businessmen have always prided themselves on. As one computer expert said, "Perhaps for the first time in man's history, he can afford to think truly long thoughts." Except for the computer industry itself, however, such thoughts have so far not been too apparent in the record of General Motors and other giants. The visions I have read about have been chiefly of more speed, efficiency, material abundance—the kind of values that computers can process.

Meanwhile the computers cannot determine the human costs of the latest wonders. These have certainly meant hardship for a great many people, even if statistically they are a small fraction of the working force and sooner or later are reabsorbed into the economy. No one knows how many factory workers, office workers, statisticians, middle managers, and "platoons of engineers" have lost their job. No one can measure the human costs in the loss of status and skills, the problem of learning new skills or unlearning old ones in mid-career, enforced moves out of a familiar environment, the tensions of uncertainty about the future. Displaced men are no happier because automated equipment makes unnecessary

any measure of their productivity, removes the "human element."
And little has been done about the problems of steady dislocation.
Management has been concerned mainly with the economic benefits of automation or cybernation, the savings in labor costs;
though these have seldom resulted in lower prices for consumers.
(In our economy it appears that every advance in "efficiency"
results in higher prices.) Government has barely begun to face the
problem of retraining and relocating workers. Schools continue to
give students an obsolescent training. Cybernation may not proceed as rapidly as many now hope, or fear, but in any event our
society is scarcely prepared to deal adequately with even the
present rate of change.

So let us pause to review the startling changes in the work
habits of Americans in this century. One statistician estimates that
whereas in 1900 eighteen out of twenty worked with their hands,
ten of them on the farm, by 1965 only five of twenty earned their
living by manual work, one of them on the farm. Agriculture has
had the most anomalous fate because of improved technology. By
contrast with the Soviet Union, which Khrushchev boasted would
soon surpass us, fewer and fewer farmers have been producing
more and more by the year, but their success has been an embar-
rassing national problem: instead of being rewarded by increasing
prosperity, many have been having a harder time—especially the
small farmers on family-sized farms, the type long glorified in
American tradition. Unionized industrial workers have fared
much better ever since the New Deal saved "capitalism" by
violating its tradition. Today the unions are worrying over automation, but not over the problem of what already has been called
"cyberculture," to baptize the "new age of abundance and
leisure." They have been demanding a shorter work week, which
sooner or later will get still shorter; so one may wonder what
workers will do with all their leisure. (Many say they would take
on an extra job.) American schools are not educating the young
to make a more fruitful, satisfying use of leisure than do the
millions of people who depend on TV. They also continue to
inculcate the archaic gospel of work that has survived all these
changes. Hence most Americans consider immoral such proposals
as a guaranteed minimum income for all Americans whether they
work or not, even though the unemployed are helping to maintain

the "fluid" labor market that businessmen want, big corporations working on government contracts enjoy a handsome guaranteed income, and big farmers (like Senator Eastland) are paid many thousands by the government for not growing things.

To this whole question I shall return in later chapters. Here I shall conclude with a brief look at a body of professionals who have long been concerned with the major interest of modern man —the economists. As advisers to both the big corporations and government, they have in recent times become much more important because they influence the planning of both that distinguishes post-industrial society, and gives an antique air to Hayek's once popular book *The Road to Serfdom*.

The classical economists of the past defined their subject broadly as "the science of wealth," more specifically as "the mechanics of utility and self-interest." In effect they studied the economy apart from government and society as the creation of "economic man," and so made cardinal the principle of the free, self-regulating market. With a few exceptions, they largely ignored the plain truth that economic man was a product of their particular society, because they accepted the materialistic values of this society and the reigning interests of business. Their idealized concept of the market led to a quite unrealistic account even of the economy. As Robert Heilbroner wrote in *The Worldly Philosophers*, "One might read such leading texts as John Bates Clark's *Distribution of Wealth* and never know that America was a land of millionaires; one might peruse F. H. Taussig's *Economics* and never come across a rigged stock market." After the Great Depression, however, the "science of wealth" began to grow more sophisticated under the influence of Lord Keynes, who stressed what government could do to help the economy. And since World War II, the alliance of government and industry and the rising importance of economists in government have made more meaningful the original name of their subject—political economy.

Now that most economists have come around to Keynes's way of thinking, laymen may welcome their general agreement that their forebears before the war operated on mistaken theories. But there remains the possibility that current theory may be mistaken too, which might make the rest of us uncomfortable in the knowledge of our dependence on fallible expertise. In particular there is still a surprising lag in much of their thinking, as some economists themselves have noted. Recently Heilbroner observed how "extraordinary" it was that "the two most important economists of fullblown capitalism, Alfred Marshall and Lord Keynes, have virtually nothing to say about the impact of technology on the economic system," the commanding reality of our day. Many if not most economists still appear to have a naive faith in the free, self-regulating market; only lately have they begun to acknowledge openly that technology and the giant corporations have undermined this beautiful theoretical fiction. Another economist, Walter Adams, complains that his academic colleagues are neglecting the real problems of our time, such as the concentration of immense corporate power, poverty in the midst of affluence, the city ghettos, and the growing gap between the rich and the poor nations. Engrossed in model-building and mathematical-econometric methods, they "have tended to ask themselves questions that can be analyzed with their new techniques rather than finding techniques to deal with the questions they *ought* to ask."[6] For me the key word here is the forbidden word *ought*. Although economists have become influential advisers on public policy, they still tend to neglect the problem of human values.

This neglect is easy to understand. Like other social scientists, economists avoid value judgments on principle as unscientific or "fuzzy." They may say that their business as specialists is simply to understand how an economic system works, at most to suggest ways of making it work better; and since the policies they recommended have helped the country to ward off serious depressions for a generation, and after the Eisenhower administration to accelerate the growth of prosperity, we might all be grateful to them. Yet they are generally unphilosophical and so apt to be self-deceived.

Sidney S. Alexander put a blunt question to his fellow economists: "Is there a professional point of view or bias among economists about social or political goals?" The answer is clearly

---

6. Norbert Wiener suggested an ironic footnote on the mathematics up-to-date economists and other model-builders now use so proudly. A mathematician himself, he remarked that they were using the mathematics of 1850, which was basically obsolete.

Yes. They are naturally biased in favor of "stability," but with a few exceptions (such as Kenneth Galbraith) they also like continued economic growth, a still greater abundance of material goods, however trivial. Alexander pointed out that in spite of their scientific airs they unconsciously accept without question or test one basic norm, "sometimes referred to as the pig principle" —"if you like something, more is better."[7] In technical terms, they have a passion for "maximizing." Likewise economists generally appear to accept without serious question the standard defense of dubious products by big corporations, that they are only giving the public what it wants; in their self-regulating market the consumer is always sovereign, no matter how much he is duped because of his unavoidable ignorance of the quality of most of the things he buys. It would seem that economists have not studied advertising. In general, they have had little to say about ethical values—the ethics that hold a society together, provide the trust that makes economic exchange possible, determine the public responsibilities of the big corporations, and are too often abused by them and their advertisers, on the traditional assumption that any evil effects of their operations were none of their business.

And so a last word about a still more important profession in our technological society, engineering. Adolf J. Ackerman, a consulting engineer, wrote that it was in a "major crisis" today because of the common disregard of the professional oath "to serve the public interest above all others." He protested against the silence of engineers about obvious violations of their ethical code, and the prevailing tendency in their professional journals to avoid controversial issues, give no space to the voice of dissent. An editor of *The Institute of Electrical and Electronics Engineers* agreed, and spelled out some specific examples that might frighten laymen:

---

7. I am quoting from a symposium of philosophers and economists on *Human Values and Economic Policy*, published in a volume edited by Sidney Hook (1967). Milton Friedman also agreed with the philosophers that economists tended to avoid the basic issues of the value-judgments that necessarily enter into both private and public policy.

It was not engineers who mounted the assault on pollution, but surely it was engineers who first knew what was happening to our water and to our air. It was not engineers who forced the program of safety in automobile design; but who was more aware of the problem? Who better than the engineers involved knew of the dangers in the Apollo spacecraft design and could have issued more effective warnings? Who can better police extravagant proposals for government contracts than the engineers, and indeed, who can better tell us of the deficiencies and dangers in atomic power plants than the engineers who design them?

The kindest explanation of their silence is that they are employed by corporations, whose executives have never had to take a professional oath "to serve the public interest above all others."

11

Language

In spite of all the wonders produced by modern technology, language remains the most marvelous instrument ever devised by man. It is impossible to exaggerate our debt to the nameless prehistoric ancestors who somehow, over long ages, evolved this elaborate system of verbal symbols, or complicated noises. Without it man could never have developed his brain-power, carried on his distinctive activities, transmitted all the lore he was accumulating, ordered his complex social life. Quite simply, he could never have realized his humanity. And now language has become all the more important because of the wonders of modern technology. Communication is infinitely more rapid, far-flung, and busy than ever before. Day and night, language is constantly at work ordering the manifold affairs of our society. So one might wonder about a question few students of technology have asked: How is language faring under such extensive, varied use?

The answer is another basic anomaly. Modern technology, governed by the ideals of efficiency and rationality, rests upon the machine, which has acquired an admirable precision, and upon elaborate organization, which requires clarity in communication. With science, it has helped to make "definite" one of the key terms of our age. As characteristically our society is now training students in "communication skills," the fashionable new technical name for what used to be called speaking and writing. It has also organized the new science of linguistics, which has developed so rapidly that we know far more about the nature, structure, and operation of language than men ever did in the past, and so might hope to use it more efficiently. Yet language has suffered grievously in our society—and suffered most conspicuously in the qualities valued for our technological purposes. Never before has it been so limp, bloated, muddy, and imprecise as it now is in much common usage. By the rulers of society it has often been used most efficiently for irrational purposes. For commercial purposes it has been systematically corrupted and debased. For ostensibly scientific purposes it has been overlaid with technical jargon that gives only the illusion of clarity and precision. And at best our many diverse specialists are finding it ever harder to communicate to the rest of us, or with one another. Briefly, the state of language reflects all the basic disorders in our world. Too often we don't speak the same language; or when we do, it is too likely to be an uncivilized language.

Since I propose to dwell on the worst, let me say at once that we of course still have a great deal of fresh, lively, effective speaking and writing. Apart from all the varieties of serious literature, we may rejoice in much excellent journalism, including a much needed kind—the reporting of technical news, such as developments in science, for the layman. Many professional men meet a difficult demand on specialists today, that they be able to make themselves clear to others; there is much good shop talk. In the universities millions of students have the opportunity of learning to speak, write, and read more intelligently, and if most of them achieve at best a limited mastery of language, the minority who do well are at least a large number, much greater than in past societies. Those who want to inform themselves about the goings-on in the many specialized fields of interest may therefore be irked by how much lumpish, dreary prose they have to put up with; but with patience they can hope to find out what they want to know.

Still, they need much patience (as I did in specialized reading for the purposes of this book). My concern remains the too prevalent disorders and diseases of our common language, whose quality is an index to our culture, the more important because all our thought and feeling are conditioned by language. Finally my concern is the unprecedented difficulties of communication, which are another key to the human predicament today.

Our story begins tiresomely with the uses of literacy, stimulated by the growth of industrialism. This tended to accentuate the worst qualities of popular speech, such as triteness and coarseness, while depriving it of the homely vigor and color that often enlivens folk dialect. Half-educated people are especially addicted to the crude stereotypes that make unnecessary any effort of thought, and to a fondness for novelty that only makes livid the pale cast of staleness in popular language—stale humor, stale fancy, stale metaphor. Popular expressions borrowed from science and technology meet the same fate; the "acid test" always brings out cliché. In general, popular language at its worst is suited to a technological society chiefly as an illustration of the stereotypes about a standardized, mechanized, dehumanized society, all the mass-men whose sensibilities have been dulled, and who have no style of their own because no personal identity.

Such too familiar tendencies were amplified by the rise of the new kind of popular culture, manufactured by professionals for a mass market. Its producers throve by debasing language. They reduced simple sentiment to sentimentality, simple thought to cliché, judgment to stereotypes of black or white. Grays appeared only in the uniform qualities of style, no less when it was lurid, as in the sensational press. All the worst tendencies of the mass media have involved a contempt both for the audience and for a language once known as our mother tongue. Here I will add only that possibly honorable efforts to simplify the immensely complicated issues of our society also indicate a low opinion of the ordinary American's ability to read. Tabloid editorials, for example, break up their thought into a series of little paragraphs made up of a sentence or so, implying that readers are incapable of following a mature, coherent argument, have to be spoonfed one little idea at a time.

Advertising notoriously flourishes on similar attitudes. The perfect ad, according to the experts in motivational research, is one that employs a minimum of language, never a sentence with a dependent clause. Presumably it achieves the ideal of catching attention without arousing thought or powers of judgment. One routine technique is the incessant repetition of slogans, or a kind of brainwashing. Another is meaningless superlatives, like "super" and "de luxe." Since the experts in motivational research maintain that these add value to a product because they give the consumer more satisfaction, I suppose only eccentrics would be pleased as I once was to buy a can of ripe olives labeled simply "very large," and to find them larger than another brand labeled "mammoth." Similarly I am less depressed by all the "sensational" dramas, "epics," and "colossal" spectacles produced by Hollywood than by the extravagant blurbs provided by popular reviewers. Although they exercise some ingenuity in finding new superlatives for all the mediocre movies, they emphasize the prevailing lack of discrimination—the discrimination especially needed in view of the superabundance of alleged goods.

More obviously dangerous is the corruption of language by mass propaganda for political purposes. We all know how Hitler and Stalin featured villains or devils, such as Jews and capitalist-imperialists, and how they exploited the immense power of words, the frightening truth that men are most willing to fight and die for a word, especially if its meaning is vague. They also featured what George Orwell called "doublethink" and "newspeak": words meant the opposite of what they used to mean. Thus "equality" meant subservience to an elite, "true freedom" meant obedience without rights, "people's democracies" were those that had no democracy, etc. One recalls the saying of Confucius: "When words lose their meaning, the people will lose their liberty." But then Americans might consider how words are used by our own national leaders, for instance about our frightening war in Vietnam.

We need not waste time over the official communiqués of our military, which are notoriously untrustworthy. (I recall one "smashing victory" in which the Vietcong overran one of our posts: our victory was that we had not only killed a lot of them but succeeded in evacuating most of our men, though at a loss of quite a few helicopters and bombers.) More troublesome was the rhetoric of the Johnson administration, just because its leaders were no doubt basically sincere. In defending the "honor" of America, we were committed to "liberating" the South Vietnamese, protecting the "free world" against Communist aggression, safeguarding the interests of "democracy," whereas for years we had been defending in fact a series of corrupt, undemocratic, unpopular Vietnamese governments, just as we had supported a number of dictators in other countries; to our policy-makers "democratic" meant in effect simply "anti-Communist." A Buddhist party opposing the government in South Vietnam was suspect because it declared "We want democracy, freedom and peace for the Vietnamese people"—what the Vietcong said they wanted. But these are old linguistic habits. More troublesome was the official use of the language of technology, an abstract language efficiently calculated to neutralize or conceal all moral issues, obscure human realities, beginning with the simple truth that millions of human beings were suffering from our defense of our national honor and the cause of freedom.[1]

This started with our "escalation" of the war, which meant bombing North Vietnam. Presently we were using "anti-person-nel" bombs, cannisters containing thousands of pieces of sharp steel, enough to kill a city block of civilians. In South Vietnam we were busy with the "pacification" of areas suspected of harboring Vietcong, or destroying what our generals called the "political infrastructure" of guerrillas. Our methods included dropping napalm, or "incinder-jell," on villages and the "defoliation" of some areas. A colonel in charge of a "search and clear" operation was most thorough: "We are developing a new style of jungle warfare here—we remove the jungle. . . . We're going to dry up the water. We're going to remove the people, and make the area in which the guerrillas have hitherto lived literally unlivable." By thus "freeing people from Communist control," we produced hundreds of thousands of wretched refugees, and achieved most efficiently what Tacitus said of the Romans: "When they make a wilderness, they call it peace." It was regrettable that our efforts at pacification were often thwarted by "terrorists," i.e., Vietcong who attacked our military installations or who killed civilians as callously as our bombers did. The Vietcong were never human beings but simply guerrillas, as merciless as their weapons, whereas American soldiers were clean-cut, wholesome "boys" who were never associated with their frightful weapons.

The jargon of the military may go unnoticed because it is so

---

1. For most of the specific examples that follow I am indebted to an article by a colleague, Kenneth R. Johnston, on "The Rhetoric Surrounding the Vietnamese War." To "surrounding" I would prefer "disguising."

common in other activities. In government it is ridiculed as "gobbledegook," but its routine forms are almost immune to criticism because useful for bureaucratic purposes. The habitual use of the impersonal and the passive voice ("It is advised that") is a way of blurring or evading responsibility; such phrases as "under active consideration" are convenient excuses for inaction; and the most absurd folderol may still inculcate a respect for bureaucratic forms, give an illusion of superiority over underlings still prone to speaking plain English. "Business English" has likewise inherited some prolix forms, but businessmen have taken as well to considerable technical jargon that is scarcely crisp, pointed, or strictly businesslike, and fails (in their own terms) to "pinpoint" or "highlight" their meaning. Elsewhere I have noted their fondness for a term that irritates me, "reaction" instead of "opinion"; if this describes accurately enough much alleged thought or opinion, human beings are none the less capable of "responding" instead of merely reacting, and often do stop to think. But the term has become standard in everyday speech, including the speech of the educated, as have many other pseudo-technical terms. All readers (I hope) find some of them irritating. Even so almost all of us have come to use many of them because they are standard. In any case they clutter up the language with more needless synonyms and faded metaphors, they blur instead of sharpen meanings, they encourage lazy thinking, and they make it harder for people to speak freshly and directly out of their own concrete experience. Perhaps "green" can no longer be the word for the life of technological man, but one must hope that it need not be so gray as his language too often suggests.

In the sciences the use of technical language is of course quite proper. Physicists have to use it because they are always dealing with quantitative abstractions from experience, and now a sub-atomic world that cannot even be pictured. With them laymen need only to guard against what Whitehead called "the fallacy of misplaced concreteness"—the assumption that their abstractions give us the literal, essential truth about the physical world, all that really matters; whereas what matters most for our daily living purposes is our concrete experience of the qualities and the particularities of the natural world, perceptions and sensations no less real and valuable because they have been stigmatized as "merely" subjective. In the sciences of man and society technical language is likewise proper, at least to begin with. These scientists want neutral terms that denote clearly and unequivocally, that are not blurred by the multiple meanings of familiar words, and that are free from the emotional and imaginative connotations that enrich meaning but may also cloud and confuse it. With them we have to remember that their abstractions are not the whole truth or the essential truth either, and that for our living purposes we have to translate them back into simpler English.

In this effort, however, we run into the serious reasons for complaint. Often the language of social or behavioral scientists is simply a clumsy, ponderous statement of an elementary idea. ("Chimpanzee selection of larger sizes of food pieces in a direct choice situation is mediated by visual size perception.") Their writings are swamped by a barbarous technical jargon that is not at all precise, but only pretentious, spurious, or strictly unscientific.

A parody that made the rounds is the 23rd Psalm, "The Lord is my shepherd," as it might be translated by a committee of sociologists:

The Lord is my external-internal integrative mechanism,
I shall not be deprived of gratifications for my viscerogenic
    hungers or my need-dispositions,
He motivates me to orient myself toward a nonsocial object
    with affective significance,
He positions me in a nondecisional situation,
He maximizes my adjustment.

The parody is fair because every one of these terms may be found in the writings of sociologists. Their own works are rich in as ludicrous examples: "The Golden Rule is another codification of considerations which should govern our choice of actions, lest we end by sub-optimizing in terms of our inter-personal objectives."[2] In the security of their professional jargon they seem utterly

---

2. In humanity I again refrain from naming the authors of the examples I use. There is no point in singling them out from the crowd, and the disease is so common that readers can take my word for it that I am making up nothing.

impervious to ridicule. They seem unaware of all their pseudo-concepts and circumlocutions, as in this definition of socialization: it means "the enculturation of persons and their assimilation into groups so that they become active members of society." "Enculturation" was left undefined. Or lest I seem too harsh, as a cranky English professor, I cite Pitirim Sorokin, who wrote chapters on the "speech disorders" of his fellow sociologists. Among other things, he complained of the "blind transference of terms from the natural sciences," the "ponderously obscure description of platitudes," and the "introduction of many neologisms which hinder precise communication." He offered this example: "The strength of the valence on the percept of, say, a given food will be a function of the general need-push for food activated by the given controlling matrix plus the degree of cathexis to the particular variety of food as determined by the shape of the generalization fork." Since I find it impossible to translate this one, I merely remark that scientists are supposed to be aspiring to clarity.

Needless to add, some eminent sociologists write well, demonstrating that it is quite possible to write about society scientifically, yet preserve a respect for the English language. Nor are such speech disorders by any means confined to sociology—they are found all over the various behavioral sciences. From a political scientist I have cited a mild example: "The property of being pro or anti, favorable or unfavorable, and so forth is inevitably found in attitudes polarized by a two-party system." (This I can translate: "American voters are always for or against things.") The blight on language is worse because so many of these scientists feel obliged to publish whether they like to write or not, inasmuch as publication is necessary for professional status; so they naturally write in the jargon approved by their profession. And I have labored the point because the blight is on all kinds of professional studies, and therefore to some extent on the language of all college-educated people.

Thus everywhere one encounters areas and levels, dimensions and frames of reference, activation, motivation, orientation, and so on, through the many "processes"—the educational process, the planning process, the thinking process, even to the disease process. Personal relations are always "interpersonal," although such relations can only be between people.[3] The children of the poor are "underprivileged," slums are "disadvantaged areas." Behind these terms there is no doubt a humane impulse, which also appears in "underdeveloped" countries, but in any case they swell the flood of technical jargon. Whatever the impulse, the effect on style is the same. It is often simply wordy: "on the college level" means only "in college." It sounds pretentious: "finalize" is no better than "finish," motives are no clearer or deeper for being called "motivation." It is always woolly, smothering simple ideas under abstractions. At a time when we badly need an antidote to all the abstractions that technology keeps piling up, and to the spreading tendency to obscure human realities, the habit of jargon amounts to a failure of mind. We live by language, and it is becoming ever harder to write simple, clean, honest English.

Among the worst offenders are the educationists—the people who train and supervise our teachers, who in turn train our children. Fifty years ago, Jacques Barzun commented, they borrowed the term "focus" from photography, "and we have not had a clear statement from them since." By now they have developed an elaborate methodology with jargon to match. To cite a few random Ph.D. dissertations, one inquired into the "job motivations" of deans of students; by sending a questionnaire to some hundreds of deans, this diligent researcher discovered that the "greatest motivational influence upon career choice" was their interest in students. A thesis on "Selected Problem Areas Within the New Social Studies" focused in part on the "inquiry problem area," in particular "inquiry motivation." Another on "The Persuasive Program in Adult Education" took up such "functions" as "activating the persuasive process," "selecting, analyzing, and recruiting persuadees," "transmitting the program message," and "evaluating the persuasive event." As Ph.D.s these people will

3. The Faculty Council of my university solemnly debated a proposal to permit men and women students to visit one another in their rooms in the residence halls. Every speaker referred to this as "intervisitation." I confess to a wayward hope that the students would settle down to copulation, or "interpenetration."

take their place among the leaders of the profession, or of the "educational process."

It is little comfort either that very few people will ever read such dissertations, which are now being ground out by the thousands. So too with the thousands of professional journals: they are read by very few but fellow specialists, with what pleasure I can only conjecture. The possible profit they get brings up a deeper problem, the peculiar difficulties of communication in our kind of society. Specialists in any one "area" read only their own professional journals. They do not have time to look into the many other journals, on dozens of different subjects, but anyway they would find most of their articles as hard to read as the general reader does, many of them as incomprehensible. Addicts of technical jargon like only their own variety; if not repelled by other learned varieties, they are unwilling to take the time or trouble to master them. And behind all the needless jargon lies the plain necessity of a great deal of technical language in the many different sciences. This can be read with understanding only by specialists; the rest of us have to wait until it is translated or popularized. Much significant work being done—even fascinating work—is simply not available to the great majority of educated people.

In this chapter a pertinent example is the new, fast-growing science of linguistics. Some years ago Alfred Kroeber, the late dean of American anthropologists, remarked that they envied its rigor and elegance, but added wistfully that linguists concentrated on the structure of language, not the problems of meaning that interested anthropologists; while recognizing that language had meaning, they seemed to regard it as "a kind of impurity," a little of which unfortunately had to be admitted. Now I gather that linguists have considerably broadened and humanized their study of language. They are cultivating a field called sociolinguistics, linking it with sociology. They are exploring connections with psychology and philosophy; Noam Chomsky believes that linguistics is moving toward a higher synthesis that will illuminate some of the profoundest problems of the human mind and the essential nature of man. It is accordingly contributing to many other disciplines—philology and literary studies, anthropology, political science, the history of ideas, and logic. It is also helping out with the practical problems of the new multilingual African and Asian countries. Altogether, it seems to me an exceptionally fruitful study.

But most of this I have only "gathered," at second or third hand. Linguistics has its own highly technical language— phonemes, morphemes, homophony, algorithm, pansemiotic, etc. —that I have not had the time to master. When I tried to read a book on the new "generative" or "transformational" grammar that I heard was the chief excitement in the field, I was discouraged by pages full of diagrams and formulas. I know best the value of linguistics in the teaching of English, particularly through studies of usage and "standard" English, but about this I have some reservations. While linguists have helped to combat schoolmarm English by making clear that correctness is determined by usage, not by the arbitrary commandments of grammarians or lexicographers, they have tended to condone any usage, no matter how lazy or slovenly. Teachers who want students to write mature sentences might be dismayed by one authority: "From the standpoint of speech a sentence could be scientifically defined as *any stretch of utterance between breath intakes.*" Jacques Barzun cites a "scientific" definition of reading offered educators: "A processing skill of symbolic reasoning sustained by the interfacilitation of an intricate hierarchy of substrata factors that have been mobilized as a psychological working system and pressed into service in accordance with the purpose of the reader." To my knowledge, linguists have been little concerned about either the diseases of technical jargon or the peculiar difficulties of communication today, I suppose because their science throws no light on these problems.

The formulas designed to elucidate transformational grammar point to another difficulty. While for specialists such mathematical apparatus no doubt makes for clarity and economy, it is often an obstacle for the general reader. Today it is found in many books on subjects that formerly were less technical, notably philosophy. It discouraged my own inquiries into symbolic logic. Likewise I bogged down in "game theory," lately popular in social science.[4] The language of mathematics, superb for its purposes, also gives increasing difficulty to nonspecialists; it appears that modern forms of it can no longer be translated into ordinary language. For such reasons most of us are cut off from the most exciting frontiers of science today, biochemistry and biophysics.

Robert Oppenheimer summed up this whole problem most eloquently in a lecture on "Prospects in the Arts and Sciences," one of a series of lectures entitled *Man's Right to Knowledge* (Columbia University Press). This right, he pointed out, has exacted a heavy price in human values:

> The frontiers of science are separated now by long years of study, by specialized vocabularies, arts, techniques, and knowledge from the common heritage even of a most civilized society; and anyone working at the frontier of such science is in that sense a very long way from home, a long way too from the practical arts that were its matrix and origin, as indeed they were of what we today call art.
>
> The specialization of science is an inevitable accompaniment of progress; yet it is full of dangers, and it is cruelly wasteful, since so much that is beautiful and enlightening is cut off from most of the world.

Oppenheimer went on to consider the comparable plight of the artist today, from which he perhaps suffers more, and his fellow men too:

> The artist depends on a common sensibility and culture, on a common meaning of symbols, on a community of experience and common ways of describing and interpreting it. He need not write for everyone or paint or play for everyone. But his audience must be man; it must be man, and not a specialized set of experts among his fellows. Today that is very difficult. Often the artist has an aching sense of great loneliness, for the community to which he addresses himself

---

4. This I gather was no serious loss. In the one effort to apply game theory to political science that I struggled with, the author built up a mountain of mathematical apparatus out of which finally emerged the tiniest mouse of a tentative conclusion.

is largely not there; the traditions and the culture, the symbols and the history, the myths and the common experience, which it is his function to illuminate, to harmonize, and to portray, have been dissolved in a changing world.

Much the same problem confronts all scholars, specialists in either the arts or the sciences. They work in separate professional communities, villages with only thin paths to one another, and to the great world around them. The superhighways—the mass media—do not help, Oppenheimer added; the little art and science they purvey is lost in the clamor of news and entertainment. All this is not wholly new, since the great world rarely paid much heed to what was going on in its intellectual villages, but the problem is much more serious and difficult in a world that technology has at once brought together and splintered. Oppenheimer concluded:

> Never before today has the integrity of the intimate, the detailed, the true art, the integrity of craftsmanship and the preservation of the familiar, of the humorous and the beautiful stood in more massive contrast to the vastness of life, the greatness of the globe, the otherness of people, the otherness of ways, and the all-encompassing dark.
>
> This is a world in which each of us, knowing his limitations, knowing the evils of superficiality and the terrors of fatigue, will have to cling to what is close to him, to what he knows, to what he can do, to his friends and his tradition and his love, lest he be dissolved in a universal confusion and know nothing and love nothing. . . .
>
> This cannot be an easy life. We shall have a rugged time of it to keep our minds open and to keep them deep, to keep our sense of beauty and our ability to make it, and our occasional ability to see it in places remote and strange and unfamiliar; we shall have a rugged time of it, all of us, in keeping these gardens in our villages, in keeping open the manifold, intricate, casual paths, to keep these flourishing in a great, open, windy world; but this, as I see it, is the condition of man; and in this condition we can help, because we can love, one another.

If loving one another is too much to ask, Oppenheimer on another occasion suggested a simpler way of easing our plight: we can have one another to dinner.

Finally the problem is one of education, an intimately related subject considered in the next chapter. At this point I should add that a study of the humanities will not necessarily restore a respect for the mother tongue. Their scholarly language is also often technical, dense with abstractions, and many of the articles in their professional journals are little more readable for nonspecialists. Although one cannot expect scholars to write a distinguished prose, style has become suspect to many of them, smacking of popularization. Even literary studies are generally addressed to specialists instead of cultivated general readers.

Sophisticated literary critics have condoned another kind of contemporary abuse of language, the flaunting of the once tabooed four-letter words. The taboo was irrational, of course, and both ludicrous and dangerous when it was extended to all plain references to sex. (Oldsters can remember the days when the word "syphilis" could not be mentioned in newspapers; the euphemism for it was "social disease.") It was primitive in its implication that mere names were evil, or could produce evil effects, inasmuch as people were always permitted to use other words to refer to the same experience or organ. Nevertheless the taboo against the four-letter words had some humane uses. Like blasphemy, it preserved their psychological values as a means of letting off steam on occasion. It even preserved a kind of respect for them; their earthiness and lusty humor could be appreciated when one met them in Chaucer and Rabelais, not on every page of novels. It lent some charm to pornography itself. To me the classics of pornography are pretty monotonous in their endless variations on a single theme, but now *Fanny Hill* can seem refreshing, even delightful, because it achieves its erotic effects without using any naughty words at all—its language is ultra-refined.

By contrast contemporary pornography is simply another debasement of language. The many cheap paperbacks put out for the drugstore and newsstand trade, all exploiting violence and sadism with sex, are perhaps the clearest evidence that people are being dehumanized in our mass society. A mere glance through some of them is enough to discover how uniformly unimaginative and banal they are; all might have been written by the same hack writer, to the same formula. With serious novelists "pornography" is not clearly the word for exploitation of the sordid or sadistic facts of life, inasmuch as they declare a serious artistic purpose, such as a disgust with modern life. But even those who are not repelled by their obsession with the disgusting may complain of their disrespect for the language, including the erotic vocabulary. The new freedom is too ostentatious, too seldom employed with either artistic restraint or imaginativeness. It is becoming another convention, reducing the lusty four-letter words to clichés, or the opposite of freedom. It is depriving our most intimate experience of dignity, diminishing the capacity for genuine sentiment. It is another symptom of the violence and the barbarism of our time.

Hence it suggests some reservations about the much-publicized thesis of Marshall McLuhan in *Understanding Media.* According to him, "the medium is the message," and the new all-embracing sensory medium of television—visual, auditory, tactile —is solving the basic problem of communication. While it is "cool," it nevertheless evokes depth-involvement and depth-commitment, and it is tribalizing men again, unifying mankind, bringing the whole world together in a "global village" because everywhere it has the same profound, beneficent effects on people, especially the young. It has also doomed print-oriented "typographical man," dating from the invention of the printing press; in the electronics age now dawning written language is fast losing its importance. So it would appear that all the problems I have worried over in this chapter are not really serious after all.

Now, McLuhan's book is a familiar type today—original, suggestive, stimulating, and brash, glib, and half-baked. It struck me as a variation of the old story of a bright young man going overboard with a fancy new half-truth. In insisting that the medium alone is the message, he entirely ignores the "content" of TV programs, and has only scorn for the literary people who complain of the irresponsible violence and vulgarity of the staple content, the incessant, blatant message of the commercials. He offers no proof at all that television is having such wholesome effects as he claims for it—effects that are plainly belied by the rising dissension, disorder, and violence in our supposedly "cool" culture. Since in his view any criticism of his thesis only proves that one is a "linear," print-hounded dodo, I think it fair to say that he himself keeps on writing books, extremely repetitious books, crammed with learned references suggesting that he is an uncommonly bookish man.

Yet I say this to preserve the measure of truth in his works, which his exasperating manner may obscure. However uncertain the effects of the new media, they are surely making some difference. In particular McLuhan had illuminating things to say in his earlier work, *The Gutenberg Galaxy*, about the consequences of the invention of the printing press. Although characteristically he went overboard in trying to make out that it was the main cause of almost all the major historical events thereafter, print culture did deeply affect mentality—our mentality today. It established the widespread habit of silent, private reading that now seems natural, but that had been known to few but scholars. (St. Augustine wrote of how surprised he was when he saw St. Ambrose reading without moving his lips.) The separation of eye and ear that made readers more visual-minded, disposed to believe that "seeing is believing," also had something to do with the growing dissociation of sensibility, the sharper dualism of heart and head, mind and matter. Likewise habitual reading led to habits of detachment, suspended judgment, the separation of thought and action, and so stimulated the growth of critical, skeptical attitudes. In producing the "detribalized" man, it compensated by stimulating as well the growth of individualism, notions about private or personal rights that most of us cherish. And so on, through much else that now seems natural or even necessary. "Far from wishing to belittle the Gutenberg mechanical culture," wrote McLuhan at this stage, "it seems to me that we must now work very hard to retain its achieved values." It was only when he grew bedazzled by our "electronic" culture that he scorned those of us who still cling to these values.

A much less sensational fellow pioneer, Walter J. Ong, is much saner in his studies of the shifting sensorium, which our technological society has made us more aware of. He not only fully respects the values of print culture but recognizes that it is very much alive, and necessarily so. A new medium never simply replaces the old; old states of mind stubbornly persist. In *The Presence of the Word* Father Ong offers a comprehensive, balanced survey of the impact of technology on the uses of language that serves much better to round out this chapter.

For the sake of perspective again, it is helpful to consider oral cultures such as Homer's Greece, in which only a few men were literate, primarily for commercial purposes, and hearing was believing. Today we can more fully realize why we cannot think and feel as they did. Among other things, men were capable of what now seem prodigious feats of memory; they had to be because none of their learning and literature was stored in books, the minstrels who recited Homer had to carry the whole epic in their head. Men had no history as we know it, only myth, legend, and epic, from which our scholars have worked laboriously to disentangle some historical fact. They had none of our passion for the "facts," since they had no way of either recording them or looking them up. Similarly they lacked our studied means of causal analysis and abstract knowledge. "Truth" as they conceived it was more intimate and alive; learning and understanding were a pleasure, not "work." For such reasons—not by inborn nature or conscious choice—men were more highly imaginative. But if we incline to envy them we should realize that their world, which they so little understood, was not a friendly one. Ong notes that their culture generated much hostility because when trouble came they tended to think that "some*body* had *done* something." They were not deeply religious either because the somebody was often a god, whose doings were capricious and unpredictable. Homer never told the Greeks to love the gods.

As for our own age, technology has affected language much more ambiguously than McLuhan suggests. So far from lessening the importance of print, newspapers, magazines, and paperbacks have vastly increased the consumption of it. Lettered words have become more ubiquitous than ever before; we meet them all day long on street signs, road signs, billboards, and at night in neon. And though people have perhaps grown more visual-minded, judging by all the pictures, comics, and ads, they are also more oral and aural than were Homer's Greeks, who endured—or enjoyed—

more silence. I would call our age the Age of Talk—endless talk, in committees, on platforms, in classrooms, over the telephone, over the air waves; the new electronic media fill the day with talk. People may be less impressed by talk than those in less literate countries, such as the Cubans who can listen to harangues by Castro lasting hours, but they listen to a great deal more of it. Ong emphasizes that *"sound unites groups of living beings as nothing else does."* While by its nature it perishes the instant it lives, the sound world has "depth, dimension, fullness" such as the visual cannot achieve.

Then others might add that modern technology enabled Hitler to rise to power: by his broadcast harangues he united millions of Germans behind his irrational, inhuman creed. Television may also "tribalize" people in ugly ways. There remain all the obvious ways in which the mass media have abused the power of the word, debased language, and dehumanized our world. The talk too may suffer from the same disorders, in an age that has scarcely cultivated the art of conversation. But since I have dwelt chiefly on the worst, at the end I might point to some humane tendencies of modern technology that Ong chooses to emphasize.

Both print and picture have obviously heightened the awareness that we live in One World. This by no means necessarily makes people so neighborly or deeply "involved" as McLuhan has it in his talk about the global village created by TV, but at least they have a more vivid sense of a world full of human beings than past societies had or could have; for the long run it is the only hope for the efforts at better understanding and better human relations indicated by such euphemisms as "underdeveloped" countries. At home a mass society is not so impersonal or atomized as its critics have it, but has retained some sense of community as people have learned to maintain decent relations with many different kinds of other people in stores and offices, or through chance acquaintance—relations that may be more pleasant because they are casual, lacking "involvement." (Who wants to be deeply involved with the pleasant postman, milkman, cleaning woman, tradespeople, and many others one meets often?) Ong remarks that verbally the world of technology and commerce is relatively peaceful, more so than Homer's Greece or the medieval world;

verbal hostilities are most apparent in traditional activities, military, political, and intellectual. He adds that if there is little "love" in all this, except for some saints love has always been reserved for intimate relations with a few. ("There is no indication in the parable of the Good Samaritan that after his rescue a close friendship between rescuer and rescued flowered.") And as the telephone and automobile enable people to keep in touch with distant friends, maintain more close personal relationships, so our technological society provides ampler facilities for having one another in to dinner.

12

Higher Education

On education, especially in the universities, I would dwell at some length if only because I have been engaged in it all my life; but the main reason is again that the "knowledge industry" has become the biggest industry in America. Whether or not the university will emerge as the primary institution in our technocratic society, it remains our best hope for managing such a society wisely and humanely. It is molding the prospective leaders in both business and government, the specialists serving them, the thinkers in all fields, and the large educated class that should be influential in setting national goals and standards. More than ever before "the ancient and universal company of scholars" must provide the intellectual authority and leadership that we cannot expect from either government or the church. For the long run it may be considered our only hope. And lest I seem biased as a professional educator, let me add at once that I am not at all complacent about the state of higher education in America, including my own efforts. At best education is a long, slow, uncertain process, but today it is much more complicated than ever, and hardly a radiant hope.

To begin with, our educational tradition differs significantly from European tradition. The first full-fledged democracy with universal suffrage, the United States was also the first to provide free public schools for all, well before the industrial transformation of the country.[1] In Europe the growth of industrialism had

---

1. The aristocratic, slave-owning South remained an exception. At the end of the nineteenth century no Southern state made education

much more to do with the spread of public education, but even so secondary schools and universities were still reserved almost exclusively for the upper classes down to the World War; few children of the working class enjoyed the privilege of a higher education. By contrast the American high school was from the outset a democratic institution, open to all even if most poor youngsters had to go to work instead. Likewise there were many more colleges than in Europe, however poor, and they included the distinctive state university with free or minimal tuition. During the Civil War Congress assured the flourishing of these universities by passing the Morrill Act, which gave them large grants of public land. (In time their domain would about equal the size of Belgium.) The state universities also accentuated the practical, vocational bent of higher education in America, again for democratic reasons. Whereas Oxford and Cambridge students were mostly young gentlemen with a secure, respectable berth awaiting them, many college students in America were poor boys working their way through, and most of them always had to keep in mind the necessity of making a living when they graduated.

Hence there was a marked difference in the response to the growth of industrialism. As aristocratic institutions, European universities were typically cloistered communities of scholars; the scientific and industrial revolutions had alike taken place outside them, with little or no contribution from them. Although German universities began making more of the study of science early in the nineteenth century, Oxford and Cambridge remained conservative, devoted to their traditional classical curriculum. As late as 1880 Thomas Huxley was still on the defensive when he pleaded the necessity of science in higher education. Despite its industrial leadership Britain was slow as well in introducing technological education, and so lost this leadership; it long had only mechanics' institutes, attended by workers instead of potential managers. American colleges were also slow in catching up with the Germans in science, but they were much more responsive to the practical, professional needs of a society devoted to business, even in farming. Thus the Morrill Act required the

---

compulsory, fewer than half the children went to school, and only a very small minority finished grade school.

state universities to set up agricultural and mechanical colleges to supplement the liberal arts. The Massachusetts Institute of Technology began its distinguished career in 1865 as a place "intended for those who seek administrative positions in business," which required "familiarity with scientific methods and processes." Cornell University, founded at about the same time, proclaimed the ideal of offering students instruction in all subjects, including technical ones. John Purdue, founder of the technical university named after him, was more exclusive, stipulating that his money should revert to his heirs if ever a Greek or Latin book were found in its library. Everywhere the study of the classics in the universities dwindled as students turned to more practical subjects.

This trend was clearly suited to a nation on its way to becoming the greatest industrial power in the world. American colleges and universities were pleased, however, to call themselves "institutions of higher learning"; so one might wonder how high or learned was much of the instruction they offered. Advocates of a liberal education might condone the backwardness of mossy Oxford and Cambridge in clinging to the traditional ideals of a university. Thorstein Veblen had such ideals in mind when in 1918 he wrote of the alleged higher learning in America:

> The greater number of these state schools are not, or are not yet, universities except in name. These establishments have been founded, commonly, with a professed utilitarian purpose, and have started out with professional training as their chief avowed aim.

By now the state universities more clearly deserve the name, and almost all are at least aspiring to it; yet we are a long way from what was once thought of as academic home. We may rub our eyes as we recall what John Stuart Mill took for granted a century ago:

> The proper function of a University in national education is tolerably well understood. At least there is a tolerably general agreement about what a University is not. It is not a place of professional education. Universities are not intended to teach the knowledge required to fit men for some

special mode of gaining their livelihood. It is very right that there should be public facilities for the study of the professions. . . . But these things are not part of what every generation owes to the next, as that on which its civilization and worth will principally depend.

Now, there is no longer the slightest question that a vast deal of professional education is what our kind of civilization depends on. The question is whether the universities are upholding the values on which the "worth" of a civilization depends. Are they giving all their students a sound basic liberal education? Are these students aware of what a "liberal" education once meant—one befitting a "free" man, not merely a practical training in the "servile arts"? What are their notions of "worth"? Such questions have been forced more insistently by the educational explosion since World War II.

This explosion has sent hordes of students to college—now nearly half of those of college age—thereby enormously increasing the size of many universities, especially the state universities. One might expect the quality of education to suffer from this mass influx, but actually standards of admission have gone up—so much so that one hears laments over the pressures on our poor teenagers, even though they still study much less than European students in secondary schools have to. The immediate problem is mere size. The big university, now called a multiversity because of its many thousands of students and vast assortment of specialized courses in different schools, is not at all a cloistered institution, of course; it is sometimes described as an educational supermarket or super-service station for society. Neither is it a community of scholars; the faculty and students of its diverse departments may get together in commencement halls or football stadiums, as its beery alumni do on reunions, but for the most part they go their separate ways, scattered all over a sprawling campus. As Clark Kerr wrote, "The university is so many things to so many different people that it must, of necessity, be partially at war with itself." Its elaborate organization may look like educational anarchy.

These obvious complaints call for some qualification. The multiversity is a unique institution that offers the no less obvious compensations of variety, a wealth of opportunity for the cultivation of all kinds of interests. It is much more cosmopolitan than the traditional university, especially because of the new institutes or departments of international studies—Russian and East European, West European, East Asian, African, Latin American, etc.[2] By contrast the small college is much more of a community, but like small towns it is apt to be provincial and smug. The more heterogeneous students in the big universities give little sign of pining for more community, beyond the congenial groups they form. Many prefer the advantages of privacy, moving out of residence halls to rooms or apartments off campus. Independent types are freer to experiment with designs both for living and for learning.

As I see it, there can be no ideal college, or ideal size for one. In any case higher education in America is pluralistic, much more diversified than it is in Europe. Many private colleges and universities are maintaining something of their own traditions. Although the innumerable small colleges are mostly mediocre, many are good, some excellent. They include purely liberal arts colleges, a kind lacking in the Soviet Union, Germany, and France. A more distinctive American novelty is hundreds of new, fast-growing junior colleges, which siphon off students of limited capacities, ambitions, or financial means. Adult education, now attracting millions of people, is spreading the helpful idea that education is a continuing process, not something "completed" when the student gets a degree. Taken together, the two thousand diverse institutions have been contributing to a great deal of experimentation in programs, a constant ferment, or a simple restlessness that is much less common on the more dignified, or the stodgier, academic scene in Europe. The big foundations—another distinctively American institution—have been subsidizing much of the innovation. For my present purposes I feel both heartened and dismayed by all the qualifications called for in the familiar generalizations about the means and ends of higher education in

---

2. My own university, which once was as provincial as most Mid-Western universities, has won international fame for its linguistics departments. It has over a hundred students majoring in Uralic and Altaic studies alone—languages, I confess, I had not heard of before.

America, the necessity of continually resorting to some equivalent of "on the other hand"; but anyway readers are well advised to keep on hand some pinches of salt, and maybe of pepper too.

On that other hand the big American universities now dominate the academic scene much more than they did in the past. They are making it harder for the smaller private colleges to recruit and retain an able faculty, or simply to get the funds they need to meet the spiraling costs of higher education in an inflated society. They themselves keep getting still bigger. If a few think of setting a maximum enrollment of 25,000 students, most appear content to grow indefinitely, even to take pride in their size, though like American government they typically expand without plan. Their growth is slowed up chiefly by the proliferation of university extension centers and new state universities, but these in turn are growing so fast that some are already almost as big as their elders. The expectation of still more millions of students in the years ahead raises the prospect of a land full of knowledge factories, processing the higher learning by mass production methods.

In the big universities students naturally get little individual attention, since classes have been growing larger too. Although the relative shortage of teachers is being relieved somewhat by the so-called "technological revolution in education," this may raise other misgivings. Beginning with audio-visual aids, technology is contributing a large battery of hardware and programmed materials. It includes, I have read, "computers for educational data processing and as controls of teaching systems, instructional systems themselves, etc."; again the magic word is "systems." In the schools teaching machines are speeding up "educational automation," which is already producing test-grading machines too. Professors are being put into cans of videotaped lectures, carefully tested by educational psychologists, who refer to them as cans of "teaching behavior." One specialist concludes: "The educational future will belong to those who can grasp the significance of instructional technology." Old-fashioned teachers may no doubt be too leery of this significance, as of all the new hardware, and too unhappy over the prospect of ending in cans; a professor who lectures to several hundred students might as well address thousands on television. But certainly mass production is tending to make education more impersonal, to sacrifice spontaneity and imaginativeness, and to reduce the higher learning to the kind of proficiency that can be tested or graded by machines—not to mention the significance of all the horrendous jargon that "instructional technology" is producing.

Among other penalties of bigness are the basic problems raised by all the large organizations that distinguish a technological society. The administration of a multiversity is perforce a bureaucracy. Like all bureaucracies, it tends to proliferate, to produce a growing army of vice-presidents, directors, deans, assistant deans, secretaries, typists, file clerks—altogether a much larger ratio of administrative personnel than in European universities. Forever busy with paper-work, it passes on more work of this kind to its faculty—people who used to be only teachers and scholars—and takes much more of their time by all the committees it breeds; whenever a question is raised, reads the new administrative law, appoint a committee.[3] It may give too little attention to the problems of education proper because it is primarily concerned with the day-by-day business of administration—keeping records, housing and feeding students, processing applicants, maintaining and enlarging the "plant," providing more parking space, and so forth. Students at Berkeley have protested that as far as the administration goes they are only cards in a computing machine. (Their signs read: "I am a human being, do not fold, spindle, or mutilate.")

Hence the president of a big university has an awesome job that seldom inspires awe. Although the institution is now run more like a corporation, he needs to have a much rarer combination of qualities than is required of executives in corporations. Ideally he should have a proper appreciation of scholarship, some understanding of the claims of the many diverse departments in

---

3. I remember fondly the old days when I was an instructor in a university and received only an occasional notice from the administration, had nothing directly to do with deans, turned in no reports except student grades, and sat on no committees. Today the official mimeographed materials I find in my office mailbox run to well over a hundred pages a month.

his university, more humility about all he cannot know, a concern for the welfare of both the faculty and the students, and a devotion to the ideals of academic freedom; but he is also expected to be good at administration, at money-raising, at public speaking, and at public relations with alumni, legislators, and parents or voters, most of whom have no passion for academic freedom, regarding it as only a nuisance they have to put up with in order to get the otherwise valuable service provided by the university. In practice he usually spends much less than half his time on strictly educational matters. Often he is chiefly a promoter, valued more for his skill at public relations than for any intellectual authority.

A particular complication of the problem of the big university is an apparent boon, the vast sums of public money it gets for research and development. One result has been the increasing devotion of the faculty to research, at some notorious expense of teaching. The university has been described as a "hotel keeper for transient scholars and projects," shopping for research grants, competing in the new game of "grantsmanship." Another by-product of the knowledge industry has been much more emphasis on the needs of society, immediately of government—so much more that Clark Kerr describes it as the greatest change in the university, which has become "a prime instrument of the national purpose." This issue has been confused somewhat by traditional attitudes. Conservatives see in federal aid a threat of government control of the universities, while liberals have worried over the powers of boards of trustees, made up largely of businessmen, who appoint the presidents of the universities.[4] Actually the universities are freer today than most were in the past. The federal government has taken pains not to exert direct control, trustees interfere less often and less directly than they used to, and the presidents they choose are less domineering than many once were. The real problem is that the universities and their research men

---

4. One anomaly of education in democratic America is that its universities have long been run much less democratically than most European universities, whose faculties typically elect their own deans or rectors, make all the regulations, and in general run the universities to suit themselves—though not necessarily to suit their students.

have in general been all too willing to respond to the demands made on them from outside. For the sake of public grants they have carried on all the research the government wanted, much of it for military purposes, and have therefore got involved in some unseemly projects. (A scandalous example was Project Camelot, in which a flock of social scientists worked for the Army on the problem of how to smother or subvert the revolutionary aspirations of the masses of wretchedly poor Latin Americans.) Likewise the universities have been serving the interests of business much more directly and assiduously than in the past.

To be sure, American universities—above all the state universities—unquestionably must meet the demands of their society. The question is whether they are meeting the demands on their own terms. Traditionally these terms have included some independent idea of what society needed—not merely what it wanted. The universities were supposed to be centers for the disinterested pursuit of truth, an ideal not too popular in America. They have developed a still less popular tradition of freedom not only of inquiry but of criticism and dissent. In commencement oratory they habitually proclaim their dedication to such lofty aims as providing both moral and intellectual leadership, maintaining a primary concern with civilized values, serving the highest interests of humanity. In a society devoted to material wealth and power they might reasonably be expected to serve as the conscience of America, a reservoir of idealism. And as self-governing institutions they could insist on their own terms. Government and industry need them even more than they need government and business support, and today they can always count on many more than enough applicants for admission.

Many of their faculty have in fact been insisting on their traditional terms, especially the need of criticism and dissent. They have been a major source of the social criticism that may be considered a primary function of the university today. One reason they are insisting, however, is that many of their fellows engaged in research, and most administrators, appear willing to play along with the powers that be. Often these men keep silent on controversial issues, out of either prudence or indifference. They may appear morally insensitive because the interests they serve are so often callous or venal. The universities themselves look venal in their acceptance of commercialized sports, the routine hypocrisy of recruiting their "amateur" athletes by giving them a free college education—a bribe not offered students gifted merely with brains. With some distinguished exceptions, the universities of America have officially not been providing a technological society with a high order of moral and intellectual leadership.

In particular federal aid has underscored a basic problem, or whole set of problems, by the invariable emphasis on the "national interest," in terms of military power and economic growth instead of the personal development of the students. Because of this almost all the billions spent on research and development have gone to specialists in science and technology, or to the development of what is called "scientific manpower"—a term that might make these specialists shudder more than they seem to. A small fraction has gone to the humanities; there is no public demand for literary or philosophical manpower. The official custodians of the national interest seldom act as if the liberal arts were really vital to this interest. When in 1965 the National Foundation on the Arts and the Humanities was finally established, to match the National Science Foundation, the President requested only a few million dollars for it. Congress can always be counted on to slash his request, in the interests of economy. A Republican leader who wanted to cut out the appropriation entirely protested that "we got along pretty well in this country for a century or so without spending money for culture and the humanities."

Anyone concerned with civilized values may be less sure about how well we have been getting along. Traditionally the humanities have served as the custodians of these values, the ends of the good life. As Coleridge said, "We must be men in order to be citizens," and a liberal education is supposed to develop men, whole men— not merely manpower. Hence people engaged in it have been much alarmed by the decline of the humanities in the universities. As a dean of Columbia University, Jacques Barzun declared flatly that liberal education in America is doomed. Even in staid England the alarm inspired a collection of essays entitled *Crisis in the Humanities.*

Now, I doubt that the humanities are doomed, or again that

the problem is grave and urgent enough to deserve the fashionable name of "crisis." The decline has been only relative, by comparison with the growth of technical studies, and it has been considerably exaggerated because of the usual nostalgia for the good old days, when they supposedly dominated higher education. In fact there never was such a golden age in America, at least in the state universities. In absolute terms, indeed, the humanities are flourishing as never before. Their departments too have greatly expanded because they attract a respectable share of the increasing horde of students; they are getting more financial support than they did in the past, even in terms of expenditure per capita; they are turning out proportionately more graduates with advanced degrees, some of whom are aided by government fellowships; they are getting more attention in technical schools too; and their staff is much better paid than formerly. The chairman of my own English department administers an annual budget of well over a million dollars—manages a much bigger business than mere English professors dreamed of before the war.

Still, he is not exuberant about this business. He knows that many a single professor in the sciences commands an annual budget of $100,000 or more. Apart from the usual administrative headaches, he can count on more whenever a financial pinch comes, because of state legislators always inclined to be economy-minded, for English is typically among the first departments to go begging. As it is popularly conceived, the national interest does not require that college graduates be able to read, write, and talk well, or have a real command of their language. Most graduates could not meet a seemingly elementary requirement of higher education, that they be able to speak good sense in good English about fundamental matters of general concern, including such questions as what the national interest really is or ought to be.

So let us consider C. P. Snow's celebrated, much abused lecture "The Two Cultures," the sciences and the humanities. His thesis was that these cultures, alike essential to an adequate education today, are unfortunately separated by mutual incomprehension, to some extent even by hostility. Critics were quick to point out that there are more than two cultures—one could easily make out two dozen—and that there is as little communication within them as between them. Sir Charles was more vulnerable because of his complacence over science and technology. He took for granted the greater importance and value of science, dwelling chiefly on the faults of literary people; he said very little about the values of literature, the reasons why scientists ought to be better acquainted with it. But after such qualifications his basic thesis seems to me clearly valid. Most specialists in science do not have an adequate grounding in the humanities; most specialists in the humanities do not know enough about the fundamentals of science and technology. And the current stress on research and development is widening and deepening the gulf between them.

In a following lecture Snow recognized the social sciences as a "third culture," a potential bridge between the other two. Ideally they could serve as a mediator because their subject is man and society. Actually, however, they are generally taught and practiced as sciences, not as humanistic studies, and if anything they have been tending to widen the gulf, arouse more suspicion or hostility among literary people, for reasons I suggested previously. Similarly with the study of engineering. Eric Ashby has said that it could unite science and humanism because it is inseparable from men and communities, concerned with the application of science to their needs. In fact most students of technology acquire very little humanistic culture, and they tend to be more indifferent to it than are students of the natural sciences.[5] And technical or vocational studies have been attracting a larger proportion of the students flooding our colleges than have either the humanities or the natural sciences. They might be grouped in another separate culture, distinctive in its incomprehension of both the other two. Too many of their graduates are cultural illiterates.

The best excuse for such narrowness, all the practical training that students need to fit them for their profession, may now look antiquated. As long ago as 1897 John Dewey pointed out a peculiar difficulty of education today:

---

5. I could cite sociological studies to support this generalization, and many others in this chapter, but prefer not to clutter up the discussion with scholarly paraphernalia. Although my generalizations are necessarily rough, readers should know that they are not merely personal impressions or prejudices.

> The only possible adjustment which we can give to
> the child under existing conditions is that which arises
> through putting him in complete possession of all his powers.
> With the advent of democracy and modern industrial condi-
> tions, it is impossible to foretell definitely just what civiliza-
> tion will be twenty years from now. Hence it is impossible
> to prepare the child for any precise set of conditions.

Since Dewey wrote, the much faster pace of technological change
has made his comment still more pertinent, above all for college
students. For years engineers and other technical students who
went into business have never used much of the supposedly
practical training they got, but today much of it is strictly
obsolescent, in process of being outmoded by the latest innovations. Many technical students are being prepared thoroughly,
meticulously, for a "set of conditions" that is ceasing to exist.
They may find themselves in not only a rut but the wrong rut. In
this view, as Robert Hutchins argues, a liberal education may be
considered the most practical kind. The humanities in particular
are supposed to put students in a fuller possession of all their
powers.

But we therefore need to look more closely into their claims,
which traditionalists may too easily take for granted. Students in
the humanities also concentrate on specialized subject matter, and
while it will not be simply superseded as scientific theories and
engineering techniques often are, one may ask what purpose is
served by all the knowledge or "book learning" they acquire. It is
clearly useful for students who are going on to become teachers
or scholars in their subject, but what of the many more who are
going out into the world? Specifically, the guardians of the
humanities might ponder some pointed questions asked by a
responsible business leader who wanted to know more about their
value. Just why are they considered so important? "Speaking quite
practically, what can the humanities do for me, for my family, for
my business, for my community? . . . Do they make people better?
Do they make people happier? Do they make people more
capable? How do you know?"

He could read many resounding answers. To cite a few

random examples, the humanities preserve our rich cultural heritage; they are "the very soul of our culture," the heart of the "civilizing process"; they elevate the spirit by giving us "a sense of man's innate worth and of his infinite capacities;" they further our understanding "of such enduring values as justice, freedom, virtue, beauty, and truth"; they are a "storehouse of wisdom" that enables us to make better judgments; they endow us with purpose and "concern for man's ultimate destiny"; they "help men to live more fully and creatively and to expand their dignity, self-direction, and freedom"; and scientists themselves are wont to say that without a knowledge of the humanities "science would lack vision, inspiration, and purpose." Or the businessman might ponder Cardinal Newman's classic statement of the ideal values of a liberal education:

> It aims at raising the intellectual tone of society, at cultivating the public mind, at purifying the national taste, at supplying true principles to popular enthusiasm and fixed aims to popular aspiration, at giving enlargement and sobriety to the ideas of the age, at facilitating the exercise of political power, and refining the intercourse of private life. It is the education which gives a man a clear conscious view of his own opinions and judgments, a truth in developing them, an eloquence in expressing them, and a force in urging them. It teaches him to see things as they are, to go right to the point, to disentangle a skein of thought, to detect what is sophistical, and to discard what is irrelevant. It prepares him to fill any post with credit, and to master any subject with facility. It shows him how to accommodate himself to others, how to throw himself into their state of mind, how to bring before them his own, how to influence them, how to come to an understanding with them, how to bear with them. He is at home in any society, he has common ground with every class; he knows when to speak and when to be silent; he is able to converse, he is able to listen; he can ask a question pertinently, and gain a lesson seasonably, when he has nothing to impart himself; he is ever ready, yet never in the way; he is a pleasant companion, and a comrade you can depend upon; he knows when to be serious and when to trifle, and he has a sure tact which enables him to trifle

with gracefulness and to be serious with effect. He has the repose of a mind which lives in itself, while it lives in the world, and which has resources for its happiness at home when it cannot go abroad. He has a gift which serves him in public, and supports him in retirement, without which good fortune is but vulgar, and with which failure and disappointment have a charm. The art which tends to make a man all this is in the object which it pursues as useful as the art of wealth or the art of health, though it is less susceptible of method, and less tangible, less certain, less complete in its result.

Still, Newman paid this tribute at a time when Oxford and Cambridge were offering a routine classical education that now looks pretty sterile. The important question remains unanswered: Do the humanities today actually achieve these high goals, or come anywhere near them? As the businessman asked, "*How do you know?*"

The honest answer is that we don't know. I doubt that we can know, since the lofty objectives ascribed to the humanities do not lend themselves to measurement. As it is, we go on giving examinations that test chiefly a student's knowledge of a particular subject, scarcely his wisdom, vision, creativity, humanity, soulfulness, or what not. We do not know either what kind of subject matter would help most in attaining the lofty objectives, but go on teaching all the traditional subjects in more or less traditional ways. As an English teacher I am disposed to believe that a proper command of language and literature could contribute most to self-realization, or broadly to a fuller realization of both our common humanity and one's personal identity, but the history of literary studies offers no convincing evidence that they make men more humane, and recent history gives more reason for doubt; Dr. Goebbels was only one of the many German doctors of letters who welcomed Hitler. I am sure only that many students specializing in the humanities achieve at most a mastery of a particular subject, or rather of some subdivision of it. Their studies have made them more capable in some respects, and we must hope on the whole better people, but we must wonder how much they can contribute to their community, or whether they could answer effectively the questions asked by the sympathetic business leader.

The root trouble, in the humanities as in the sciences, is specialization. Both teachers and students obviously have to specialize in order to master any subject; we all know the value of specialization, in the fuller understanding and appreciation that can come only as one gets more deeply into a subject. But it has become increasingly narrow, intensive, and often crabbed because of the demands of a highly professionalized society, and lately the knowledge explosion.[6] Students find it ever harder to realize one important value of a college education—an opportunity to explore, cultivate various possible interests, discover the uses of intellectual pleasure for leisure, or even discover what they like most. The specialization grows more cramping as students advance. They may never get a philosophical grasp of their whole subject, much less of its relations to other subjects, or to the means and ends of a technological society. Hence it is by no means certain that students of the humanities either get a sound liberal education, or come into complete possession of their powers. A specialist in any one of their subcultures is not necessarily broader, wiser, freer, or more humane than specialists in the sciences.

One consequence is the incomprehension of science deplored by C. P. Snow. Although all students of the humanities have had a course or so in some science, in which they picked up some technical knowledge together with probably fallacious notions about "the" scientific method, most know little about the history, philosophy, and sociology of science, or about how and why it has revolutionized thought and life. They have no clear idea of what questions science can and cannot answer, why its answers are always partial and dubitable, why its triumphant advance may make the humanities all the more important, or why these are not

---

6. The new subject of the history of technology—another potential bridge between the "two cultures"—has already produced scores of specialized volumes, such as *A History of Tin Mining and Smelting in Cornwall* and *The Cornish Beam Engine*. The quarterly *Technology and Culture*, which reviews two or three dozen books in each issue, exemplifies the dangers of specialization. The reviews are all about technology, have much less to say about culture, and rarely dwell on problems of human values.

living merely on left-overs.[7] They have a legitimate excuse in that introductory courses in science are usually taught as technical courses, a basis for more advanced work, not as humanistic studies for nonspecialists, or even as introductions to the fundamental question—the nature of scientific inquiry. But students of literature in particular may share the traditional attitude reflected in T. S. Eliot's *Notes Toward the Definition of Culture*, in which he completely ignored science. Many literary people still think of it as something apart from true culture, not a major branch of culture, or one really essential to a liberal education.

Another consequence of specialization already noted is that students of the humanities are generally at home only in their own subculture, and have little command of Western cultural tradition as a whole. Majors in literature may be poorly grounded in history or philosophy, or both, and vice versa. Again this is not primarily their own fault. They are kept busy meeting the requirements for their degree, have too little time to explore other subjects, and might well be staggered anyway by the host of important, disconnected subjects in the modern curriculum. In this fragmentation of learning, at any rate, some fundamentals get overlooked. Nowhere, in high school or college, are students given a basic training in how to read, sensitively and intelligently, the common forms of writing, from newspapers through articles and treatises to fiction and poetry. While they are indoctrinated in the values of their own specialty, they are not trained to think about the basic problems of values, in the light of both the humanities and the sciences. From long experience I know that most seniors are nonplussed by such elementary questions as what does it mean to be a civilized person? What are the essential values of civilization? What are its costs?

A particular reason why the study of the humanities may not achieve its professed lofty aims is that it too has been deeply influenced by science, through the spread of comparable modes of research. In America this began with the veneration of the

---

7. I speak with assurance out of my own experience, as one who for years has offered a course in Literature and Science, always to a small class. Such courses are not required for majors in English.

German Ph.D., which prescribed that the doctoral dissertation should be a contribution to knowledge. Students doing research in the sciences could expect to make such a contribution, but in the humanities only the most brilliant could hope to do so; the rest confined themselves to factual studies, often of matters of such minor importance that nobody had bothered to look into them before, and so got less training in the creative uses of mind than scientists get. In our century the spell of scientific methods, or the appearance thereof, has deepened.

Thus history is now often classed with the social sciences instead of the humanities. In aspiring to be objective and impersonal, scholars may pride themselves on dispelling the old idea that history is also an art by research monographs written with an apparently studied avoidance of style. Similarly government, once a humanistic subject, is now called political science; specialists in it work diligently to support this honorific title, no less because they cannot agree on what their science is or ought to be about. Philosophy has been dominated by the schools of logical positivism and linguistic analysis, alike concentrating on methodical analysis; in their obsessive fear of "meaningless" questions or "nonsense," they discredited not only metaphysics but ethics. In literature, where the German Ph.D. long dictated confinement to historical and biographical studies, the so-called "new critics" brought back literary criticism proper, but reduced it to a formal analysis, a highly professional performance uncontaminated by "subjectivity." Their initial effort to rule out all reference to history illustrated a common tendency of specialists in the humanities to seek autonomy and "purity" in an effort to maintain their traditional authority, or achieve a prestige comparable to that of science.

At the same time, the humanities establishment is prone to tendencies that undermine the prestige it seeks. Since it is concerned with preserving our cultural heritage, it is typically turned toward the past. So it should be in an age enamored of modernity, lacking a deep sense of the past or proper respect for it; American students in particular need more understanding of the past, more appreciation of its great works, and a fuller sense of not only how much we owe to it but how much of it is alive in the present, for better or worse. But too often the establishment lacks a deep enough concern for other needs of the present. Conventional scholars continue to maintain a detachment from contemporary issues; in an age of revolutionary change both their tranquil interests and their frequent hostilities with fellow scholars may too often seem irrelevant. Scientists may seem more humane than professional humanists because they are much more concerned about the problems of the present and the future, like the population explosion.

For such reasons one may feel less unhappy over Washington's treatment of the new national foundation of the humanities as a stepchild, giving it only a few millions instead of billions of dollars to play with. Early reports of the projects under consideration suggest that the foundation may be inclined to support chiefly conventional scholarship. This has an absolute value for specialists, ultimately a real value for the rest of us too, since it enables us to read the past with a finer, fuller understanding; it has given us far more knowledge of our whole cultural heritage than any previous society ever had. Yet this is to say that for all but the specialists scholarship is valuable only as a means to some end, and scholars may lose sight of the end. Many are not scholars in the ideal sense proposed by Emerson, Man Thinking, but intellectual celibates. In any case there is scarcely a crying need for more learned articles, for we already have a great many professional journals, more than enough to dignify the humanities by a kind of knowledge explosion of their own. If the explosion has made little noise to speak of, it has not generated much heat or light either outside of scholarly circles. The most apparent result has been a swelling flood of articles, too often inspired only by the need of publication for promotion. From much of this printed matter a nonprofessional might never guess that the humanities are "the very soul of our culture."

More typical of our technological society, however, is another establishment—the professional educationists. In the many teachers' colleges and schools or departments of education they train our prospective teachers, principals, and superintendents, who in the high schools are supposed to lay the basis of a liberal education. By and large they are not known for either their own excellence as teachers or their dedication to the ideal objectives of the humanities. For some time now they have been dedicated rather to the proposition that pedagogy too is a science. They have been carrying on hundreds or thousands of research projects, dignified by measurements or statistics, concerning all manner of "processes"—developmental, educative, etc.—and efforts to "evaluate" methods. These presumably have some value for pedagogic purposes, but mostly seem mechanical or trivial by humanistic standards, of little help in inculcating the values of the humanities, in which teaching still appears to be more of an art. At best the educationists have been turning out teachers well trained in theory and techniques, but too often lacking a sound grasp of their subject.[8]

Once more, however, I have been saying something like the worst about our educational and humanities establishments, and saying it only in order to emphasize more the values of a sound liberal education. There are of course many enlightened educators, "progressive" in the best sense. Likewise there are many stimulating teachers of the humanities who help students to realize their capacities for the pursuit of truth, beauty, and goodness. There are more than enough of them to warrant the hope that in practice the humanities are to some extent achieving their ideal objectives. That we can never be sure to what extent is no reason for disparaging them, only better reason for repeating that values other than formal knowledge and technical proficiency cannot be measured precisely. It is now time to dwell on the enormous difficulties faced by all educators in the effort to prepare students for life in a technological society and a revolutionary age.

Let us begin with an elementary objective on which everybody seems to agree, that our universities should turn out good citizens. Everybody might agree too that their graduates should accordingly have some understanding of their society, be able to listen and read intelligently and to think independently, and

---

8. No subject has suffered more than English, the most fundamental one, taken by all students. About half of the English teachers in our schools did not even major in the subject. Judging by the way they too often write, educationists share the common assumption that anybody can teach English, which in small high schools may be taught by barely literate football or basketball coaches.

have a responsible concern for human values, which in American democracy include moral values like justice and equal rights. They should be able to deal intelligently with questions about both what we *can* do and what we *ought* to do. For these purposes a purely scientific or technical education will not suffice; apart from such common deficiencies as a considerable ignorance of history and government and a limited command of English, technically trained students are likely to have an excessive faith in scientific method or technique, too little awareness of the complexity of social and political problems, too little understanding of questions about what *ought* to be done. Neither will an education in the humanities alone suffice for enlightened, responsible citizenship, since so many of our problems are technical and it gives too little idea of what we *can* do. The plain trouble is that most college graduates—whatever their specialty—have too limited an understanding of our technological society for potential leaders. Very few, in either the sciences or the humanities, have studied the impact of science and technology on society and culture, the source of our imperious problems. But how, then, to give them all an adequate understanding of their changing world? How to prepare them to make the best of this world?

Since most educators even in technical schools agree, at least in principle, on the need of some liberal education, let us consider first the humanities. Students majoring in science and technical or vocational studies have time to take only a few courses in the humanities, and so have a particular need of good basic courses, ideally a coherent program instead of a miscellany. What should such a program consist of? Surely these students need a better acquaintance with Western tradition, as the source of not only our cultural values but our institutions, the kind of people we are and how we got this way. It follows that they need to know some history; so what would be a good basic course in it? To the common answers, American history or a survey of Western civilization, the common objections are that the one is likely to be too provincial, the other too superficial. Other subjects too have obvious claims—literature, philosophy, foreign languages, and the fine arts, the latter as a supplement to an education that is otherwise purely verbal. Since we are perforce living in One World, the study of anthropology would seem especially important, to give some sense of the nature and the power of culture. The rise of the whole non-Western world also makes desirable some acquaintance with its radically different traditions, in particular the historic civilizations of China, India, and Islam. But to any proposal of a comprehensive program for technical students it may always be objected that a study in depth of some one subject would be preferable, or that better than any prescribed program, which students are always inclined to balk at, would be giving them their head, allowing them to explore or cultivate their own particular interests, and so to come into fuller possession of their own powers. They might even be allowed to design some courses (as students have done in an experimental program at San Francisco State).

Similar questions may be raised about any program in science for students of the humanities. They need to know something about the history, philosophy, and logic of science in order to grasp its fundamentals, but presumably they ought to know something too about the major ideas in the various sciences, the many revolutionary theories. Physics, long regarded as the most fundamental science, has further claims because of the revolution in modern physics and the world-shaking applications of nuclear physics; then physicists add that nonprofessional students need to know calculus in order to understand their science. Chemistry may speak for itself because of its prominence in modern technology. Biology would demand attention if only because of the theory of evolution, one of the key ideas of the last century, but it has become more important because in recent years it has produced the most revolutionary discoveries. There remain the obvious claims of psychology and the social sciences. Since the objective is a better understanding of a technological society that conditions all studies, by its influence on all interests, one could argue for a basic course in the history of modern technology. If, that is, there were not as good arguments for dozens of other courses, and the unhappy students were not already oppressed by the extent of their ignorance and the impossibility of doing much about it.

To all these questions perhaps the best answer is courses embracing a number of subjects, focusing on the relationships lost sight of in the compartmentalization of knowledge. These have grown respectable under the technical name of "interdisciplinary" courses (another term that makes me shudder), but even so they can serve humane purposes. Specifically, they might be designed to bridge the gap between the "two cultures," for example by a study of the theory of evolution and its influence on religion, philosophy, history, literature, social and political thought, anthropology, Freudian psychology, and almost everything under the intellectual sun. Another way of bringing various disciplines together is by focusing on contemporary problems. Despite their tradition of facing toward the past, teachers of the humanities have been yielding to the insistent claims of the present in a revolutionary age, giving much more attention to modern literature, history, and philosophy, and some have been showing interest in programs centered on problems that call for the cooperation of various specialists, the union of techniques and a concern for human values, or of know-how and know-what. One obvious example is the deterioration of the American metropolis. A study of it might begin with the history of the city, and perhaps a novel about modern city life, and then move on to what sociologists, political scientists, economists, engineers, architects, and city planners have to say about means and ends. Only then the problem is to find teachers willing and able to teach such courses.[9]

Today teachers may be most troubled by the final consideration—the interests of the students themselves. While our society

---

9. Some years ago M.I.T. gave all its students in science and engineering what impressed me as an admirable basic course in the humanities, offered by its department of English and History. The first semester was devoted to a study in depth of ancient Athens, its history, government, economy, social life, religion, philosophy, art, drama, and so on; the second semester was devoted to a similar study of Renaissance Florence, in our own civilization; and at the end of the year the students wrote a long paper on their hometown in the light of these studies. The course was dropped because both English and history teachers had to do considerable reading outside their field, and neither wanted to take so much time off from their own research. And the most conscientious teachers may resist the growing demand for "generalists" because of their devotion to standards of excellence. They may ask: Can there be excellence "in general"? This book might serve as a red flag.

appears to be serving its youth handsomely, it is most concerned with its own immediate needs rather than theirs. The universities have provoked a growing dissatisfaction on the campus by their too faithful, uncritical service of the official "national purpose." The upshot was something new in the history of American education—the dramatic student revolt. Only this too has become an increasingly complex issue.

Paul Goodman suggests that the best advice for earnest students is still that offered by Prince Kropotkin long ago: "Think about the kind of world you want to live and work in. What do you need to know to help build that world? Demand that your teachers teach you that." This might appear to be a quite reasonable demand. Even teachers willing to heed it may object, however, that whatever students think they want is not necessarily what they need or what is best for them. But the chief difficulty arises when one asks what in fact do they want. The radical demonstrators who made the headlines obscured this difficulty by their apparent certainty—they want "student power." Tacitly they assume that they would exercise this power for their own purposes, whereas whatever power students win would naturally be exerted in the interests of the majority; student government is everywhere at least nominally democratic. (At one big university the aroused students demanded—and got—the right to park their cars on an already congested campus.) The radicals forget what they are well aware of when they meet hostility on the campus: most American students are complacent.

Ten years ago Paul Jacobs reported that the great majority of college students were "gloriously contented" and also "unabashedly self-centered," prepared to discharge their routine duties as citizens, but otherwise politically irresponsible and often politically illiterate. Today students are more disposed to complain about this or that, in particular restrictions on their social life; but most remain basically content with their education and their prospects. The kind of world they want to live and work in is pretty much America as it is, affluent, comfortable, and conservative. Studies have shown that most leave college with essentially the same view of themselves and their society that they entered with. At best they have grown more tolerant because they are confident of their prospects, which financially are indeed far better than they ever were in the past. Many can put up with nonconformists because their own conformity is not due to insecurity or anxiety.

The complacent majority still includes many collegiate types, drawn by the traditional attractions of "college life"—fraternities and sororities, week-end parties, football, and all the other forms of rahrah. When not anti-intellectual, they at best have no keen intellectual interests; they are content with the social and economic values of a college degree. With them may be lumped the types that sociologists have called "ritualists," who come to college simply because it is the thing to do and are distinguished by their lack of commitment to anything, even rahrah. They might be pitied because of the pressures put on them by their middle-class parents, but also might remind us that so far from being oppressed, American students are the most pampered students in the world. At any rate, teachers need not worry much about either of these types, or ask them what they think they need to know. If they clutter up the campus and act as a drag in the classroom, they give little reason for feeling any particular responsibility to them.[10]

More troublesome are the considerably larger number of students, especially in the technical schools, who have serious vocational interests. They force the basic problems of education in a technological society. Typically the vocationalists tend to resent the courses their deans require of them for the sake of some liberal education; they regard the humanities as a nuisance to be got rid of as quickly and easily as possible, so that they may get on with their specialty. Content to be "one-dimensional" men, they are otherwise usually satisfied with their education. They might well complain of it, inasmuch as corporation executives often say it is less serviceable than a broad basic education would be, pointing out that the corporations will retrain them for

10. The sales talk of too many administrators recalls what Robert Hutchins wrote about "collegiate life" thirty years ago. "Undoubtedly, fine associations, fine buildings, green grass, good food, and exercise are excellent things for anybody. You will note that they are exactly what is advertised by every resort hotel."

their own particular purposes; but even so they prefer to stick to their specialty. They know what President Johnson emphasized in urging business support of higher education, that a college degree is a good investment, enabling a graduate to earn on the average $300,000 more in a lifetime than those who have only a grade-school education. And as might be expected, they score consistently low in tests of cultural sophistication, social conscience, and political liberalism. Similarly few technical institutions stand out for independence of thought on the campus, opportunity to hear controversial speakers, or ardor over academic freedom, civil rights, or any burning issues. In this respect teachers' colleges are even more conservative, suggesting again their devotion to technical training more than to humanistic learning.

Today the most serious problems are raised by the many professionalists, in general among the most capable students, who are bent on going on to graduate school and joining the new "meritocracy." They have become so numerous that they have brought about what Christopher Jencks and David Riesman call the "academic revolution." For the graduate schools have profoundly affected all higher education. They exert great pressure on the colleges beneath them, whose reputation now depends largely on how many of their students are admitted into graduate schools and how well they do there. They intensify the drive to specialization, the impatience of students to get done as soon as possible with other required subjects. By their requirements for admission they heighten the emphasis on formal examinations, grades, and credits—all the trappings of academic "performance" —and make it harder for gifted students with independent interests who want a good education, not merely good grades. They are stirring increasing complaints even from conscientious performers, who feel they have no choice but to become either playboys or grinds.

As Jencks and Riesman point out, the professionalists most plainly belie the popular beliefs about how democratic higher education in America has become—so many of its young people going to college and now getting advanced degrees. "Coming of age in America can be a race for the top," they write. "It is seldom a sprint, however, in which victory goes to the naturally gifted or enthusiastic. It is a marathon, in which victory goes to those who train the longest and care the most." Hence the children of the upper-middle class have an immense advantage, in not only financial means but training and incentive. They have been prepared from childhood on by parents, usually themselves college-educated, who take for granted that they will go to college, and they are also impelled by the common feeling in America that much more painful than the frustration of ambition to go up the social scale is the fate of those who go down it. By contrast only the most ambitious young people from the lower class have sufficient incentive to enter the marathon. Handicapped by a poor background and poor training as well as little money, the great majority are easily discouraged in what hopes of college they may have; only 7 per cent of the college students come from the bottom quarter of the population. And now a bachelor's degree is no longer enough for those aspiring to professional status.

In view of all such tendencies, one may appreciate more the ardent young radicals on the campus, the novelty of the student revolt. The militants can indeed be pretty obnoxious, even to those who sympathize with their indignation over the war in Vietnam and the treatment of the blacks. Although most of them are very bright and have keen intellectual interests, they can seem anti-intellectual because of their contempt for the normal academic decencies, their fierce intolerance (especially of mere liberals), and the violence of their language, sometimes their behavior.[11] They can also seem unrealistic in their impatience to make over America into the kind of society they want, the hopeless call of some for a revolution among whites too, and meanwhile their increasing addiction to a lawlessness that is antagonizing not only the public but the liberals on campus. But otherwise they can at least serve as a healthy antidote to the political apathy and unreasoned conservatism of too many American students. Although they flaunt the trappings of "alienation," such as long hair, they can be called good Americans because of their implicit

---

11. The *Berkeley Barb* has called for guerrilla bands to sweep through college campuses "burning books, busting up classrooms, and freeing our brothers from the prison of the university."

optimism, their faith that something can be done about all social evils. The liberals they despise may forgive the lack of ideology that distinguishes them from European and Latin-American students, who have behind them a long tradition of student radicalism such as the United States never developed. "Anyone under twenty who is not a radical does not have a heart," runs an old foreign saying; "anyone over forty who still is one does not have a head." The militants on the campus at least have a heart, and they may keep their head better because few are converts to Communism or any other radical ideology.

In any case, the militants are far from being so ineffectual as they might seem because of the tiny minority they represent. Most active in some of the most distinguished universities, which set the tone for higher education, they have won concessions that are influencing other institutions. They have enlisted much faculty support because their attacks on mulish administrations initiated much needed reforms on the campus. Above all, they may help to bring about a more wholesome "academic revolution" because they have stirred a much larger number of earnest, thoughtful students who incline to deplore their tactics but sympathize with their indignation and their ardor, and also support their more reasonable demands.

The prospect of such a revolution may seem less bright when one considers Prince Kropotkin's advice to students. Except for their call for social justice, especially for the blacks, the young radicals are seldom specific about the kind of world they want to live and work in. While demanding student power, including more voice in determining academic policy, they are usually vaguer about what they need to know for their purposes, just what kind of education the university should offer. Chiefly they want more attention to contemporary problems, and for this reason are likely to be impatient with most courses in the humanities, or study of an American past they see no use for because, as Christopher Lasch observed, it has little to teach them except the common failure of radical movements, the lack of a revolutionary tradition; but otherwise they have little to say about the basic problems of education in a technological society. They ignore the claims of all the students who want to be good professionals, technicians, researchers—claims that the universities cannot afford to ignore in our kind of world. Nor are the many thoughtful, more moderate students who earnestly want to help build a better world much clearer about the kind of education the university should offer. They too feel that their education is not "relevant" enough, but when asked relevant to what—to immediate burning issues or to broader interests than the subject they are specializing in—their answers are usually vague or confused. If as individuals they think they know what they need to know, as a group their interests are so varied that they are of little help to teachers aspiring to draw up a sounder curriculum, with an eye at once to breadth, depth, and balance. I suppose they make clearest that there can be no one ideal curriculum, any more than the big American university can hope to be an ideal community of scholars such as Robert Hutchins still dreams of.

Nevertheless I should say that these students are our best bet for the future, the best reason for believing that the revolt on the campus may make a real and lasting difference, and not merely because they are still willing to listen to those of us past the unspeakable age of thirty. They are taking a much more active, enlightened interest in national and world affairs than students used to (for example in my own generation). They do know that they want somehow to be more actively "involved," as so many were happy to be in the campaign of Senator Eugene McCarthy. They want to be "committed" to some positive cause beyond private, selfish interests or more affluence, the goals of their parents; for this reason many are turning away from business. They make plainer that the student revolt is due to a real failure of the universities to satisfy the needs of thoughtful students. They are looking for an idealism they too seldom find, asking for more devotion to the lofty aims paraded in commencement oratory. If they are vague or inconsistent about what they need to know, they want more freedom to explore, inquire, dissent—and dissent from academic authorities too. In short, they are seriously concerned about human values and the good life.

So the question is whether the universities can be expected to respond to their needs. Many are granting student demands for more power over their social life, which are obviously reasonable if students are to be treated as adults instead of "college boys," given the rights of other free citizens at a time when almost half the population is under 25. Some universities are taking steps to grant them as well some voice in educational policy. But apart from the nice question how much power students should have over their teachers, the immediate difficulty here is not merely the arbitrary rule of administrations, inasmuch as most presidents and boards of trustees are now careful to heed the wishes of their faculty. The difficulty is the vested interests of the faculty. They set the style of the university, the tone in its classrooms. In this respect the government of the university is highly decentralized, each department largely determining its own policies. While the faculty are inclined to be politically liberal (except in schools of business, engineering, and education), they are more inclined to be academically conservative. Wishing to go on turning out scholars like themselves, most professors support the drive to increasing specialization, the requirement of the Ph.D. for admission to the brotherhood, the rule of "publish or perish." They are likely to resist proposals for radical innovation in the curriculum, in particular any that would reduce the importance of their own specialty in the academic scheme. Would-be reformers in the big universities may despair of the problem of overhauling so vast a curriculum in the face of the conservatism of their colleagues, who may recall the ironic observation of E. M. Cornford—"Nothing should ever be done for the first time."

More difficult is another issue raised by the student revolt—the demand, backed by some of the faculty, that the universities serve as agents of major social change. The revolt has at least stirred them up enough to make them reconsider their social role, their possible service as the conscience of America. Their faculty represent a considerable reservoir of moral and intellectual leadership, lively concern for social justice and personal freedom; they remain a primary source of the critical judgment badly needed by an affluent society. The academic scene as a whole remains much livelier than the scene in Europe, or less academic in the depressing sense, and more open to change. Under the recent pressure the universities have begun, for example, to rush in courses on the history, literature, and culture of the blacks. Yet they can hardly be expected to act as deliberate agents of anything like radical social change—least of all the state universities, whose administration has to cater to majority interests and keep a prudent eye on legislators and taxpayers for sordid financial reasons. And these bring up still another difficulty.

The costs of higher education have been soaring steadily because, unlike mass production in industry, its managers can do little to increase its economic efficiency, or the "productivity" of professors, except by increasing the size of classes in already over-crowded buildings. Clark Kerr estimates that the colleges and universities, which are now spending some $20 billion a year, will in less than a decade have to spend more than twice as much, and that to keep them solvent the federal government will have to triple its present contribution of $5 billion. His estimates are possibly excessive, since one may question whether the universities actually need all they want; they want new chairs, a Ph.D. in every classroom, and other such academic luxuries primarily for the sake of prestige. But in any case their prospects of getting such support would seem to be pretty poor, given the current drive to reduce government expenditures on everything but the wants of the Pentagon. And their prospects are poorer because of another recent development, a growing public hostility to the universities. Surveys have revealed that more than two-thirds of the public complain of not only the rising dissent on the campus but their promotion of social and scientific change, and of a meritocracy that people consider fundamentally undemocratic. Briefly, the American people think the universities are already much too active as agents of social change.

At the end I return to the enormous difficulties posed by the basic question: What kind of men do we need for our new kind of society? First of all they need to be men, whole men, free men, and so need something like a liberal education for a full life. Most of them need as well some professional training to fit them for the innumerable specialized jobs in our kind of society. The universities must give them this training; but at the same time they should hope to give all students a better understanding of their society and its new horizons, make them good citizens with a proper concern for civilized values, help them to make satisfying use of the increasing abundance and leisure, develop them as independent persons with a capacity for both the firm convictions and the adaptability needed in a changing world, and prepare them for responsible leadership in a future that the experts in forecasting tell us will be still fuller of problems. Again I should say that there can be no ideal curriculum for all these purposes, or even no satisfactory way of determining priorities. The most that can be said is that in a growing awareness of the problems we are making some progress in preparing students for life tomorrow instead of life the day before yesterday. This is no reason at all for complacence.

13

The Natural Environment

In all the talk about "underdeveloped" countries these days, Americans may forget that our own country is woefully "overdeveloped." For years we have been not only squandering our natural resources but fouling our nest. Long intent on mastering nature, man has never shown less respect for it than have modern Americans; and by our abuse of it we have disregarded our own human interests too, losing control over our environment. Ex-Secretary of the Interior Udall summed up the outcome of the American Dream:

> This nation leads the world in wealth and power, but also leads in the degradation of the human habitat. We have the most automobiles and the worst junkyards. We are the most mobile people on earth and we endure the worst congestion. We produce the most energy and have the foulest air. Our factories pour out more products and our rivers carry the heaviest loads of pollution. We have the most goods to sell and the most unsightly signs to advertise their worth.

Technology also provides the means for repairing much of the damage, but this would call for a prolonged national effort on an immense scale, and first of all a public realization that what we do in the next few years will affect the country for many years to come. As it is, there is little such realization. Except for some alarm over the smogbound cities, problems of the environment constitute what Udall called a "quiet crisis," one recognized by relatively few people. About no set of problems is there less reason for optimism.

So I begin this gloomy chapter with the usual conscientious reservations. We know surprisingly little about the effects of the natural environment on people, but do know that all through history they have proved capable of adapting themselves to many different kinds of environment. We are not certain even about how much people are suffering physically from the pollution of the air; they might get used to it as they have to many disagreeable conditions. Least of all can we define an "optimum" environment, befitting the wealthiest nation on earth. Yet there is no question that our environment has been deteriorating in various ways, and that it has some bad effects on people, however immeasurable. The bacteriologist René Dubos is certain that the constantly increasing need for hospitalization and medical care, due in part to mere affluence, is due as well to the kind of environment Americans now live in. Just because we know so little about its effects, others fear that such contamination as all the carbon gas we produce may turn out to be disastrous. And at least some psychological effects are plain enough. Smog is clearly a nuisance, as are congestion, noise, and filth, the up to five pounds of garbage generated daily per person. What Americans have done to "God's own country" seems worse when one considers spiritual needs, including the satisfaction of the esthetic sense. Although most Americans have evidently got used to all the ugliness around them, people who live in beautiful or just pleasant surroundings know they make a real difference.

Let us consider a typical example of popular attitudes. There are still a few rivers in America, or stretches of river, that are unpolluted and undefiled. In the summer of 1968 Congress considered the Scenic Rivers bill, which would protect six of these immediately and on a couple dozen others bar dams and other projects for eight years while they were being studied. Included in the latter was a stretch of the Susquehanna River. Unfortunately, a series of dams for this stretch had already been projected by the Army Corps of Engineers, than which no bureaucracy in Washington is more ambitious, single-minded, short-sighted, and thick-skinned. Opposition to the bill was led by Congressmen from the region, backed by local business interests eager to attract industry. Always pork-minded, the House shouted down the bill in a session attended by fewer than half its members. Later the House reconsidered, but meanwhile no wave of public indignation swept the country. It is a safe bet that most newspapers did not bother to report the incident, which most readers would have skipped over anyhow on their way to the comics or the sports page.

In the same session the House emasculated a bill to save the remaining redwood forests; this had naturally been opposed by the logging companies exploiting the forests. A little earlier it had passed the Federal Highway Act, adding 3,000 miles to the interstate highway system, this at a time when it was drastically cutting appropriations for welfare purposes in the interests of economy; it economized only by eliminating provisions for parks and wildlife sites, and by cutting out 90 per cent of the minimal funds proposed for highway beautification. The billboard lobby had been fighting the infant beautification program.

In all this the House was faithfully serving the economic interests that make esthetic and other human interests secondary when not negligible. The trouble with the redwood forests was precisely that they were as "priceless" as lovers of them said: they could not be priced on the market—except in the form of lumber. With other programs in the public interest the trouble is that the costs can be estimated. One authority, for instance, has said that to make all our rivers and streams pure again might cost $20 billion a year, an expenditure he was sure most people would think ridiculous. Actually it would amount to less than half the current annual increase in the gross national product, but one can hardly imagine either Congress appropriating it for such a purpose or public opinion demanding it. Surveys have shown that Americans are more willing to pay for exploring outer space than for improving the environment on earth. For economic purposes, at any rate, polluted rivers are not a critical problem.

The problems may look still more hopeless as one considers them more closely. To begin with, it might seem relatively easy to prevent further desecration of the environment: the government has only to clamp down on the "selfish" business interests that have been blasting, defiling, and polluting it. These interests are powerful, however, and they do not seem simply selfish to the many people they employ and the many local communities dependent on them. The "public" is not something apart from them but has always been implicated in their activities. But much more difficult is the problem of repairing the damage, restoring the environment. This would require vast expenditures, much greater than most people realize, and if by unlikely chance Americans were willing to spend enough to do a thorough job, their leaders would face further problems. The public has many other needs, such as education, health, and housing; even a country with a GNP of $900 billion cannot spend unlimited amounts on all of them; and how does one determine priorities? Who is to decide how much should be allocated to problems of the environment, and on what basis?

Economists have been of little help in this matter. They naturally think in terms of quantities, market values, dollar costs —economic considerations that can never be ignored, of course, but can never be sufficient either. My main concern here is the *quality* of our environment, human values that cannot be priced; choices must finally be made by value judgments. Nathaniel Wollman has accordingly proposed a "new economics of resources" to deal with these problems: it would recognize that the cost of doing nothing about a deteriorating environment is itself high, and "by developing new methods of analysis" would show us how to incorporate values "into a measured system." One who hopes that economists will make some such effort may still doubt, however, that they can measure noneconomic values, or develop methods even so roughly scientific as those of the "old" economics. They would also run into the possible need of a "new politics" too. Both in Congress and in the country at large only a minority is really concerned about the environment.

Hence our hopes must rest on this minority, specifically on its efforts to enlighten and arouse the apathetic majority. In the end, again, our hope must lie in education—the excuse for this chapter. It is a good enough excuse because there has at least been a growing interest in the problems of the environment, both professional and public. One promising sign is the new science of ecology, dealing with the relations of living organisms to their environment, and now embracing the as new science of demography.

Its promise is naturally still clouded. It has inevitably bred more specialists, who on all problems are likely to say we need first of all much more research; laymen might say that the problems are too urgent to wait for specialists to complete their labors. Much of the research is likely to take the form of quantitative studies, in an effort to attain the respectability of physics and chemistry, and so again to neglect problems of quality or values. The new science has also bred the usual technical vocabulary, including terms like "ecotone" and "ecosystem" that are not so precise as they sound, and that encourage the usual jargon.[1] Nevertheless ecologists have produced valuable pioneering studies, such as William Vogt's on human population that awakened thinkers to this problem. They are also encouraging efforts at synthesis, among them studies in "bio-politics," suggesting a fruitful cooperation with the social sciences. All such studies are being encouraged by grants of millions to universities by the Ford Foundation. Chiefly ecologists have revealed the problems man has created, the dangers he now faces, but this is of course what we most need to know.

The conservation movement, a forerunner of ecology that got under way toward the end of the last century, was at first concerned mainly about such natural resources as forests, and the most popular theme ever since has been irretrievable losses of resources. There used to be much alarm over dwindling supplies of coal and oil, with dire predictions about how soon they would be exhausted. Then the alarm shifted to soil erosion, dramatized by Paul Sears in *Deserts on the March*. Lately there has been more talk about a looming scarcity of water. Specialists could no doubt add other candidates for alarm, perhaps including our forests again because of the recent "paper explosion" due to duplicating machines. (By 1965 these were producing about ten billion copies of one thing or another, and were expected to turn out seventy

---

1. One specialist has presented a "two-dimensional conceptual model, based on levels of integration and points of view, as an aid in orienting knowledge with relevance to the ecological crisis." Humanists might still not rejoice when the writer added that their point of view would probably have more influence than that of natural scientists on the "conceptual environment" that is "critical in achieving ecological homeostasis."

billion within a few years.) Today, however, there appears to be a widespread faith that technology can solve the problems it has created. Plenty of good enough land is still available, with the help of big dams for irrigation, and rising crop yields assure an abundant food supply. Instead of conserving coal, in a declining industry, the Department of the Interior is spending millions on developing new uses for it. With new technical devices explorers keep finding new sources of mineral ore, offshore deposits of oil, more resources in the ocean. Technology is discovering or creating substitute materials and exploring new sources of energy. I suppose it might solve the water problem too by such methods as purification, rainfall by weather control, and the desalting of sea water.

Yet in a long view—the prospect of a completely industrialized world—there is still plenty of reason for worry. Harrison Brown points out that the time is approaching when we will have to get along on the lowest grades of metals, and before many hundreds of years man will have consumed all the ores that it took nature many millions of years to lay down. Technology will no doubt keep us going, he adds, if need be on ordinary rock, but we will have to put more and more of it into the system, and like the Red Queen run faster and faster to keep in the same place. As it is, right now we use about ten tons of steel per person to keep America going. Americans might be appalled by the huge wastefulness, which grows worse by the day as more throw-away bottles, cans, and containers enter the market. So the dumpheaps and junkyards keep growing, while beer cans litter the landscape. On all counts, the world of the future bids fair to be a much less attractive one to live in. Add the steady emission of poisonous elements, such as the radioactive gases and wastes produced by nuclear power plants, and it promises to be much more dangerous too. The ecologist Paul Shephard asked: "Who would want to live in a world which is just not quite fatal?"

Meanwhile we have a big enough national problem in simply combating the increasing pollution. Most Americans, I take it, would welcome clear water and fresh air again, and would even be willing to pay something for what economists used to call "free goods." (I write as one who remembers the days when people could still swim and fish in the Hudson River, and who has a vivid idea of how many more goods there are that money cannot buy in our affluent society.) But even apart from the many billions they would now have to pay for them, this problem too is not at all easy. Industries that have been polluting rivers by long established practice regard it as a right, in accordance with a law of business or nature; the country will have to learn to accept much more regulation. This will have to be nation-wide regulation, moreover, because if only some cities and states impose it, industries could legitimately complain that their competitive position would suffer; too many small cities and poor states eager to attract industry are willing to give it a free hand. Government itself has been directly responsible for much of the pollution as cities dumped sewage into rivers to hold costs down, in deference to their taxpayers. And by now we have poisoned much more than rivers. Big as it is, Lake Erie is already a dead lake and Lake Michigan is fighting for its life. Harbors and coastal waters everywhere have been fouled. The fouling of beaches by oil tankers and drillers has only dramatized man's ability to pollute the ocean too, which is being enhanced by plans to build 500,000-ton tankers; all over the world marine life has been harmed enough by residues from oil-burning ships to stir some demand for international control. Countless thousands of the chemicals we produce also find their way into the sea, as do all kinds of other human waste. Now we face proposals to dump into it surplus poison gases and radioactive wastes.[2]

The pollution of the air has been getting worse, no less because it directly affects people by damage to property if not health. Inasmuch as economists are thought to be more realistic than ecologists, we might consider a study of the problem by Azriel Teller, an adherent of the "old" economics. In the name of "economic rationality," he dismisses as "specious reasoning" the

2. Sir Bernard Lovell is worried by the pollution even of outer space. It already contains well over a thousand fragments of burned-out rockets and spacecraft, "earth-circling junk." In the debate over the ABM's authorized by President Nixon, George Kistiakowsky noted a further danger, that the computer that would automatically make the hair-trigger decision in case of apparent attack is "really a very stupid thing," and would have to be programmed to distinguish between a warhead and all these pieces of space junk.

suggestion that "it would appear to be wise to reduce general air pollution of all types insofar as possible," for the costs would be prohibitive. Anyway, "No individual has the endowed right to obtain clear air free, simply because it is a necessity." Since this is a collective good, society must "simulate the market for air," and it can do this only "if it can estimate the demand and supply schedules for clean air." Teller makes some effort to measure the demand by estimating "the cost of the physical and psychic damage that results from different levels of air pollutions," and then estimating the costs of abatement. He advocates a "selective abatement," in which each city would decide for itself "the air-quality standards it needs"; the federal government should conduct research into methods of abatement, but should not establish national air-quality standards. Except for some doubts about measurements of "psychic damage," I suppose all this is indeed practical by American standards and about the most we can hope for. So long as Americans insist on the sovereign rights of their gasoline chariots, the main source of the pollution, they no doubt deserve the kind of air they have to breathe.

Outside the city the natural environment has likewise worsened, to some extent irretrievably. Technology cannot replace or restore all that has been destroyed, such as the venerable redwood forests.[3] Powerful new techniques have also created new problems by unexpected secondary effects on the environment. In *The Silent Spring* Rachel Carson made pesticides the notorious example. Chemical companies then rushed to the defense of DDT and their other insect-killers, which have indeed much improved crop yields and nearly eliminated such diseases as malaria in some regions, but the public controversies that followed were pretty disgraceful. The commercial producers of the pesticides had plainly ignored the deplorable side-effects, the killing of large numbers of birds and fish, possibly some animals too; only now are they testing and screening their products some-

---

3. Since Americans are so conscious of dollar costs, Californians should know that they paid heavily for the destruction of the redwoods by logging companies. The loss of topsoil in the forest regions led to devastating floods, and the costs of repairing the damage were charged to the taxpayers, not the logging companies.

what more carefully. Meanwhile DDT has entered living tissue everywhere, including many things people eat. Similarly marine life has suffered from synthetic detergents, which came into wide domestic use before it was realized that they resisted bacterial decomposition. They still end in the rivers.

The polluted rivers in which fish can no longer live suggest an incidental irony, that Americans now have better fishing rods and fewer fish to catch. It points to a crueler paradox, that while modern man has been learning so much more about life he has been destroying animal life on an unparalleled scale. As the buffalo herds were slaughtered and the carrier pigeon was exterminated, just for sport, so for various reasons many other species are threatened with extinction. The U.S. Fish and Wildlife Service has identified 78 species that are in danger, as well as some dozens of others that have become rare enough to be nearing the danger point. Among them, perhaps appropriately, is the national symbol—the bald eagle. With up-to-date mechanized equipment, the whaling industry is also threatening to extinguish the whale, sending Moby Dick into limbo with the bald eagle. (The one gratifying thought is that the industry would thereby extinguish itself, join the mythical ostrich with its head in the sand.) Ecologists are concerned about what all such heedless slaughter is doing to the ecological balance, but nature lovers might be unhappy for simpler reasons.

The common talk about how modern man has become alienated from nature may make one wonder how much we really need natural beauty. Millions of slum-dwellers have got along with very little greenery and wild life, while millions more of city-dwellers have been content with some trees, a patch of lawn or garden, here and there a little park. Few people appreciate the incredible richness of nature—the millions of species, the billions of billions of plants and animals. Most would be surprised (as I was) by the news that an acre of pasture contains more than a billion insects, mainly a kind called springtails. Yet almost all people do get some pleasure from nature. Considering how much poorer life would be were there no flowers, trees, birds, or other animals, one might find more credible the belief of the entomologist Howard Evans that "a world of trees, flowers, and wild creatures is needed to nourish human attributes now in short supply: awe, compassion, reflectiveness, the brotherhood we often talk about but rarely practice." There are not many such Wordsworthians today; but at least city people still like open country enough to have made it more valuable, as uncrowded space for them to escape to—the "wide open spaces" that once helped to create the American spirit. Only they thereby create another problem.

Most of America is still open country, of course, and some of it has been set aside for public recreation, which can now be regarded as a "right" of industrial man. All told the country has many millions of acres in national forests and parks. As the population grows both larger and denser, however, and enjoys longer vacations, much more space will be needed for recreation. A dozen years ago the director of the National Park Service estimated that eighty million Americans would visit the parks in 1966, and accordingly quadrupled expenditures, but even so his estimate proved too low. Like beaches, the popular parks are already often more crowded than the cities people escaped. "It is a measure of our troubles," wrote Robert Patterson, "that an American will soon have to buy a reserved ticket for a seat in the woods; if he will not accept that fact, he might not have a woods to sit in."

Nature lovers may not rejoice either in all the cabins, lodges, food concessions, and other buildings in the parks. The automobile must be brought into them too, provided with more parking lots and camp sites to increase the congestion. Some conservationists complain that our parks and preserves should not be merely picnicking places, designed to "provide shallow amusement for bored and boring people"; but most Americans will otherwise. Granted that they have a clear right to their picnicking, the litter they too often leave intimates that they have little interest in preserving beauty and quiet, what they might be seeking in their escape from the cities. And ugliness has come to many once beautiful regions, such as Lake Tahoe, through the creation of "motorscape." The approaches are crowded with gas stations, motels, lunch stands, souvenir shops, and other garish accessories of the tourist trade, with a wilderness of signs, all hailed by billboards for miles in advance.[4] The highways themselves betray the same disrespect of the natural environment in the interests of economy. Highway engineers are ready to invade or destroy any scenery in their way in order to enable automobiles to go faster and farther, now with the help of computers that plot the route of new highways with a complete disregard for scenery. (A remarkable exception proves the point—a highway engineer who for several weeks walked a thirty-mile stretch in Maine in order to decide how to take advantage of the terrain for the sake of interest and natural beauty.)

Granted that the human costs of the "planned deterioration" in the environment cannot be measured, or government expenditures on it allocated "rationally," it seems to me that many of the costs are nevertheless plain enough to justify considerable expenditure, and the burden of proof ought always to lie on those who are degrading the environment. But efforts to combat the evils meet political obstacles. A conservationist association wishing to preserve some of the public domain from a corporation seeking to exploit it will find the law hostile if it tries directly to influence legislation. The corporation can deduct from its taxable income the money it spends on lobbying; the association cannot spend tax-deductible donations for the same political purpose—and it cannot hope to get the money it needs unless donations are tax-deductible. Even the growing concern within the government over the various problems of the environment raises familiar difficulties. Too many agencies share the concern to make possible coordinated effort, and none has the authority needed to plan and carry through long-range programs. It appears that we can expect no more than some timid piecemeal programs.

The best hope remains the recent changes in attitude and policy. The government has made some starts, such as the Wilderness Act, which reserved 9 million acres of the public domain for

---

4. Tourists might be reminded of Ogden Nash's poem:

> I think that I shall never see
> A billboard lovely as a tree.
> Indeed, unless the billboards fall
> I'll never see a tree at all.

wildlife and recreation, and the Clean Water Restoration Act. CBS has offered television programs on our many abuses of the natural environment that gave an even grimmer account than I have here. Private associations are working to preserve or improve the environment, with growing public support. Even a few aroused citizens can get surprising results. Frank Tysen, a specialist in urban ecology who is crusading against the "ugly-makers," tells of heartening examples of what can still be accomplished in America by ordinary people. Thus two or three citizens took on a big oil company that was preparing to put up a gas station of the usual box design on a shaded corner of their city, at the expense of a couple of fine old trees; they demanded that the company spare the trees and design a station suited to the location, and they succeeded in stirring up enough support from the townspeople to induce it to give in. Now all the major oil companies employ architects to design gas stations for communities that demand some style.

Tysen grants, however, that on the whole the "ugly-makers" are still having their own way. The vested interests are too powerful, public resistance is too sporadic or feeble. And the prospects of improving the environment look poorer when the whole problem is viewed in the light of a crisis that is not at all "quiet"—the population explosion. The statistics ought to be familiar by now. Whereas in the first sixteen centuries of the Christian era the estimated human population on our planet about doubled, to reach half a billion, it has grown to more than three billion, and at the rate it is growing it would double by the year 2000, double again in another thirty-odd years. Billions of human beings will be competing for resources not yet adequate to sustain them all decently.

In the United States, as in Europe, this problem still does not look like a "crisis." It was not long ago that Americans were worrying instead over a declining birth rate, at least among the educated. Even if they continue to increase at the rate that has carried them past 200 million, amid public rejoicing, they can count on getting plenty to eat and having enough resources to sustain their rising standard of living. Now the birth rate has been falling off again and may decline more, but in any case this is not a problem that we are heedlessly bequeathing our children: the young will decide how many children they want. Yet the rising population is aggravating the problems of the environment, threatening to worsen its quality. The prospects are that soil, water, and air will be more polluted unless big programs are carried through. Less open space will be near at hand, less land available for recreation, unless again there is a sustained effort by government. There will be less beauty, less peace and quiet. It will be harder to maintain a truly human life by the standards of the many, including René Dubos, who regard quiet, privacy, and some open space as not luxuries but vital necessities. "They will be in short supply," Dubos predicts, "long before there is a critical shortage of the materials and forces that keep the human machine going and industry expanding."

In Latin America and the non-Western world the population explosion is unquestionably a crisis by any standards. Here, where population is growing fastest, people cannot count on the most elementary requirements for a decent living. The great majority of them very poor, they are no longer passive and resigned to their lot as their ancestors were over the centuries, but have got some idea of all the goods Americans and Europeans enjoy, and want more of them than their fathers did. To meet their rising expectations, modest though these still are by American standards, will therefore require a much greater and faster increase in the production of food, energy, and minerals. Here again apostles of modern technology are confident that it can satisfy all these needs. Scientific agriculture, for example, can take care of the food problem, with the help of the unexploited resources of the ocean and chemical compounds; though to Americans enjoying their steaks and fried chicken the diet of the future looks unappetizing, one enthusiast estimates that it would be possible to produce enough food to support 50 billion human beings. But even if, with Western help, the have-not nations succeed in averting mass starvation, the plain trouble remains that their economic growth has barely been keeping up with their rising population. Efforts to educate their people have not kept up with it. There are more illiterate people on earth today than ever before.

Birth control, the obvious means of controlling the population explosion, meets as obvious difficulties. All through history, to be sure, people have practiced it by various methods—superstitious practices, abstinence, *coitus interruptus*, abortion, and infanticide. Today some of the non-Western countries are trying to disseminate the necessary knowledge and contraceptive devices, and their people are growing more receptive to the idea; so I assume it is impossible to forecast the world's birth rate in the year 2000. But meanwhile the problem is of course very different in Latin America and the non-Western countries than in the United States, where the vast majority (including most Catholics) practice birth control. The majority of their people lack even a rudimentary knowledge of contraceptive methods, and most ignorant are the impoverished, illiterate peasant masses who most need the knowledge. Everywhere efforts at birth control run into peasant inertia or meet resistance for various social, cultural, moral, and religious reasons. As serious are the economic difficulties: government has to provide the necessary pills, since most of the people are too poor to afford them. In short, as Aldous Huxley observed, death control is cheaper and easier than birth control. So population continues to grow, while at least a billion people remain undernourished. It raises the hard question whether population controls should be enforced by the government, and people no longer allowed to go on breeding as they please.

For America too the population explosion in the rest of the world has possibilities explosive enough to produce unmistakable crises. Already fearful of the power of Red China, it has to reckon with the probability that there will be a billion Chinese before the end of the century; sooner or later it will have to recognize their nation, even if a billion Chinese can be as wrong as fifty million Frenchmen. It will also have to deal with increasingly restless peoples all over Latin America, Africa, and Asia. Peoples poorly fed, housed, and educated, if literate at all, can hardly be expected to admire affluent America. Envy of it will lead more naturally to a growing resentment of its wealth and power, as is already apparent in most of these countries, and it may be dangerously aggravated by racist feelings, since their people are mostly colored. And America has been doing very little about these explosive possibilities. Surveys of attitudes toward government programs indicate that by far the least popular program is foreign aid. Too little aware of their own deteriorating environment, Americans know much less about world ecology and most of them couldn't care less. Congress is as indifferent about efforts to deal with the basic ecological problems of mankind, such as the International Biological Program, in which 55 nations are cooperating. "Mankind is living incredibly dangerously," a scientist told it in supporting a request of only $5 million for the American contribution to the Program; whereupon Congress granted only a third of the amount.

One may accordingly be more grateful for a by-product of modern technology, the business of systematic forecasting, which makes clearer the need of doing much more about the problems looming up. Among the more optimistic specialists in this business is Joseph L. Fisher, president of Resources for the Future, Inc. (As usual, I am not taking seriously the many popular prophets of a future radiant with technological wonders, or the many scientists who are automatically optimistic, in the belief that science can do almost anything.) From preliminary world studies he and his colleagues concluded, not too resonantly, that "the population-resources outlook is not without elements of hope." The United Nations, for example, has been making some effort to deal with the problems of world ecology by providing various kinds of technical aid services. Fisher banks in particular on the adaptability of man, "the extraordinary capacity of human beings and institutions to react to emerging problems in new and constructive ways." If "reaction" may be the precise term for too much of their behavior, a great deal of significant change has in fact been taking place all over the non-Western world, affecting even the most backward regions. However slow or uncertain their "progress," it looks more remarkable in a historical perspective, for the capacity Fisher speaks of has seldom been conspicuous before this era of rapid change.

I would stress still more than he the uncertainties of the future, the impossibility of forecasting with assurance. But I would also stress the absolute need of efforts at forecasting, so that we may anticipate problems instead of merely "reacting" to them, as Americans have to pollution. If the experts often seem too confident about their estimates, it would be worse if they stopped out of scientific fear of being wrong. And if their estimates often seem exaggerated, so much the better: we might then prepare for the worst. There is no hope unless we face up to the alarming possibility that in a generation our much abused planet will be crowded with six or seven billion human beings, the more congested because so many of them will be living, or trying to live, in cities.

14

The Social Environment:
The City

"We shape our buildings," said Winston Churchill, "and then they shape us." Now that the city has become the shaping environment of most Americans, and has much plainer effects on them than does the natural environment, it would call for special attention in any event, but a particular reason is the growing alarm over the "urban crisis." Modern technology has been blighting the big cities, and what is in worst shape is precisely their heart. The center of an amorphous region called the "metropolitan area," this grows more congested even though for years people have been fleeing it by the hundreds of thousands. Thereby they have helped to bring on the most obvious crisis, the state of the Negro ghettos. By now the cost of restoring the American city, making it a fit place to live, is commonly estimated at $100 billion. If the outlook for it too is "not without elements of hope," I propose to dwell chiefly on what is wrong with it for the usual reason, that most Americans still do not regard its condition as really critical, and there is no immediate prospect of anything like an adequate national effort to restore it. For my text I take a statement by Adlai Stevenson:

> I pray that the imagination we unlock for defense and arms and outer space may yet be unlocked as well for grace and beauty in our daily lives. As an economy, we need it. As a society, we shall perish without it.

If so, the chances are we perish.

Now, the modern American metropolis is necessarily very different from the great cities of the past, lacking their tradition, and it cannot hope to recapture the kind of antique charms that American tourists appreciate in Europe. But since there still are many city-lovers, let us consider what the city has meant and might still mean, and why all through history ambitious people were drawn to it. "City air makes free," ran a medieval saying. It freed men from not only the serfdom or common oppression of the peasantry but their inveterate conservatism or inertia. The city was what it had been since its beginnings, the center of both commerce and culture, the essential means and ends of civilization. It attracted men by its variety, color, and stir, with always something going on. The medieval town was congested and unsanitary, of course, but it offered ample compensation. It had a rich civic life, centered in its market-place and its squares, where its citizens could make merry or get solemn together on their many holidays and festivals. As the European cities grew, they maintained much of this civic tradition in their piazzas, platzes, or places. People who were well off continued to live by preference in or near the center of the city, which was never merely a business district.

With the rise of the new industrial towns came the horrors I have already mentioned, the foulest environment ever created by man. The late Victorian cities, however, were no longer the "mere manheaps, machine warrens" that critics have called them. By this time they not only were more sanitary but in rivalry had developed civic pride. They proudly adorned themselves with public buildings, mostly in deplorable styles, yet exhibiting aspirations to carry on the old tradition of civility and culture; many of their buildings have aged with more dignity than most modern building is likely to. Meanwhile the old cities of Europe were still maintaining their civic traditions, as they would to this day. Their prosperous people mostly still prefer to live in or near the center, and are not fleeing to the suburbs in droves. Besides pleasing old quarters, they have many avenues in which it is a pleasure to stroll, as it is in few American avenues except for window-shoppers. Hence American tourists are pleased to recover the use of their legs in Europe. At home they go downtown on business or on shopping expeditions, rarely if ever for the pleasure of strolling or relaxing. In only a few quarters can they find the "atmosphere" they pine for, and this soon becomes commercialized, as in Greenwich Village, or else is reserved for the well-to-do, as in Georgetown, Washington.

Hence they force the question that city planners have begun, rather tardily, to ask: What is or should be the *purpose* of the city? In America its apparent purpose, one remarked, is "sustaining the population, enabling it to consume." Efforts at urban renewal have suffered from the lack of any clear goal beyond immediate pragmatic purposes, merely tinkering or "reacting" to problems instead of trying to create a suitable environment. Some planners are accordingly talking of providing on a larger scale for diversity, novelty, and openness. Others, with an eye to the slums, stress the need of creating a more democratic urban environment. A few even define the goal as "urbanity"—a necessarily vague term, but referring to civilized values that include Stevenson's "grace and beauty." In general, the more imaginative planners are saying that the purpose of the city is to serve the needs of people—and not merely their economic needs.[1] It is the purpose that the cities of the past served unconsciously, to promote the growth of people, enable them to realize more of their potentialities. For city planners the problem is complicated by the growing feeling that they ought to consider more the wishes of the people, especially the poor, make more effort to understand them, and get them to participate in the task of improving the quality of the environment that shapes their lives. The trouble remains that most Americans have little idea of what city life can be at its best, as limited an awareness of their own higher needs and potentialities.

Nevertheless the big American city still fulfills some of the historic functions of the city.[2] A center of commerce, it is also a

---

1. I am drawing here on recent reports of the American Institute of Planners, which held a two-year consultation on the environment we should try to build over the next fifty years. The reports have been published by Indiana University Press in three volumes, *Environment for Man, Environment and Policy,* and *Environment and Change,* edited by William R. Ewald, Jr.

2. In generalizing I am perforce neglecting the considerable diversity of American cities: Washington, a spacious political capital that

center of culture, the sophistication that enables people consciously to take it or leave it. Containing the principal theaters, concert halls, art galleries, museums, auditoriums, and the like, it differs in this respect from the cities of the past in offering a greater variety of cultural fare, to go with the still greater variety of material goods in its department stores; technology has brought both kinds of goods into it from all over the world. Otherwise the modern city offers the traditional attractions of animation and diversity. As one who fell in love with New York in my youth, and prowled all over Manhattan, I learned to discount the traditional complaints about the coldness and soullessness of the big city, and the modern complaints about its standardized, atomized masses. In its slums I found plenty of squalor, but also real "neighborhoods," animated communities with some tradition and culture of their own.

Yet for the same reason I am more keenly aware how woefully New York has deteriorated, and how critical its condition is today. In this respect it is like almost all the big cities in America. They are not beautiful, of course—no industrial city has ever been; they were not built to "human scale," with an eye to openness and spaciousness, grace and amenity. But neither are they functionally rational or efficient, as one might expect a city to be in a technological society. Congested, blighted, and polluted, they make some authorities doubt that the central city is worth saving or restoring. Its many poor people are simply unable to enjoy all the professional and cultural services it affords, and business could be dispersed to outlying regions, where corporations have been moving anyway, since computers have made it possible to establish headquarters almost anywhere. Meanwhile, however, the city remains

---

has almost no industry but is far from being a cultural capital; Los Angeles, a product of the automobile, described by somebody as "seven suburbs in search of a city"; New Orleans, Boston, and Philadelphia, which retain relics of their historic tradition; Detroit, whose heart is the deadest by night; Chicago, long a symbol of the violence of America; San Francisco, considered the most attractive city in America by most visitors; New York, whose Manhattan remains the most spectacular and world-famous; and so on. But one can generalize readily because of common tendencies everywhere, and above all the common problems.

the chief center of business, professional, and cultural activity, and the millions of slum-dwellers have no place to go, inasmuch as the suburbs are bent on keeping them out. I assume that the urban crisis is here to stay for years to come.

The crisis is worse because it has deep roots in the American past. Like New York, most of the big cities are still contending with the tradition of favoritism, corruption, and graft that earned municipal government in America the reputation of being the worst in the world. They are contending as well with the commercial tradition I touched on earlier. With a few farsighted exceptions, such as the setting aside of Central Park in New York, the rapid growth of the cities was governed by the aim of growth and bigness for its own sake—not the growth of its people—or more specifically for the sake of business and real estate values, the interests of speculators. Lewis Mumford has cited as a classic example the statement by the engineer of the Public Service Commission of New York: "Every transit line that brings people to Manhattan adds to its real estate values." The transit lines also added to its congestion when skyrocketing real estate values made skyscrapers profitable, despite the loss in economy because so much space must be given to elevator shafts. The haphazard growth of the metropolis was rational by the standards of the market, but irrational for civic purposes. While speculators made fortunes, the booming American city characteristically gave back little in the way of civic amenities, such as arcades, promenades, sidewalk cafes, and squares where people might enjoy satisfying human relations instead of superficial or fretful contacts. Given all the crowding, noise, dirt, and pollution today, Wendy Buehr has observed that our wonder-working technology enables a head of lettuce to cross the continent and arrive at market in better condition than the customer who walks a few blocks to buy it.

The recent growth of the suburb was likewise governed by the purposes of private contractors and realtors, with little or no public control over the uses of the land or provision for community life. One result was carefully segregated white communities, heavily subsidized by government with highways, sewers, water, and other services, all to the benefit especially of investors. Another result was the spread of commercial slop on the approaches to the city, now aptly called "slurbs." Although modern technology had made the suburbs possible by electric power lines and rapid transport, it was not until the 1950s, with the immense expansion of the automobile industry, that the "exploding metropolis" took its place with all the other explosions. Today almost a third of the American people live in suburbs, representing the greatest migration in history over so short a period. New York (which I shall continue to use as a prime example) has become the center of a metropolitan sprawl of more than sixteen million people over more than twenty counties. And though Manhattan has by now lost about half the population it had in 1905, this has not meant more peace, quiet, and room in the center. A million or so people still work in it, since it remains a great business center for the whole nation, and many of those from the suburbs come in by car. In other big cities, which have more available space than the island of Manhattan, most suburbanites drive in to work. Hence room has to be found for all the cars to park as well as to move. In Los Angeles two-thirds of what is charitably called its heart is taken up by provisions for the automobile—streets, freeways, garages, parking lots, gas and repair stations, dealers' lots, etc. Again the villain is the "Sacred Cow of the American Way of Life."[3]

Transport by the most untamed animal in the country is by far the most highly subsidized by government. While the federal government spends billions on highways, cities devote more of their budget to caring for the automobile than to any other civic need, and more and more highways are built to bring it into the already congested centers. The results have been ludicrous. Trucks built to go sixty miles an hour often move more slowly in New York than did the old horse-drawn wagons. According to one authority, it costs more to move an orange from the West Side of Manhattan to the East Side than to move it from Florida to New York. Taxis on many crosstown streets may barely keep up with pedestrians. As for the pedestrian, who was once regarded as a

---

3. Lest I seem obsessed by the automobile, I should add that I too am pleased to own one. We all know its advantages. My simple point remains that most Americans appear not to realize how high a price they pay for their cars.

human being, a planning report for Los Angeles described him as "the largest single obstacle to free traffic movement." To traffic engineers he may look most human when defined as "a man who has two cars, one being driven by his wife, the other by one of his children." In some suburbs he has actually been stopped by the police as a suspicious character. Even in Europe, where many people still like to use their legs, traffic engineers may consider him a mere nuisance. At an international conference some suggested that pedestrians be required to queue up and cross the streets only in small trickles, while others wanted simply to eliminate sidewalks, the habitat of the creatures. All no doubt would have been shocked by the solution of Julius Caesar, who relieved the congestion in Rome by banning private wheeled traffic in the daytime.

To Americans it makes no difference either that the private car is the most uneconomical method of transport, especially when they drive one to a car. A National Highway Users Conference denounced the subversives who were proposing to substitute various kinds of public transportation in metropolitan areas. In New York Robert Moses, who once confessed that traffic control was one of "our few failures," nevertheless continued working most indefatigably to bring more automobiles into the city. When citizens of Staten Island proposed an alternative route to a new parkway to be bulldozed through the heart of its green belt, Moses denounced it as "impractical and visionary" because it would cost more. (He also sees red at the mention of "regional planning"; his name for people trying to improve the civic environment is "long-haired planners.") As a result of such "planned chaos," Lewis Mumford wrote, "the time is approaching in many cities when there will be every facility for moving about the city and no possible reason for going there."

For years Mumford has also crusaded valiantly, and vainly, against the multiplication of big office buildings in midtown Manhattan. These have plainly increased the congestion that is bad for business too, but the businessmen who invested in them thought only of the immediate profit for themselves. As one told Mumford, " 'Money' is not interested in looking further ahead than the next five years." Most of the new buildings are shoddily built because tax clauses for property depreciation make shoddiness profitable; it does not pay to build for keeps when a building may be written off for tax purposes in forty-five years. Hence the new buildings will have to be replaced before long, only the congestion will last indefinitely. Profits have also called for tearing down much sturdier old buildings. In New York a conspicuous example was the demolition of Penn Station, a survivor of its age of Roman splendor, to make way for commercial buildings and a new sports arena. (Grand Central had already been cheapened by commercial litter.) No architectural masterpiece, Penn Station had none the less handled space grandly, much more generously than contemporary architects are allowed to do by their business sponsors; by its nine acres of columned, vaulted concourse it had earned its reputation as a monumental landmark. Yet no serious public effort was made to preserve it. Its destruction epitomized the commercial values that are ruining the American city. In an editorial the *New York Times* described it as the shame of the city:

> Any city gets what it admires, will pay for, and, ultimately, deserves. Even when we had Penn Station, we couldn't afford to keep it clean. We want and deserve tin-can architecture in a tin-horn culture. And we will probably be judged not by the monuments we build but by those we have destroyed.

About modern architecture I shall say more in a chapter on traditional culture. Here the main point is that most city architecture is not a matter of high culture, only of business; often the architect merely translates into blueprints the orders of his clients. The occasional striking building, like Lever House or the Seagram building, that stands out in the jungle of routine or pretentious buildings emphasizes the urban chaos. Peter Blake reminds us that architecture in all the great periods of the past was a force for order and unity as well as beauty and civilization; a colorful chaos can be charming, but only within a system of architectural order. He laments that municipal authorities have as little sense of style or order: New York "commissioned its last decent building in

1812"—its old City Hall ("still standing—unless they sold the site to some developer"). Altogether, the most public of the arts is symptomatic of the sickness of the American city. It reflects a basic disorder that too often manages at the same time to be colorless and dull. Chaos has become 100 per cent American in "googie architecture"—hot-dog stands shaped like hot dogs, ice-cream stands shaped like ice-cream cones, and other such efforts to make standardized commercial enterprises look like something special.

Housing in the big cities offers modern comforts and conveniences, but otherwise it is nothing to boast of, especially in view of an architect's remark that there is no longer any economic reason why every family should not live in a spacious dwelling, with a view and sunlight. The trend to smaller apartments marks an incidental irony noted by city-lovers: while Americans boast of the "conquest of space" they have ever less space to live in. Slums have been spreading over former middle-class districts, including some once dignified streets, which thereupon begin to deteriorate; landlords find it as unprofitable to keep them up as builders do to provide decent housing for the poor. Slum clearance projects have mostly been unimaginative, adding only some patches of grass between dull slabs of apartments; while often destroying communities, they fail to provide for the human needs of the poor.[4] Private housing developments for the middle class are usually as dull. Among the grimmest is Stuyvesant Town in New York, built by the Metropolitan Life Insurance Company to house 24,000 people; its collection of regimented apartment houses is not at all a "town," lacking a school, library, theater, community room, or any provision for community life. A much more common spectacle all over the land is the many square miles of monotonous middle-class residential districts, lacking the charm of many old European streets with houses in uniform style because they

---

4. Jane Jacobs reports the comment of a tenant on a piece of rectangular lawn in a housing project in Harlem: "Nobody cared what we wanted when they built this place. They threw down our houses and pushed us here and pushed our friends somewhere else. We don't have a place around here to get a cup of coffee or a newspaper even, or borrow fifty cents. But the big men come and look at that grass and say, 'Isn't it wonderful! Now the poor have everything!'"

have little or no real style. New residential districts for the well-to-do near the center of town, as on New York's East Side, might lessen envy of the rich by suggesting that they too are sacrificed to real estate values. The elegant big apartment houses occupy every square foot of space on the block that our lax laws allow, and only a few of their occupants enjoy sunlight or pleasing views. Together with the sunlamps that make do for sunlight, the narrow horizons might help to explain why Americans have too little sense of new horizons or new frontiers.[5]

The suburbs offer varying but only partial compensations for the loss of what the city can mean. Many of the older ones are settled communities, spacious and well-shaded, really attractive refuges from the city. More typical is the large, growing "gray belt," the many drab housing developments that offer little but a bit of lawn and relatively fresh air (if they are far enough from the main highways). Although encouraging neighborliness, they too lack the animation and spontaneity of neighborhoods. They house mostly transients, bound on and up, and are doomed to decay before they can mellow; like as not they will end as semi-slums, inspiring belated programs for "suburban renewal." Meanwhile they hardly bring "grace and beauty" into the lives of their inhabitants. Lacking the cultural opportunities afforded by the city, or a really urbane environment, most people in suburbia spend the evenings staring at the same TV programs. Studies of life in it, sociological or fictional, seldom give an agreeable idea of how it shapes people. It is not only growing more homogeneous or uniform but celebrating homogenization.

Although the state of the suburbs is well short of a "crisis," indirectly they have contributed heavily to the urban crisis because the flight to them has drastically lowered the potential tax income of the city. Almost all who left were middle to upper-class people; the newcomers to the central city have been mostly poor people, in particular Negroes, who cost it much more than they can pay in taxes. But since government must deal with this

---

5. Sunlight, incidentally, is another once "free good" that is getting scarcer. An authority on atmosphere has said that because of the effluents from the fast increasing number of jets it is possible that the next generation will never see the sun.

crisis, let us first consider the immense confusion in the government of the metropolitan area created by the suburbs.

This spreading area has blurred the whole idea of the "city" I have been talking about. Although the city still has official limits, necessary for geographical and political purposes, these have become meaningless for important economic and social purposes. "Instead of a tight corporate community providing definite services and functions for its own membership," writes York Willbern, "the city is now a somewhat fuzzy and indefinite association, supplying a mixed and indeterminate range of services to an amorphous and indeterminate constituency." As it may sell water to suburbs, so it is involved in various services over the whole metropolitan area—public transit systems, sewage disposal, flood control, etc. The area is made up of cities, counties, and townships, each with its own government agencies. (One tireless researcher found some fourteen hundred such local governments in the New York area.) Given this jumble, one can hardly hope for the rational, efficient coordination that is supposedly a hallmark of modern technology. The idea of home rule that created the jumble, and that is still dear to most Americans, has been largely antiquated; local government seldom inspires the civic pride it usually flaunts. While civic leadership becomes much more difficult, the ideal of community also loses much of its meaning. Thus a man may live in one community, work in another, belong to a country club or have his oldest friends in still another, drive through the jurisdictions of half a dozen on his way to work or play, and have his eye on a classier suburb he hopes to move to. Such limited and confused loyalties naturally weaken civic spirit. Voters are most easily aroused by proposals to increase expenditures, which suburban towns often vote down even when they are for better schools.

Within the city proper, government is always handicapped because its authority is limited by the state, whose unrepresentative legislatures have traditionally been more devoted to rural interests, usually hostile to the big city. Today its most obvious difficulty is raising enough money. While it has to spend much more on roads, schools, welfare, and health, it still has to depend primarily on property taxes, in recent years supplemented by local sales taxes. Businessmen in the city do not welcome either. Hence

New York, the wealthiest city of them all, recently found itself bankrupt; but when it proposed an increased tax on stock market transactions, the Exchange at once threatened to move out of the city. As all Americans ought to know, but don't, most other big cities are in as bad financial shape. At the same time their politics too has become relatively meaningless. Election campaigns rarely offer any programs except promises of more honesty and efficiency in providing the necessary services, which take up almost all the budget; and the promises are always dubious because their government is so disorganized. They have too many bureaucratic agencies with overlapping or conflicting interests, make little if any provision for coordinated planning and policy-making. Efforts at urban renewal, for instance, may require the cooperation of more than fifty agencies, and may be further hampered by dozens of different building code regulations.

At best, the problems of the cities are simply too much for them. In losing the clear identity as a corporate community they once had, they have lost as well their self-sufficiency. Their increasing dependence on Washington is by no means due merely to federal encroachment or bureaucratic lust for power, as conservatives protest; they have nowhere else to look for the billions they need. And apart from the common failure of the states to come to their aid, they have a good excuse in that their gravest problems, such as poverty, unemployment, and race riots, are strictly national, not local problems. Thus while the twenty largest cities have lost more than a million white people to the suburbs since 1960, they have had to provide for more than three million black people who moved in. These were mostly from the South, where they had been given an especially poor education and little opportunity to acquire skills; since hundreds of thousands of them have been unable to find jobs, they have increased enormously the costs of welfare services. Granted all the inadequacies of municipal government in America, it has had to grapple with problems for which it was not responsible. Even in its extravagant concessions to the automobile it was only responding to the demands of Americans everywhere.

Its growing dependence on Washington also recalls the disorganization of American government as a whole. With no problem is it more ill-equipped to deal efficiently than that of urban renewal. One can hardly hope for adequate measures when there is no central authority, municipal, state, or federal, to deal with the problem, to decide priorities, or to provide for the necessary coordination. In the city itself there are only port authorities, transit authorities, highway and housing commissions, and the fifty other agencies concerned, all with different ideas about priorities. For that matter, city planners cannot agree on any master plan for the city, and as specialists usually confine themselves to a particular problem. One who has seen the handsome pedestrian mall in the heart of Rotterdam, which the Nazis had bombed out in their wanton attack on Holland, might think that only a thorough bombing would make possible the restoration of the heart of the American city. Students of the city have remarked that one reason for the attractiveness of San Francisco is that it had had the advantage of a devastating earthquake.

One might therefore be surprised that the urban crisis is not more desperate than it is. At least the big cities are still very much alive, far from stagnant. They go on providing their basic services and trying to do more about public welfare. The majority of Americans who do not worry much over the condition of their city, except for black riots, have some excuse: after all, they can always count on subways running, water coming out when they turn the tap, firemen responding promptly to all calls, and so forth. It is chiefly professionals who are alarmed by the urban crisis. Still, this too is a hopeful sign. There has been much study of the problems of the city in recent years, enough survey and research to christen yet another new science, "metropology." With it has come much effort at city planning, a new profession that is already served by several dozen accredited schools and by institutes in some of the best universities. The federal government is now supporting both research and the efforts that have come out of it. However inadequate, its efforts represent another significant change, for the federal battle for urban renewal began only in 1962; the year before this government was doing nothing at all about it, and Senator Dirksen and other politicians laughed off the floor its advocates, dismissing them as esthetes.[6] Perhaps the

6. During the Eisenhower administration, about the only "crisis" the government was aware of was the Cold War. New Dealers and

chief benefit of the federal government's effort to help the cities has been the spreading realization that their problems are national problems. National debates over public policy may clarify possible means and ends obscured by the utter confusion of government in the metropolitan areas.

In this effort to make out a silver lining, I should add that the piecemeal efforts at least hold some promise. In the suburbs whole communities have been planned; Radburn, New Jersey, partially indicated what could be done to build a "town for the motor age," with neighborhoods planned as units. Within the city there has been much scattered imaginative response to the problems of both planning and building, enough to substantiate the feeling that our technology has no more exciting opportunity than the restoration of the city. Victor Gruen Associates, for example, drew up a plan for a pedestrian mall in downtown Fort Worth, with sidewalk arcades, varied exhibits, outdoor cafes, flower beds, and bandstands to encourage street concerts and dances. Gruen writes that it is quite simple and cheap to create such malls: one has only to take the main shopping street and place a sign at each end, CLOSED TO AUTOMOBILES. In *The Heart of Our Cities*, which is chiefly an eloquent account of all that is wrong with them, he ends on an optimistic note: "I believe we now stand at the brink of an epoch during which progress toward the improvement of the urban environment will take place at an ever-accelerating rate."

This, however, might sound like the automatic optimism of the American spirit, which has never learned enough from experience and is always disposed to the notion that we are standing on the brink of something bigger and better. If efforts to improve the city are progressing at a faster rate, there is still a very long way to go, usually in the face of opposition from chambers of commerce, politicians, realtors, and other vested interests, not to mention taxpayers. (Fort Worth turned down Gruen's plan.) A number of cities have indeed built monumental civic centers, quite impressive on a first visit, but few people return to them for pleasure; they are not vital centers, only sideshows, collections of monu-

Fair Dealers were still too busy fighting for policies born of the Great Depression to take up the problems created by an affluent society. Adlai Stevenson was among the first to publicize them in his campaign of 1956, but only to get badly beaten.

ments segregated from the rest of the city. Little more vital are most of the cultural centers, such as Lincoln Center in New York; they too are segregated, reserved for special occasions for a few people. Most schools of city planning are concerned primarily with administration and economics, not the needs of human beings. Victor Gruen reports that when he lectured at various schools on creative planning, the students were utterly bewildered and their teachers distressed because he was confusing or distracting them.

Considered as a national problem, the urban crisis warrants little more optimism. The increasing attention the national government is giving the cities remains far short of their needs. The federal programs available to them may seem impressive when a mere catalogue drawn up by the Office of Economic Opportunity ran to several hundred pages, but this represents a relatively small expenditure, as well as the usual bureaucratic confusion. Congress is far more generous to the Department of Agriculture than to that of Housing and Urban Affairs, which has become far more important; since 1960 the government has spent more than fifteen times as much on agriculture. The idea of spending $100 billion to save the American city still looks like a pipe dream, which could be realized only if the military defense program were drastically reduced; and one may doubt whether even then the needs of the city would be given a similar priority. Jane Jacobs dismisses the whole idea most scornfully:

> But look what we have built with the first several billions: Low-income projects that become worse centers of delinquency, vandalism and general social hopelessness than the slums they were supposed to replace. Middle-income housing projects which are truly marvels of dullness and regimentation, sealed against any buoyancy or vitality of city life. Luxury housing projects that mitigate their inanity, or try to, with a vapid vulgarity. Cultural centers that are unable to support a good bookstore. Civic centers that are avoided by everyone but bums, who have fewer choices of loitering place than others. Commercial centers that are lack-luster imitations of standardized suburban chain-store shopping. Promenades that go from no place to nowhere and

have no promenaders. Expressways that eviscerate great cities.
This is not the rebuilding of cities. This is the sacking of
cities.

Although no longer a city-lover (except when in Europe), I
think this may be too harsh a verdict. Perhaps we should be more
grateful for small blessings: an unmistakable improvement in
school buildings; here and there attractive housing developments,
even attractive shopping centers; the real convenience of the
expressways (since almost all of us often ride in automobiles);
and always some urban oases, dignified old streets or quarters that
have survived the widespread demolition. Except for slum-dwellers, life in the big cities is not simply hell. Many people continue
to enjoy it, many more to be basically content with it; and even
slum-dwellers generally prefer it to the impoverished life they led
before moving to the city. As for the shortcomings of the city
planners, they might be pardoned in view of the novelty of their
profession, still suffering from growing pains, and in particular
the difficulty of planning at once for the rich diversity and the
order demanded by city-lovers. (They could never satisfy, for
instance, both Lewis Mumford and Jane Jacobs, who do not see
eye to eye on what the city ideally should be.) Always they have
the excuse that most Americans are not at all clear about the kind
of city they want or might like.

Yet the fact remains that the big city is still deteriorating.
Efforts at urban renewal so far have aided chiefly middle-income
families, the beneficiaries of almost all the reforms since the New
Deal. The poor have not only benefited much less but often
suffered from slum-clearance projects. Moved out of their homes,
they were provided with no housing they could afford but had to
seek quarters in other slums; urban renewal programs have
destroyed more housing than they have built. And poverty has
been increasing in the cities. Unemployment, which during the
Great Depression affected all classes, is now concentrated among
the poor. For them the chief reform has been welfare benefits, but
these are generally demoralizing. Legislators and supervisors have
robbed the poor of self-respect by insisting on policing their
domestic life, to make sure they are entitled to their pittance and

are not earning any money on the side. Still maintaining its tradition of backwardness in social legislation, the United States remains the only industrial democracy that has no family allowance, just as it spends a lower proportion of its GNP on social welfare measures than do the other Western democracies.

In particular there remains the plainest crisis—the revolt in the spreading black core of the cities. Modern technology had much to do with this through farm machinery, such as cotton-pickers, that eliminated the jobs of Negroes in the South, and so stimulated their migration to the Northern cities. The affluent society that enabled more white people to move to the suburbs also helps to explain why the blacks have at last rebelled against their degradation, after a century of submission; the commercials on their second-hand TV are daily reminders of all the goods enjoyed by the whites, the opportunities they are denied in their kind of reservation in the ghettos. The root cause of the crisis, however, is the obvious one—racial discrimination. As the President's Commission on Civil Disorders reported of the ghetto, "White institutions created it, white institutions maintain it, and white society condones it." Discrimination is most pronounced in the suburbs. Here the many who left the city where they make their living have turned their back on its problems, and as voters and taxpayers are mostly hostile to expensive state and national efforts to deal with them. In the election campaign of 1968 politicians expressed far more concern over maintaining "law and order" than getting at the causes of racial violence; for Richard Nixon the "forgotten Americans" who touched his heart were such types as the prosperous suburbanites. Not many legislators betray an awareness that in a decade or so black people will be a majority of the population in most of the big American cities, as they already are in the nation's capital, and their ghettos will be a still bigger and more cancerous core unless costly programs soon get under way.

In this atmosphere there is little immediate prospect of an adequate effort to provide decent housing, schools, and jobs for the blacks. The necessity of cooperation by business raises further difficulties. The construction industry is among the most backward technologically, ridden by archaic practices that steadily increase costs while denying jobs to the blacks. Facing a national need of millions of new housing units, and entrusted with the "developing" of land in the slums turned over to them, builders have not developed such methods of mass production as Henry Ford did to turn out cheap cars, or as the Soviets have done in housing. City-lovers might object that such methods could produce only bigger Levittowns, not healthy neighborhoods or vital city centers, but builders have shown little enterprise either in such civic efforts. Business leaders who recognize their responsibility in helping the government to solve the problems of the city, occasionally even admitting that private enterprise was initially responsible for the whole mess, rarely suggest a bold new strategy, comprehensive programs for dealing with the problems, or immediately the need of reappraising the "philosophy" of business, the primary obligation of corporations to their stockholders.[7]

Hence I again assume that any large-scale programs will have to be organized and financed by the national government, as some business leaders themselves are now saying. And again I am not thereby declaring my faith in "socialistic" enterprise. If the government does settle down to a real effort to deal with the problems of the city and its black ghettos, no crash program will do. These problems call for not only vast expenditures but long-range planning and sustained, coordinated effort—all made possible by modern technology, even relatively easy for military purposes, only not at all easy for the many different agencies that would have to cooperate, or for welfare administrators who have to go to Congress every year hat in hand, always in an atmosphere of partisan politics and fearfulness of antagonizing taxpayers, the poor forgotten Americans.

I conclude that the most we can hope for is elementary social justice for the blacks. There is little or no prospect that the

---

7. Among the exceptions is J. Irwin Miller. Recognizing that a Negro explosion might destroy our free society, and that business is part of the Establishment, he has called on it to join the side of the "Revolution" even though it doesn't like the idea of working for social reform. "No one is going to solve the problems of our time unless he is powerful and expert and organized"—as big business is. But I still find it hard to imagine the big corporations joining any Revolution.

spreading black cores will be made over into attractive city centers, throbbing with the kind of life that would please Lewis Mumford, Jane Jacobs, or Victor Gruen. The forecasters talk instead of Megalopolis, huge metropolitan areas extending to cities once widely separated, such as Boston and Washington—an area that has been given the suitably ugly name of Boswash. Since the United States is the only advanced nation that has no policy or plan for building New Towns, even though these would be much cheaper than slum-clearance projects in cities where land is far more expensive, Megalopolis will presumably be developed by realtors and private contractors, without civic plan, while government provides more superhighways and all the other services it will need. People will have to live and work in it, but to judge by the forecasts it will be more suitable for showing off technology, or accentuating the neglect of the needs of human beings. Some planners seem happy in designing cities to be built underground or even underwater. In these technology could come wholly into its own, eliminate all the archaic green stuff that clutters up the landscape.

A word, lastly, about the big cities in the rest of the world. They are much more diversified (not to mention beyond the range of my knowledge), but their problems permit some generalization. European cities are struggling with traffic as more and more of their people too take to cars. A few, notably Stockholm, have displayed imagination both in modern residential building and in taming the automobile; others, notably Rome, sometimes suffer from traffic snarls even worse than Americans put up with. In many other countries the cities have been growing too fast to provide decent housing and employment for their new occupants. Tokyo, now the most populous city in the world, has no sewerage for most of its metropolitan area. Teeming Calcutta is unable even to house most of its people, who have to live in public streets and alleys. Mexico City, grown to more than six million, has become modern enough to produce both traffic congestion and smog, but many of the people who keep flocking to it from the countryside have to live in slums or shantytowns; like other Latin-American cities, it is busily reproducing the blight and sprawl of American cities before it has developed a decent standard of living. African countries have been following our example on a much lower standard. Other new capitals, such as Kuala Lumpur, have shiny boulevards and modern buildings that emphasize the growing sameness of cities the world over. Brasilia is an exception only in its lifelessness; tourists find more interesting the shanty-town thrown up as temporary housing for construction workers. Nowhere to my knowledge does the city of the future promise to offer the charms of the old European cities, or of New York as I knew it in my youth.

15

The Mass Media

There is no subject about which one is more likely to think wearily that all has been said, over and over, than the mass media. It appears less complicated than other subjects, for the major tendencies of the media are quite obvious, even blatant, and both their critics and their defenders ring variations on too familiar arguments. It may not even be so important a matter as is generally assumed, inasmuch as we cannot be certain how much people have been affected or changed by them. Yet they can never be ignored in a study of technology, for the "communications revolution" they represent is unquestionably significant.[1] The mass media have grown into another immense industry. They are a primary source of popular culture, now called mass culture because their audience numbers so many millions. Whether deep or not, their influence is certainly wide and incessant, day in and day out; most people not only depend on them for entertainment but get through them the news about what is going on in the nation and the world, the food for what thought they give to major problems. Political leaders in all countries, democratic or Communistic, alike take for granted their power over people. Altogether, I am inclined to think that their influence is commonly somewhat exaggerated, but never to question that they are a real power. And as always most Americans do not realize how great a power it is, or how often abused.

Actually, of course, the subject is about as complex and

---

1. Even the word "communication," today so popular, did not appear in the index of the 1895 *Encyclopædia Brittannica*.

ambiguous as any other involving modern technology. It has been oversimplified by the crude term "mass"—mass media, mass society, mass culture. As Raymond Williams wrote, "There are in fact no masses; there are only ways of seeing people as masses." We have to go on using the term because these ways are clearly relevant, but we also need to keep in mind that it is a crude term and obscures other aspects of a society and culture more heterogeneous than any in the past. The mass media—newspapers, comic books, popular literature, moving pictures, Tin Pan Alley, radio, television—are different media with different effects, which have given Marshall McLuhan a field day. They differ as well in the degree of sophistication and freedom with which they treat such popular subjects as sex. To some extent rivals, they have conflicting interests; newspapers were pleased to play up the scandal of the rigged quiz shows that embarrassed the television networks. Their products are widely varied, ranging from pure trash through mediocre work to reputable work by high cultural standards. Their variety and range are in one respect unique, that on any given day a single newspaper or broadcasting station may run the whole gamut from trash to excellent work. And call this melange mass culture, they are by no means its only source. Popular culture is still shaped as well by schools and churches, family, town, and regional tradition, and various subcultures.

Still more varied are the responses of the "mass" audience to the media. Its heterogeneous members bring to them different capacities for understanding and appreciation, and in their enjoyment of them range from passive "reaction" through mild or flagging interest to serious absorption. True, the great majority have common tastes and interests, which enable the mass media to prosper by satisfying or exploiting them, and which force us to generalize. In various ways they differ significantly from the common people in past societies, first of all in that their culture is largely a manufactured product, not a folk inheritance, and that it is no longer a working-class culture either. They are conscious "consumers" of culture, using or lapping it up. But for this reason they are not—as their critics often imply—simply fooled, deluded, or degraded by it. They know they are always being sold goods. They have some awareness that their entertainment is always make-believe, some degree of immunity to the propaganda that comes with it, some idea that you can't believe everything you read in the papers or hear on the air. Finally, the composition of the mass audience is extraordinary in a way now difficult to realize, which Edward Shils considers "the heart of the revolution of mass culture": a vast deal of the entertainment is produced for and consumed by youth, who in past societies were served chiefly by only a few teachers and toymakers. They too have become more sophisticated consumers of culture, as of all kinds of goods.

Intellectuals, the most searching critics of mass culture, have nevertheless done as much to oversimplify its issues as have its commercial producers. Many scorn all the "mere" entertainment offered by the media, forgetting that such entertainment is a real human need and the desire for it is quite normal, often shared by them in their relaxed moments. Feeling beleaguered as an unpopular minority, they also tend to forget that the majority never will develop a keen interest in ideas, or attain a state of intellectual grace, even if educated, and that our society will not therefore be doomed, any more than the great societies of the past were by the common illiteracy. In attacking the mass media intellectuals may likewise fail to come to terms with them, as all who support democracy must. Some declare that "mass society" and its media are alienating people, who they imply are basically good, but more in effect attack the "masses" the media cater to; they imply something of the fear or hatred of the "rabble" common in Europe before the rise of democracy. They might be reminded of Bertolt Brecht's suggestion to the Communist government, that since it had lost confidence in the people it should simply dissolve them and elect another. At the same time many have an excessive fear of popularization, any concession to the naturally limited capacities of people. Now that they are attacking "midcult" too, deploring that "middlebrows" are not highbrows, they may look simply snobbish, for they are scorning a great many earnest, thoughtful people who are trying to improve their minds or start out on the high road to culture—an effort that democrats ought to welcome.[2] And since critics are accustomed to dealing only

___

2. Dwight MacDonald, the most uncompromising critic of midcult, also exemplifies most clearly a kind of futility to which intellectuals are

with "good" literature, they are likely to deal crudely with popular art, which cultivated people in the past largely ignored; or when they discover that it can have merit, like jazz, they may go overboard in their enthusiasm, as some now do over the Beatles.

Yet intellectuals themselves—who are always critical enough of one another—often point out such occupational hazards. I make all these preliminary reservations chiefly in order to clear the ground for a dispassionate consideration of the legitimate complaints about the mass media, as well as to stress that the effort to see them steadily and whole makes it necessary often to say the obvious. To begin with, the media raise the same basic questions as the big corporations. They too are largely controlled by a few men, the heads of the major networks, Hollywood studios, big popular journals, and major newspapers or chains. The questions again are: For what purpose do they exercise their power? How legitimately? How responsibly? To what public effect?

Now, they are certainly not wicked men engaged in a conspiracy to debauch, oppress, or enslave the American people, as some Marxists still suggest. Insofar as they "exploit" people it is by catering to their worst tastes, giving them what they want. A Texas oil millionaire may use his broadcasting station to spread right-wing propaganda, but this only emphasizes the caution or timidity of the major networks. They differ from the controllers of the media in Fascist or Communist countries precisely in that they do not use their power primarily for political purposes, and are often criticized for their flabbiness. For their primary purpose is simply to make money. They may or may not be devoted to the

---

prone. Years ago he was an editor of the *Partisan Review* when it was Trotskyite in politics, aristocratic or avant garde in literature—exalting a kind of literature that was quite foreign to the interests of the "workers" to whom it was presumably devoted. Today the despised middlebrows still make MacDonald look schizophrenic by providing many earnest supporters of political causes he is dedicated to. In his anxiety to keep high culture pure and lofty he tends to divorce it from other vital interests, and may leave it high and dry. I should then confess that my own taste in some of the arts may be middlebrow, but add that few of us are capable of being in all the arts the "cognoscenti" for whom MacDonald wants to make high culture an exclusive preserve.

public interest, as they usually profess to be, but in any case they are concerned first and last with private profit. And because they head a big business this controlling purpose seems wholly legitimate in America. In the infancy of radio even so staunch a champion of business as Herbert Hoover said that of course so powerful an instrument for public education and public good should not be turned over to private interests, but of course it was. When television came along there was no longer any question. Although it was put under some regulations, since broadcasting stations were being given free use of air waves limited in number, the regulations have never been strictly enforced. Now that the big stations have become immensely valuable property they regard their franchise as not a public gift but a private right.

The issue here has been clouded by the traditional insistence on freedom of the press, and then of the air waves. This is indeed an essential freedom in a democracy; few Americans would want government-controlled television such as France has, not to mention Communist countries. (Even the much admired BBC barred Winston Churchill from the air for five years before the World War.) The American government has not been a serious menace to this freedom, however, except in matters of security regulations and now and then alleged "un-American" activities, which the mass media hardly go in for. When it tries to manage the news, newsmen can and do publicly protest. But the major networks have been considerably less zealous about maintaining their freedom in presenting the news. When CBS staged an interview with Premier Khrushchev and was itself surprised by the tremendous public response, it raised some interesting questions about the house rules of the networks: Who decided to stage the interview? For what reasons? Against what pressures? And why not more such programs? The answers to the last two questions were soon made clear. President Eisenhower tossed off a critical, uninformed comment, and superpatriots everywhere protested against the interview. Although many newspapers defended the right of CBS to produce the program, it practically apologized, and no other network defended it. Today the networks are likely to boast of their boldness when they venture an occasional program on a controversial subject. Much more often they defend only their right to put on as many trashy programs as they please and to work in still more commercials.

Hence they force the critical question: How responsibly do the controllers of the mass media exercise their freedom and their power? Above all in discharging their most important public function—reporting the news, informing the public about vital issues?

Concerning the press, the oldest of the media, this question has become more important since it has grown into a big business, with more chains and ever fewer competing newspapers. In 1947 a Commission on Freedom of the Press, headed by Robert Hutchins, offered some criteria for judging the press, standards of performance drawn from earnest or ceremonial pronouncements by leaders of the journalistic profession. To serve a free society well, its press should provide "a truthful, comprehensive, and intelligent account of the day's events in a context which gives them meaning," serve as "a forum for the exchange of comment and criticism," provide "a representative picture of the constituent groups in society," responsibly present and clarify "the goals and values of the society," and provide "full access to the day's intelligence." As might be expected, the press largely disdained the Commission's criticisms, but thereby it only made plainer that most American newspapers failed to come up to these standards, or anywhere near them.

Since they are ideal standards that no popular press in any country comes up to, or is ever likely to, critics should first note that in recent years publishers and editors have grown somewhat more responsible. Objective reporting, which in the last century was far from standard in American newspapers, has long since become the accepted goal in theory, and it is now less often flagrantly disregarded in practice. The *Chicago Tribune*, for example—the self-styled "world's greatest newspaper" that most newsmen agreed was among the world's worst—has become less rabidly partisan since the passing of Colonel McCormick.[3] Dicta-

---

3. Years ago I used to teach college freshmen something about how to read newspapers by asking them to buy a copy of the *Tribune* on a day coming up, and then to compare its presentation of the

torial types of publishers, such as Hearst, McCormick, and Henry Luce, appear to be on the wane. The press remains a one-party press, up to 90 per cent Republican, but most newspapers report election campaigns fairly enough, except for some partiality in headlines, largely confining their bias to the editorial page, where it has relatively little effect; the results of elections often demonstrate that publishers have considerably less power than they like to think (or than the Hutchins Commission thought, taking them at their own word). And while some of the worst tabloids have disappeared, the surviving newspapers include a number of the best ones, headed by the *New York Times*, which are about as good as any in the world. Together with the better weekly news journals they justify the comment of Leo Rosten: "Never in history has the public been offered so much, so often, in such detail, for so little."

Nevertheless the great majority of American newspapers fail to provide a comprehensive, intelligent account of the day's events "in a context which gives them meaning," or anything like "full access to the day's intelligence." The very concept of "news" in America tends to rob contemporary history of continuity and significance, for it must be recent or hot, and it gets most attention when it's the latest "flash." What happened the day before yesterday is not newsworthy; what is reported on the front page is a medley of discrete events, usually with little if any background information or interpretation. Even the better newspapers often sin in their headlines, playing up the sensational at the expense of the significant. And in this hectic atmosphere the ideal of objective reporting may itself muddy notions about what is fit to print, facilitate the manufacture of news that for years kept the wild charges of Senator Joe McCarthy on the front page. As Elmer Davis complained, "This kind of dead-pan reporting—so-and-so

---

news with that on the same day of the *New York Herald-Tribune*, another Republican paper but an honest one. The *Tribune* never failed me: in every issue there would be glaring examples of slanting and distorting in the selection, headlining, and reporting of the major news of the day. More than once it gave the front-page headline to rumored or manufactured news that was not even mentioned in the *Herald-Tribune*.

said it, and if he is lying in his teeth it is not my business to say so —may salve the conscience of the reporter (or of the editor, who has the ultimate responsibility) as to his loyalty to some obscure ideal of objectivity. But what about his loyalty to the reader?" The question is more pertinent because the press is never in fact wholly objective (any more than historians are), but typically patriotic or loyal to the national establishment. Thus for years most papers failed to report the hopeless corruption of the South Vietnamese government and army, which was known to every good reporter on the scene.

Underlying all the specific shortcomings of the press by the standards of the Hutchins Commission lies a basic complaint, stemming from the decision made by publishers more than a century ago. Confronted with a choice of making it their business chiefly either to inform or to entertain the public, they decided to go in for entertainment, and with few exceptions they have concentrated on it ever since. Hence the newspapers are filled with comics, sports, gossip, crimes, accidents, "human interest" stories, all manner of trivial, fleeting, or sensational news, while serious news trickles through the advertisements. Thoughtful readers may find some solid fare in interpretative reports, syndicated columns, and other special features, inasmuch as there are many good newsmen; but for the most part the press subordinates their interests to the more profitable business of entertainment. Hence a people daily swamped by newsprint remain poorly informed about the rest of the world they are supposedly fit to lead, and often feel bewildered or blindly frustrated by what goes on in it.

The standard defense of the newspaper publishers is that by all the entertainment they are only giving the public what it wants. While this is a tacit admission that they are not doing as good a job as they might, they can always point to the tabloid New York *Daily News*, which is bought by a million or so more people than buy the *Times*. It is never certain that publishers are only satisfying wants, not creating them, and the better newspapers indicate that it is possible to assume higher responsibilities; but I judge that on the whole the American press faithfully reflects the reigning "values and goals of the society." By the same token it is disclaiming any serious responsibility to "clarify" these values and goals. Though many an editorial declares such a responsibility, most of what the papers feature declares otherwise. Hence many readers who want more access to the day's intelligence have taken to the weekly news journals.[4]

News reporting on radio and television is in some ways much the same. It is fairly straight and objective except for the invariable tendency to play up the sensational, which is more dangerous on TV than in the press; it featured more vividly the worst excesses of peace demonstrations and Negro riots. Commentators are fairly representative except for the usual failure to represent adequately liberal or Leftish points of view; few radicals get on the air (any more than agnostics). At least the networks offer an ample number of news programs, more indeed than the public clearly demands or wants. Yet they have not been growing more responsible. Rather, they are doing a distinctly poorer job than radio used to, when such first-rate reporters and commentators as Elmer Davis and Edward R. Murrow were given fifteen uninterrupted minutes for serious discussion of the news. Now the television networks not only introduce more commercials before and after every news program but break in with still more every few minutes. Under these conditions the best newsmen cannot develop a serious topic at all fully or provide a context that gives the news meaning; the constant interruptions trivialize the gravest news. They accentuate the utter lack of discrimination that was always apparent in most programs of local stations, on which the announcement of important news would be followed by a scatter of trivial items, capped by another plug for eyewash.

In a somber speech delivered before the Radio and Television News Directors Association in 1958, Edward R. Murrow emphasized that in a time of crisis the networks were not probing the reasons why the country was in mortal danger, but chiefly were providing it with tranquilizers. A new rash of five-minute news reports only called attention to the failure to deal at all adequately with the *why* of the news. And why this rash? An

---

4. Of these Henry Luce's *Time* was a special case—a highly profitable blend of slanted news, celebration of the American Way, and periodic spiritual uplift or moral crusade, of an innocuous kind that would not disturb either its customers or its advertisers.

official explained to Murrow: "Because that seems to be the only thing we can sell." The main business of the networks was simply to sell. Murrow concluded that the future of the free commercial system of broadcasting was bleak "unless we get up off our fat surpluses and recognize that television in the main is being used to distract, delude, amuse, and insulate us." Ten years later the controllers of the system were sitting on still fatter surpluses, but the outlook was still bleaker. Fred Friendly left CBS because his superiors refused to let him telecast live some of the important Congressional hearings on Vietnam, instead running their usual daytime soap operas and situation-comedies. The main business of the big networks remained selling. The programs that sold best were show business, tailored to suit Madison Avenue men; advertisers exert far more influence in television than they do in newspapers, often dictating the programs. (A possible exception is the New York *Daily News*, which once ran a headline "Ciggies Assailed Again—Ho Hum.") And so we are brought to the subject of mass culture, in which television has succeeded Hollywood as by far the most popular manufacturer.

Now, the big networks of course offer some good programs. Their limited originality is scarcely surprising in view of the obligation they feel to fill every hour of the day from early morning to midnight; they could never command enough talent to make their dozens of daily programs fresh and imaginative. Yet the good programs make clearer how far short they fall of realizing the ideal possibilities of the new medium (not to mention the pipe dreams of Marshall McLuhan). Most of their staple wares for the popular evening hours are mediocre, while their soap operas and most of the other fillers of the daytime are simply dismal.[5] And though the industry still offers the old excuse that it

---

5. I write about the "waste land" mainly from hearsay, lacking the fortitude to engage in extensive firsthand research, but on my travels I have made a point of sampling programs. When I was marooned for several days in a Washington hotel by a curfew because of Negro riots, I thoroughly explored the possibilities of the broadcasting stations in the nation's capital, and was rewarded by very little even in the way of passable entertainment; only occasional reports on the local excitement enlivened long dreary stretches of time-killing. Everywhere my experience has confirmed the old saying that one doesn't have to drink the whole sea to find that it's salt.

is young and growing, it has not been improving with age and experience, but growing worse in this respect too. A medium that once attracted some promising playwrights has become, in Friendly's words, "a grind house for inferior Hollywood movies." The playwrights quit because they were denied the freedom to do honest work, and now even the better movies are hard to enjoy because of constant interruptions by commercials.

One might therefore wonder about the docility of the American people, who once had some reputation for standing up for their rights: no European people puts up with so many commercials throughout the livelong day—up to ten thousand a year. The apparent contentment of Americans staring at their TV brings back the old question, whether the networks are not right in maintaining that they are simply giving the public what it wants. There is considerable evidence to support the contention of Murrow and others that the people are more reasonable and mature than the networks assume. There is also plainer evidence that the people like the cheap fare they are offered, much prefer it to the serious programs sporadically ventured by the networks, and that in this respect they are no different from other peoples. (When the British, for example, were offered the same kind of show business by commercial television, the great majority turned off the more distinguished BBC programs—a choice that would surprise no one familiar with their mass circulation newspapers, which are even worse than the poorest American papers.) But the big networks are clearly not satisfying the interests of a large number of people, and the question remains whether they are not also neglecting many of the interests even of the immense audience that daily tunes in. For in always seeking the largest possible audience at the popular hours they treat people as if they literally were "mass-men."

Oscar Handlin has noted that our mass culture differs from popular culture early in the century, which was more closely tied to living traditions and dealt more directly with concrete familiar experience. Out of it came such genuine expressions of popular thought and feeling as jazz and the art of Charlie Chaplin. Handlin concluded that in improving the techniques for communication the mass media had diminished its effectiveness, producing an apathetic audience; or in view of all the signs, billboards, and newsstands, it might be said that they have helped to make communication a chaos, eliminate any semblance of genuine community. Today, however, the popularity of soap opera may make one wonder whether mass culture has not become as close to people, especially when one hears of women beginning to knit baby clothes once the heroine gets pregnant. The success of the mass media indicates that to a significant extent most Americans can be treated as mass-men. Yet they are still a heterogeneous lot, with more diverse interests and capacities than the popular TV programs indicate, and—one must hope—somewhat more intelligence and imagination.

In this hope I feel obliged to rehearse some of the tiresome complaints about TV. The most familiar is about all the violence it features; a researcher once counted almost three thousand acts of violence in a week's programs.[6] More depressing to me is the mushiness of most of the entertainment, the lack of anything like mature thought and deep feeling, anything to stimulate people. When a networks executive assured an Ethical Culture audience that radio and TV had little effect on "basic moral values," the audience answered rightly that this was precisely the trouble—the countless thousands of hours taken up by entertainment or "communication" did have little or no moral effect, because they made moral choice unnecessary. In drama all problems must be simplified, all meanings spelled out, all complexities, ambiguities, and possible misunderstanding or doubt avoided, to assure a stock response. At worst inane, at best the run of the mill is stereotyped, trivial, and irrelevant. One give-away of the standard formulas is that the dramas can be programmed by computing machines. And the daily diet of pablum is more depressing when one recalls that before the rise of the mass entertainment industry people managed to use their own wits to amuse themselves, sometimes read books. In reading they might pause to reflect, perhaps to talk back to the author. Books might illuminate life, enhance consciousness,

---

6. Defenders of TV point out that there is plenty of violence in Homer and Shakespeare, but in their drama it had some significant relation to character, plot, and theme. In TV it is mostly gratuitous, artistically meaningless.

keep people on the stretch, and make them feel more alive, the better for the experience. Now many people spend hours daily watching television—willing to watch almost anything supposed to be entertaining—it appears just in order to pass the time. Communication becomes a one-way affair, a feeding of pap to an essentially passive audience.

In providing such pap, the big networks are not positively irresponsible—they are just non-responsible. They evade the whole issue of responsibility by remaining neutral on controversial issues, taking no stand as newspapers do. Hence it is sometimes said that television has no values, because no principles. I should say rather that it respects the accepted values of an affluent society, including lip-service to conventional moral and religious values. It adheres to the Hays production code for Hollywood, that it must uphold "correct standards of life, subject only to the requirements of drama and entertainment." The ideal of its advertising sponsors is "100 per cent acceptability"—a program that won't offend any sizable group of people, except for the negligible minority of eggheads. Both its commercials and its programs inculcate the values of conformity, the goals of the status-seekers. By shaping what Edward Shils has called a "culture of consensus," television has perhaps provided some fabric for Marshall McLuhan's vision of a brave new world in which mankind will be unified, but if so it would be a consensus on tendencies that D. H. Lawrence described as "anti-life." Another word for its many mushy programs is phoney. They are drenched in fake thought, fake sentiment, fake personality, fake idealism. Sometimes, as in the rigged quiz shows, they indicate a cynical contempt for people.

To repeat, we cannot know how much television has affected people.[7] We have some evidence of the effects of the information

---

7. About Marshall McLuhan's thesis that "the medium is the message," and that it is profoundly influencing people, I have already commented in my chapter on Language. I know of no evidence that substantiates his thesis. My impression remains that people reading a melodrama, seeing it in the movies, or seeing it on TV are having basically the same experience, and that insofar as it is more vivid on TV, it makes the vulgar or violent "content" more objectionable. I should confess, however, that I am unable to understand clearly why McLuhan calls TV a "tactile" medium, perhaps because I cannot get my mind off the content of its programs.

they pick up, but much less evidence of significant changes in attitude; such changes as their disapproval of the war in Vietnam are due most plainly to the historic events themselves. Hence I would again question the common assumption that TV has a massive influence, even that it has become the major influence on people. The common complaints about its influence, moreover, are sometimes inconsistent. Thus it is charged with promoting a mindless togetherness and conformism, but also with promoting antisocial behavior, like juvenile delinquency, by all the routine violence in its programs. In general, I assume that although an occasional program may have a dramatic immediate effect (such as the panic into which thousands of people were thrown by the Orson Welles radio broadcast of "The War of the Worlds"), whatever serious, lasting effects television may have will be apparent only in the long run; and social scientists are still unable to study systematically long-range consequences. Meanwhile my guess is that it chiefly confirms attitudes already ingrained, since most people are not inclined to listen attentively to anything foreign to their interests or contrary to their own opinions or prejudices. One might wonder how they ever changed their minds—were it not that they are constantly exposed to different opinions, in the press as well as in their everyday lives.

Yet one can say confidently that on the whole television has not tended to make Americans more rational, responsible, and mature, or fit for a "Great Society." Insofar as people—especially the impressionable youth—make watching it a habit and take its drama at all seriously, it insinuates chiefly hollow or false values, cheapens the quality of life, muddies rather than clarifies their own experience, isolates them from the real world, makes them less capable of dealing realistically with their own problems. In satisfying a childish passion for the happy ending it looks less mature than the folk cultures of the past. It provides not only what Freud called a substitute gratification but a gratification inferior to what people can hope to achieve by their own effort. Hence I would say as confidently that television has not made Americans happier either—not fulfilled the one aim that could justify the immense entertainment industry. Although we can never know for certain whether we are more or less happy than our ancestors, Americans are surely not a conspicuously joyous people, and they seldom seem exhilarated by the endless entertainment they lap up.[8] Most of the make-believe has little spirit of play. As David Riesman remarked, it fails to provide a real escape either, for most of the programs and the commercials support the constant pressures to be popular and well-adjusted. Above all, it impoverishes life. It dulls the capacity of people for wonder and awe, makes them less sensitive to the exhilarating possibilities of life, makes it harder for them to realize the best in themselves—the means to lasting satisfaction instead of temporary distraction.

Mass culture is accordingly most troublesome for those who cherish democracy. For this is democratic culture, made available to all people, designed primarily to please the majority who are supposed to rule. It magnifies the perennial difficulty of reconciling the ideal of equality with ideals of excellence. Those seeking to maintain standards of excellence, in the name of the dignity and worth of man, may be most repelled by the common indignity of mass culture, most tempted to despair of the American people. And what can they do about this culture? They cannot simply ignore it and go their own way unless they are blind and deaf, for the mass media are ubiquitous and their clamor incessant. Neither can they utterly reject it unless they are content simply to despise the people. They can warn, ridicule, rail, exhort, but then they are apt to feel futile. They have been doing this for many years now, and mass culture remains as vulgar as ever, while television continues to get worse.

Still, I think their plight is not so desperate as they sometimes make out, for reasons I suggested at the outset. So far, moreover, I have dwelt chiefly on the staple fare of the mass media. There remain their better products, much reputable work, some even excellent. In the media themselves one may find considerable self-criticism, as of the venality of television. The endless search for novelty by the advertisers who sustain them has brought more variety, some tendency to play up new interests, tastes, and styles, and a rise in the general level of sophistication. Most pertinent,

---

8. I have felt this especially while sojourning on Greek islands that lacked both movies and television. The Greek way of life, though hard on the many poor, involves much more impromptu, spontaneous gaiety than the American way.

the affluent society has produced still another explosion—the "culture explosion." Millions of Americans have been discovering the possible pleasures of traditional culture, long the preserve of highbrows.

To this subject I shall return in the following chapter, with as usual less rejoicing over the culture explosion than writers in popular journals have indulged in. Here I shall merely remark that there has indeed been a significant change, and that although it has called out too much self-congratulation and huckstering, the mass media, which have been charged with inventing the fiction of a culture boom, have made solid contributions to it.[9] The better newspapers have reflected if not stimulated a rising interest in music, painting, ballet, and experimental theater. Popular magazines have published articles by the likes of Bertrand Russell, Reinhold Niebuhr, and Paul Tillich. Publishers have put out thousands of paperbacks, including many trashy ones but a fantastic number of serious, even scholarly books, which are now displayed in drugstores and newsstands. While Hollywood keeps grinding out chiefly Grade-B movies, independent producers have been turning out experimental films and hundreds of little theaters have been showing these films, together with old classics and foreign films. Radio has helped to popularize opera and especially classical music; some hundreds of FM stations broadcast chiefly such music. The big television networks have characteristically contributed least to the boom in culture, but their occasional ventures have been supplemented by independent local stations and an increasing number of educational stations. Possibly they have contributed to improvements in the movies and the press by preempting vulgar violence and reporting news so superficially; other media might best compete with most popular TV by offering better fare. All in all, the quality of the culture now being

9. In 1961 Alvin Toffler pointed out in *Fortune* magazine that Americans were spending at least $3 billion a year on culture. One had a choice of emphasis: that this was a minute fraction of the GNP, or that it was almost twice what they had spent a dozen years before and was increasing so fast that it was expected to double again before 1970. At this point I am choosing the cheerful emphasis. Toffler provided much solid evidence to support it in *The Culture Consumers* (1964), which is on the whole a balanced study.

consumed by Americans is often dubious, yet it does not come down to the mere midcult *kitsch* or supermarket culture that supercilious intellectuals have called it.

Technology too has directly contributed to the cultural boom in other ways. As it made photography a popular new art form, so it helped to popularize painting and art history by superior reproductions. To music it has given long-playing records, improved methods of recording, and high-fidelity equipment. The techniques of organization it has made standard have created arts councils, arts centers, and cultural festivals all over the land. This has meant introducing bureaucracy into art too, with all the usual complaints, but initially it has encouraged both enterprise and sustained effort. To be sure, we are still dealing with the interests of a minority, and are still far short of a renaissance. Most high culture remains alien to most Americans, just as most mass culture is alien to intellectuals; so we cannot speak of a unified national culture, a heritage shared by all Americans. But neither can we sharply, absolutely separate mass and high culture, or regard the one as simply a product of modern technology, the other as simply a victim of it.

16

Traditional Culture

In the Romantic tradition that poets were the "unacknowledged legislators" of the world, Oscar Wilde was pleased to say, "Life imitates Art far more than Art imitates Life." Offhand, no statement could be farther from the truth about literature over the last century. The criticism of industrialism and the values of the rising business class that began with the Romantics has continued to this day, swollen by more criticism of other manifestations of a technological society; but all to no avail. Society remained indifferent when not hostile to visions of life as writers wanted it, the tendencies they deplored grew stronger. Meanwhile art imitated life more obviously than ever before in realistic works, in which writers mostly portrayed modes of life they deplored or despised. Other writers tacitly admitted their ineffectuality for social purposes by turning to the new religion of art-for-art's-sake, or cultivating a kind of pure art that life could not imitate. Today a similar admission is implicit in the popular literary themes of alienation and anxiety.

Yet the story is as ever much more complicated than this. In a long view life does to some extent imitate art, which helps to set the standards of a society. Writers are never wholly ineffectual so long as they are read and studied, and that after a century we are still repeating much of their basic criticism of our society also indicates that their values and ideals are still alive, perhaps a force to be reckoned with. Technology has plainly aided literature by creating a much larger audience for it through mass education, multiplying books, and giving writers more publicity. And certainly writers have not been paralyzed by feelings of impotence or futility. The literature of the last hundred years has been distinguished by not only ample vigor but far more variety than in any previous age. Life could not possibly imitate all the diverse modes of modern art. It would be most unfortunate, indeed, if many people did imitate some of them. Readers will make different choices in writers as exhibits of the limitations, excesses, or perversities of modern literature, but all recognize that the greater writers have not been uniformly humane and wise, art does not always offer ideal designs for living.

Vastly more complicated is traditional culture as a whole—all serious art and thought, all the values of the life of the mind. The intelligentsia, roughly defined as the class that creates, cherishes, and transmits these values, have also been prone to feelings of ineffectuality because, excepting scientists, many have felt isolated, remote from the centers of economic and political power, in a society they thought had little use for them; yet they too have kept active and have been remarkably creative. A much more substantial class than ever in the past, they have also been more directly influential with the aid of technology. They have authored all the new isms in art and thought, most of which have entered the common language of educated people; some have seeped down into popular thought. If as social critics, reformers, or rebels they are at any given moment no match for vested interests, some of their ideas are sure to prove more powerful in the long run. The vested interests themselves commonly appeal to theories, such as laissez faire, originated by intellectuals in the past, and nothing is more perishable than the dogmas of ruling classes. We have only to remember that revolutions have been made by intellectuals, notably Marx and Lenin, just as the American and French revolutions proclaimed and spread ideals derived from John Locke and the thinkers of the Enlightenment.

It follows that the intelligentsia are by no means a clearly defined class with uniform interests and values. The active ones work in separate provinces, often to cross purposes. As for literature, Marx told writers that it was a form of ideology, masking the fundamental economic reality, which meant that as bourgeois they were reflecting a "false consciousness" or imitating a false social reality; their works only illustrated the very opposite of Wilde's dictum, that "it is not the consciousness of men that determines their existence, but, on the contrary, their existence determines their consciousness."[1] The historian Spengler told writers that all modern art was degenerate, only another symptom of a declining civilization. Most social scientists simply ignored them, paying no heed to the possible influence of literature or even its plain significance as an index to the sentiments, attitudes, and beliefs that help to define the social reality. (Who could understand ancient Greece without knowing anything about its literature?) These scientists merge into a very large new group, the technical intelligentsia—the countless brainworkers who are busy promoting science and technology, but are largely indifferent to the traditional values of high culture. The term "intellectual" becomes almost meaningless when Bernard Rosenberg tells us that "the mass media are swarming with intellectuals."

In short, the culture that remains when "mass culture" has been separated out is another very large and complex subject. To simplify it, I am confining myself here to traditional culture as represented by the arts, since I have already considered science and the state of the humanities. This still leaves me with a large subject, for chapters—or books—could be written on the influence of technology on each of the fine arts, chapters that I am not qualified to write; so once more I emphasize that I am only touching on some of the major developments and the issues they raise, this time more casually or informally than usual. But I also state emphatically my conviction that art is much more important than may appear in a positivistic age. As a logical positivist, Rudolf Carnap wrote that "only verifiable statements of specific empirical sciences can be legitimately conceived as knowledge of the states of nature and of matters of fact," and that "whatever therefore we take from a work of art may offer emotional or formal pleasure but cannot offer intellectual meaning." I believe that the arts do offer some such meaning in symbolic form, which

---

1. Since Marx had a high regard for literature, he was inconsistent about its uses. But the proposition I have extracted from his frequent inconsistencies provided the logic of most Marxist criticism, and of the Soviets' demand that its writers conform to "socialist realism."

is significant even if not verbal or logical (as in what Beethoven had to say in his last quartets), and that literature in particular is a significant way of knowing and dealing with reality, or the conditions of man's life.[2] Or even if the arts offer only "emotional or formal pleasure," this remains a vital need in a technological society, especially in view of the emptiness felt by many Americans. The arts might fill this emptiness more satisfactorily than golf, bridge, or mass entertainment.

In literature the most significant theme for my purposes remains the insistent social criticism, one of the plainest indications that modern culture has not been wholly dominated by technology, or that the consciousness of men is not simply determined by their existence. From the outset the criticism was highly relevant for the same reason that society at large paid little heed to it: writers were naturally concerned first and last with the human values that are my main concern, and so were as naturally sensitive to the human costs of industrial progress. Familiar as their basic criticisms have long been, we may better appreciate our debt to their original insights by going back to an early essay of Thomas Carlyle, "Signs of the Times" (1829), which clearly announced the themes of the future.

In this Carlyle introduced the term "industrialism" into English and named the new age the Mechanical Age. "Not the

---

2. It is more important because modern philosophy, as represented by logical positivism and the linguistic philosophy of Wittgenstein, is of little aid to those concerned with human values or ideals, which again I would not divorce from "intellectual meaning." Philosophy has become a highly technical study that no longer looks much like a "love of wisdom," and in any case it has not been on the forefront of thought. It lags far behind science, which the logical positivists have been pleased to analyze or serve, but in ways that scientists have largely ignored, while Wittgenstein concluded from his analysis of language and common sense that philosophy "leaves everything as it is." The trouble with his philosophy, Ernest Gellner commented, is that "it wholly obscures both the tremendous *changes* which our society has undergone, and the *choices* which it faces." Rudolf Carnap and his school are no more helpful on these changes and choices. Existentialism, another characteristically modern school, has given us *Angst*, which may be a form of wisdom, but its often desperate view of the human condition is chiefly a reflection of social changes, also of little aid in understanding them or making choices.

external and physical alone is now managed by machinery," he complained, "but the internal and spiritual also." He described prophetically an attitude that was not yet widespread, the scientific persuasion that "what cannot be investigated and understood mechanically, cannot be investigated and understood at all." In deploring the spiritual hollowness of human relations in an age lusting after Power, he anticipated the Marxist complaint he later dwelt on himself, that "Cash Payment" was the "sole nexus between man and man." And at this time he had an uncommonly balanced view, which he would lose as he soured and became contemptuous of "the masses" (another term that entered the language in this period). Declaring his faith in the "imperishable dignity" and "high vocation" of man, Carlyle wrote: "This age also is advancing. Its very unrest, its ceaseless activity, its discontent contains matter of promise."

Breadth and balance more consistently distinguished the thought of William Morris, a writer who is no longer widely read but seems to me one of the sanest, most humane critics of industrialism. Deploring the "sordid, aimless, ugly confusion" it had produced, he sought to make the traditional values of culture available to the working class. It was the province of art to set before them "the true ideal of a full and reasonable life." Tempted to despair of a civilization he hated, he was saved when he saw in its filth the seeds of "Social-Revolution," the hope of socialism. Nevertheless he remarked that most people, including too many socialists, saw only the *"machinery* of socialism"—the political machinery that would become so oppressive in Stalin's Russia. He kept his eye on the simple human ideal of enabling all men to put their best into work they liked, an ideal that socialist countries have achieved no more than capitalist ones. Himself a master of many traditional arts—stained glass, textiles, furniture, wall paper, tapestry, etc.—he lamented that only the rich could now afford hand-made products. Instead of cursing the machine, however, he valued it for all the time it saved and wealth it produced, for once it was put in its place the material abundance should make it possible to extend to all men "opportunities for enjoyable and self-rewarding work." In all this Morris was utopian, of course; but one may hope he was prophetic in a temperate letter he wrote:

I do not grudge the triumphs that the modern mind finds in having made the world (or a small corner of it) quieter and less violent, but I think that this blindness to beauty will draw down a kind of revenge one day: who knows? Years ago men's minds were full of art and the dignified shows of life, and they had but little time for justice and peace; and the vengeance on them was not the increase of the violence they did not heed, but the destruction of the art they heeded. So perhaps the gods are preparing troubles and terrors for the world (or our small corner of it) again, that it may once again become beautiful and dramatic withal; for I do not believe they will have it dull and ugly forever.

Most of the greater English writers did not share Morris's concern for the common people. Their criticism of an industrial society exposed the common limitations of literary men as social critics, immediately because of an aloofness inherited from the aristocratic tradition of culture as a possession of the few. Tending to nostalgia for the old hierarchical social order, they did not look closely into the historical record of aristocratic society, which had invariably maintained its privileges by considerable exploitation and oppression of the many poor; so they failed to perceive that industrialism was the only hope of the poor. In our own age such elitism became most explicit in T. S. Eliot, one of the literary gods of the age; a thoroughgoing traditionalist, he declared flatly that culture required a privileged class with "advantages of birth." Like some other literary gods—Yeats, D. H. Lawrence, and Ezra Pound—he betrayed Fascist inclinations.[3] In America, with its much stronger democratic tradition, elitism has flourished chiefly

---

3. Lawrence, who as the son of a coal miner had no advantages of birth, became the most impassioned, eloquent critic of industrialism and science more on the grounds of primitivism than either sympathy for the workers or devotion to tradition. F. R. Leavis, one of his foremost admirers, has managed to combine aristocratic ideals with a romantic primitivism, the myth of the happy village with a truly human environment, "right and inevitable"—a village that according to Oliver Goldsmith was already "deserted" in the eighteenth century. The greater writers have of course had no uniform or consistent philosophy.

on the revulsion against mass culture or the cult of alienation. Even the most hostile critics of democratic culture, such as Dwight MacDonald, rarely reject political democracy as Pound did; but their distress has been aggravated by a tendency to equate culture with fastidiousness (unbecoming a literary heritage that includes Chaucer, Elizabethan drama, and the eighteenth-century novel), and to make democratic culture seem more barbarous by comparing the worst of it with the best of the past—with Shakespearean drama, for example, but not with the bear-baiting that was more popular with the Elizabethans.

Another common literary theme introduced in the last century was the praise of "organic society," blessed by "wholeness," contrasted with our mechanical, external kind of society. Today this theme is commonly restated in the criticisms of "mass society" —the loss of community, the atomized city masses, the mechanical ways of life, and so forth. All this is much to the point, especially in an affluent society lacking any clear communal purpose beyond economic growth and technological advance. But it has led to a sentimental idealization of past societies as genuine organic communities, united by a common faith or purpose. If the village was such a community, the society as a whole was not clearly organic, inasmuch as the great majority of common people were outsiders, sharply separated from the upper class, who neither participated in the government nor shared in the high culture of the society. Critics of mass society and its culture overlook what Edward Shils has pointed out, that this is new just because these outsiders have been incorporated *into* the society. What distresses many literary people is precisely their ubiquitous presence, in particular their ways of pursuing happiness. These ways emphasize too that Americans do have common values and purposes, which again are what distress writers. There remain many conspicuous outsiders, above all the Negroes, and most writers want to incorporate them too (as I do); but such efforts at social justice, which likewise distinguish our society from traditional society, are no solution to the problems of democratic culture.

Modern technology, which has helped writers to appreciate more the values of community, just as the horrors of early industrialism had helped them to appreciate nature, has in other ways led to what Jacques Barzun calls "the treason of the artist." In literary criticism this appeared as the rage for structure and technique, methods of analysis, classification, and explication, with a technical vocabulary as an aid to "objectivity" and supposed precision; the critic became still another species of technician. Helpful as such criticism may be at its best, it has plainly encouraged a deal of affectation, the snobbishness of spurning natural interests and pleasures, such as the interests of the common reader, the simple enjoyment of literature, or the old-fashioned maxim of Samuel Johnson that its end is to help readers "better to enjoy life or better to endure it."[4] In fiction and drama scientism or pseudo-science helps to explain the rarity of the hero and the tragic sense of life, and then the emergence of the "anti-hero." Much contemporary art springing from a revulsion against modern life suggests no effort to restore moral or esthetic order, only a reflection of its neurosis, barbarism, or chaos. Such fashions as the theater of cruelty, the theater of the absurd, the fiction of Burroughs and Genet, beatnik poetry, and the formlessness and frantic spontaneity of Happenings are understandable, sometimes fascinating; but they are hardly admirable models for either art or life.

Of the other arts, architecture has been most plainly and permanently influenced by modern technology. To begin with, this has called for much more building than ever before, especially since the World War; in Manhattan alone well over a hundred new office buildings have contributed to a national total of a trillion dollars worth of new construction in this generation. Architects now use many new materials, such as not only steel but sheet metal, reinforced concrete, spun plastics, and glare-reducing glass. Le Corbusier in particular welcomed collaboration with engineers. He was excited by the "marvelous instrument" of the machine, which he believed would give us an "enormous, healthy

---

4. Critics have also overlooked a strange unconcern noted by Aldous Huxley. Modern poetry gives hardly an inkling of the extraordinary new visions of the cosmos in modern science, including the possibility of some forms of life on millions of other planets; whereas Dante packed his *Divine Comedy* with the "science" known to the Middle Ages, and many other poets down to Thomas Hardy at least gave signs of awareness of the cosmos revealed by science.

form of comfort," but above all would "awaken in us the joys of a maximum of individual liberty together with collective inspiration." A number of other distinguished architects have risen to their exciting opportunities—Frank Lloyd Wright, Walter Gropius, Louis Kahn, I. M. Pei, and Mies van der Rohe, to name a few. All over the land striking buildings have been going up. Lever House in Manhattan became a symbol of what modern architecture can achieve: a spacious house of glass that made effective use of color, inside and out, yet was efficiently designed for business purposes, and altogether far better planned than were the skyscrapers that had competed only for the distinction of being among the tallest buildings, or what Spengler called "swaggering in specious dimensions."[5] Lesser architects contributed to a minor artistic triumph but an important one for social purposes, the functional school building. Wright brilliantly demonstrated the new possibilities in domestic architecture (if sometimes with his mind more on the exercise of his original genius than on the practical needs of house-dwellers). As Lewis Mumford said, he created buildings "in which the modern spirit can feel at home with both nature and the machine." I suppose this is partially true even of the conventional ranch houses that have sprung up everywhere.

Architects themselves complain, however, of missed opportunities. Louis Kahn has said that most modern buildings do not look modern to him, but look more like Renaissance buildings new simply in their materials. Least original or impressive are the many new government buildings in Washington, on which none of the leading architects were employed. President Kennedy started to improve the appearance of the nation's capital by appointing a committee to redesign Pennsylvania Avenue, the "grand axis" from the White House to the Capitol, but meanwhile all the new building suggests that modern America is not good at building national monuments; a business-minded society apparently has too little sense of the human values or ideals that might inspire

---

5. In view of the still common tendency to idealize the Middle Ages one should remember, however, that the builders of its Gothic cathedrals also did some swaggering, competing in the height of their spires. St. Bernard damned their ostentation.

suitable monuments. (For its Lincoln Memorial it could do no better than adopt a pseudo-classical style.) To me, as to Lewis Mumford, especially disappointing is the Secretariat Building of the United Nations in New York: a monumental slab that does not look at all like a symbol of peace and world brotherhood, the ideals stated in the U.N. Charter, but only like another office building to house a bureaucracy, which to distant peoples, in Mumford's words, "is a stock emblem of the things they fear and hate—our slick mechanization, our awful power, our patronizing attitude toward lesser breeds who have not acquired the American way of life."

The total effect of modern architecture in American cities is clearly disappointing. Here and there an impressive building stands alone, or stands out enough to be admired as grand buildings should be, but mostly they are lost in a meaningless crowd of faceless office buildings, impressive only at night when they are all lit up. The total effect is far short of Le Corbusier's dream of a "magnificent play of masses—a grand decisive gesture." It falls as far short of the old description of architecture as "frozen music" (except perhaps for the cacophony of modern music). Lacking human scale or high human purpose, the jumble of buildings makes lovers of architecture unhappier in the knowledge that in its great periods the most public of the arts served as the permanent framework of society, a symbol of both its order and its ideals, a spiritual form, or even a representation of man's relation to the universe and the gods.

Technology, which made possible the triumphs of modern architecture (as let us remember it did of monumental architecture from its beginnings), is also a clue to the too common emptiness of its grandeur. Le Corbusier himself suggested it when he defined a house as a "machine for living." The emptiness derives even from the admirable principle that gave birth to modern architecture and rescued it from the ornateness and confusion of styles in the Victorian and the Gilded Age—the basic principle that form should follow function. The function is too often merely technical or mechanical, the form a kind of "frozen geometry"; neither finds room enough for other human functions, such as expressiveness, enrichment, or simple beauty. Likewise the fashion of smooth, polished surfaces, which can have a clean beauty on buildings standing alone or with some open space, grows monotonous when facades stretch continuously for blocks.[6] But the root trouble remains the reign of business and money values, which architecture has to serve. Hence it has become more closely allied with engineering than with art, and often provides its business patrons with little but the blueprints they want. Architects sometimes wonder gloomily whether it is possible to have a healthy architecture in a sick or corrupt society. The exciting buildings some of them design suggest that it may be possible (just as Athens went on building gloriously during the fatal Peloponnesian war its pride brought on), but presumably the future of architecture will depend on whether American society recovers a sense of high human purpose.

Painters have on the whole been less hostile to modern technology than have writers. An indirect reason, I suppose, is the prominence of painting, which in popular usage has become practically synonymous with "art"; today it has much more prestige than it did in the last century, or indeed in almost any period except the Renaissance. A plainer reason is that painters have found more use for technology in their art, which in its Renaissance heyday made much use of scientific studies. The many isms in modern painting, which come down to new techniques, have made it appear more at home in a revolutionary age than any other art. Picasso achieved his eminence by developing and varying his styles more rapidly than any other painter.[7]

---

6. Among the incidental casualties of this fashion is another ancient craft, stone-carving. In the region of Indiana where I live, which produces its well-known limestone, the industry used to employ Italian stone-carvers. Today there are only a few aged survivors, all retired; they have gone the way of the many other craftsmen displaced by machines. One producer of limestone whom I know laments that what used to be something of an art has become only another business.

7. A huge publication of all his known paintings and drawings up to 1959 has now reached its eighteenth volume—a tribute such as no artist in the past ever enjoyed while still alive. Writers hostile to modern society might be reminded that some of them have been similarly honored, as T. S. Eliot was. Volumes of critical studies of their works have appeared in their lifetime, again a tribute that had been paid to no writers in the past.

The Impressionists, who began the revolutionizing in painting, made experiments in light and color that paralleled studies being made in physics, as Seurat in particular realized. Later on Kandinsky, Delaunay, Klee, and others were excited by the discovery of a similar correspondence between their visions and the revolution in modern physics. Still others, such as Gris, Leger, and Mondrian, welcomed the new visual forms created by a technological society, and grew enthusiastic over the possibilities of an esthetic based on machine-inspired forms and such concepts as "functional" and "economical." The most enthusiastic Futurists glorified modern technology, especially "the beauty of speed," which was dissolving classical concepts of time and space; in their manifestos they proclaimed that "a racing car is more beautiful than the Victory of Samothrace." The Cubists exploited the observation of Cezanne that all forms in nature could be reduced to the mathematical forms of the sphere, the cone, and the cylinder; their technique was purely "rational." All along painters were profiting by a by-product of technology, the rise of the museum housing the world's art—a unique institution such as no past society had or could have. And the main over-all development, the growth and spread of nonrepresentational or abstract art, was clearly in line with the growth of a technological society.

About the critical question, whether modern art represents a healthy development or a degeneration, I must write diffidently. I like many paintings, do not care much for many others that are highly esteemed, but I lack clear standards of taste or judgment. (I should confess that even much of Picasso leaves me cold.) Yet I cannot simply waive the whole question either, still less dismiss it by the admirable sentiment that to understand all is to forgive all. There is obviously much pretentiousness in art criticism, sometimes smacking of a racket, often attaching a profound significance to ephemeral or eccentric work. In the art itself there is as obviously much *ersatz* and whoring for novelty. When an abstract painter remarked that Washington studios were "eighteen months behind" New York's, presumably a deplorable lag, only a fool would feel ashamed of preferring the art of Rembrandt, who was two centuries behind. In general, painters unquestionably enjoy more freedom than ever before, a freedom I am disposed to welcome; but there is no question either that it has led to a good deal of confusion. Like engineers and other technicians, they appear to feel that innovation is necessarily "progress." When almost anything appears to go, and the latest fad goes best, they might be reminded that scientists are permitted no such freedom, but held up to a rigorous discipline and definite standards. They might wonder too whether art is really taken seriously by its public.

Although judgment must finally be "subjective," inasmuch as an age that prizes objectivity is less able to agree on standards of judgment than any previous age, one can venture some relatively objective judgments about modern art as a whole. Its basic impulse is sound as an effort to create a new kind of art suited to a radically new kind of society, with new conceptions of reality. Even lovers of Rembrandt cannot expect painters to go on painting indefinitely in his style, for if they did one could complain of the "stagnation" that some critics, strangely enough, see in our age. As it is, twentieth-century art is at least a creative response, suited not only to a technologically dynamic society but to a civilization that for a thousand years has been more boldly, continuously creative than any other. For all its extravagances, it involves a healthy effort at an esthetic acceptance of machine-inspired forms. (Later generations may more easily recognize the best products of industrial design as works of art if or when they are put in museums.) In the sometimes loose talk about "significant form," artists and critics have rediscovered a truth long obscured by realistic representation, that abstract forms can have human significance, as is apparent in decorative art and architecture. In discussing "physiognomic perception," E. H. Gombrich points out that people naturally scan their environment with the question whether what they see is friendly or hostile, a "good thing" or a "bad thing," and that art teachers now try first of all to teach students to look out for the expressive character of shapes, textures, and colors.[8] Abstract art may accordingly be regarded as an emancipation even apart from the freedom it gives painters

---

8. He also mentions some related experiments in which subjects were asked about the qualities of random words, for example whether fathers were more heavy than light, boulders more serious than gay. The common answer that fathers were heavy and boulders serious may not have had strictly "intellectual meaning," but it made good sense, in that both belong to the world of potential obstacles.

from the conventional demand of recognizable representation; for those who have learned to appreciate it, it has extended and enriched the possibilities of experience. "There can be little doubt," concluded Gyorgy Kepes, "that this is an age of extraordinary vitality and promise. It calls upon artists for more than strong protest: its enormous potential for undreamed-of harmonies and rhythms demands new levels of sensibility, a new capacity for unification, a new creativity."

There can be considerable doubt, however, about not only the alleged harmonies and rhythms of much modern art but some major tendencies. The freedom of the painters has so far resulted most conspicuously in a vast confusion of fragmented visions, far from "unification"; the many different private symbols accentuate the lack of universally accepted symbols, such as the religious symbols of the past, for high artistic purposes. (The flag will hardly do for serious artists.) Another consequence has been the common failure of communication, which is no doubt due primarily to all the untrained people accustomed to looking for realism and familiar "meanings," but may be due as well to the painter's failure to render his private vision by humanly significant forms. As Gombrich warns, "We must not confuse response with understanding, expression with communication." The intellectual and imaginative reach of the ultramodern artist was open to more question when Jackson Pollock won fame by his drip method of painting, a "process" that might seem more suitable for making decorative designs than rendering a personal vision or demonstrating a "new creativity." (Picasso presumably went him one better when he announced in 1963 that he had perfected a painting machine.) And pop art, reproducing the imagery of billboards, comic strips, tin can labels, etc., seems still more mechanical. It also suggests that the avant garde is no more, but somewhere in the muddy mainstream, for it soon became the rage, again intimating that art is not really taken very seriously. Together with the commercial racket of the art galleries, the dependence of the museums on the new fashions, and the growth of an "art establishment," it led Hilton Kramer to a different conclusion about the future of art: "I have no doubt that it is going to be very like the future of life generally—tyrannized by technology and bureaucracy, rationalized by propaganda, trivialized by the mass media, and condemned to an abject dependency on the main course of society until its best instincts recoil in disgust."

The most serious complaint about modern art, to my untutored mind, is that it may reflect or reinforce the dehumanizing tendencies in a society swamped by the abstractions of science, technology, and business. It too may *reduce* life, neglect the whole man. It celebrates neither God, man, nor nature, and so might illustrate the popular theme of the "alienation" of modern man. Like contemporary literature, it gives little suggestion of the heroic, the sublime, the tragic, or on the other hand of the homely simplicities of man's life, the dignity and the pathos of mortality. The machine forms it imitates or the new forms it creates may not compensate for the loss of the rich, warm meanings men used to find in the world about them. It ignores or even disdains natural beauty, which with portraiture it leaves to academic painters; the assumption appears to be that God was a corny calendar artist. Often painters positively deform, as the "rational" Cubists did when they rendered the human face simultaneously from side and front. By contrast the Surrealists have celebrated the irrational in their highly self-conscious discovery of the unconscious. And the painters who have rebelled against the ugliness, absurdities, and horrors of modern life have typically rendered them without attempting to render as well simple beauties and delights, as Peter Breughel and Goya did. Gyorgy Kepes himself qualified his vision of the "extraordinary vitality and promise" of our age: "Rather than accept the creative challenges within the range of the visual arts, rather than learn to see a broader world, most of us, our artists included, divorce ourselves from common obligations, turn our backs on the rational, and separate man from himself, from his fellow man, and from his environment."

From my limited acquaintance with modern sculpture I judge that the main developments have been comparable to those in painting, if with the benefit of fewer isms and manifestos. Sculptors have reflected the influence of technology by making more use of metal, lately even of plastics, and producing more abstract forms, including stabiles and mobiles. A recent movement called "minimal sculpture" has moved toward industrial design, with structures of boxlike shapes and other surfaces and forms of industrial objects. Another movement called "art-and-technology" seeks a closer alliance. I have read that it aspires "to transform the visual arts into a form of electronic theatre" by "aleatory designs," both visual and aural; it is supported by a foundation, Experiments in Art and Technology, Inc., which has a membership of some 3,000 artists and engineers and is amply financed by the electronics industry. Among its products is "cybernetic sculpture." Of what I have seen of modern sculpture I again find some of it interesting or impressive, more of it not to my taste. As with painters, the "emancipation" of the sculptor seems to me too often limited by a compulsive exclusion of other natural interests or concerns. Reproductions of grotesquely attenuated or deformed figures make me think wistfully of ancient Greece and the Renaissance, when sculptors glorified the human form or at least treated face and figure with respect. My tastes are obviously old-fashioned or middlebrow; but I think it fair to say that the inspiration of most modern sculpture is not humanistic, and that if the best of it makes the most of the forms and interests of a technological society, on the whole it is not designed to offset the dehumanizing tendencies of this society. Or if I am wrong, I think it important that students of contemporary sculpture ask themselves this question.

A more genial contribution of technology, the new art of photography, had behind it a series of inventions going back centuries, such as wood-block printing, copper and steel engraving, and lithography, that Lewis Mumford applauded as one of the universal triumphs of the machine, the "democratization of the image." While the camera could reproduce scenes more accurately, men soon realized that photography could be just as selective or individualized an art as painting. By now it has become the most universal language of visual communication, enjoyed by people all over the world. Like ordinary language and everything else in a commercialized society, it has of course been put to corrupt use too. Mumford worries over an excessive democratization, the mass reproduction of images in not only books but newspapers, magazines, and ads—"images so constant, so unremitting, so insistent that for all purposes of our own we might as well be paralyzed"; they are among the favorite devices of modern propaganda. He worries over even the technical triumphs in the reproduction of great paintings, which by excessive familiarity lose the magic of the original, grow stale. I assume that most of us seldom really see the pictures hanging in our own homes, but I am more willing to pay the costs of familiarity. In particular it seems to me important that many millions of people now enjoy taking pictures, especially as tourists, for they see the world with more of a painter's eye, recover the natural pleasures in landscape, cityscape, and human figures that may be spurned too superciliously in abstract art. They might acquire more of the esthetic judgment badly needed for any hopes of improving the man-made environment in America.

Another new art that owes wholly to modern technology is the moving picture, to which it later added sound and color. Long debased by Hollywood, this has nevertheless produced some memorable films over the years, and today it is stirring more excitement than ever. One periodically hears that the novel is dead, poetry is dying, and painting is struggling to keep alive by new fads, none of which seems to me clearly true; but there is no doubt that the cinema is very much alive and has a future. While it can and does draw on the older arts, it is less burdened by tradition or annihilating comparison with the past, and as a youthful art it can more freely explore new possibilities with a less compulsive search for novelty. The spreading film festivals encourage much more boldness and honesty than do the Oscars of Hollywood.

About modern music, finally, I again know too little, like little of what I have heard of the ultra-modern, and find much writing about the "new music" as difficult as the music itself. But I venture a few observations, emboldened by a recent confession of Stravinsky: "I could not begin to distinguish music and non-music in some of the concert-hall activity I have observed of late, nor would I be confident of recognizing a new musical genius." About the latest composer hailed as a genius, his dictum would be: "Keep your hats on, gentlemen, for all I know he may be a charlatan." At the same time, Stravinsky's own career suggests the need of an open mind. Like Wagner, Schönberg, Bartok, Berg, and other now established composers, he was at first abused as unmusical and unintelligible. Today, as he says, some of the young men relegate him "to an annex of the nineteenth century."

As another antique myself, with a stronger preference for music that sings or dances, I should first repeat that music has always been indebted to technology for its instruments, and in the last two centuries for that most magnificent instrument for composers, the symphony orchestra. It has always been an "abstract" art too, yet a mode of communication whose principles and techniques have been formulated clearly enough. As for the revolution in music in our century, it began appropriately with another "crisis," called a "crisis of tonality," due to dissatisfaction with the traditional harmonic system; Schönberg responded with his twelve-tone system. In much of the music after him the jagged rhythms and the dissonances were obviously suited to modern life, which technology has made the noisiest in history. But Schönberg himself never repudiated the classical music of the past; he said it was still perfectly possible to write good music in the key of C. And it is this that distinguishes him from some contemporary composers, who announce, in Stravinsky's words, "Nothing happened before us." Of an unnamed composer "now greatly esteemed for his ability to stay an hour or so ahead of his time," Stravinsky remarked that he found "the alternation of note-clumps and silences" simply monotonous. In *Exploring Music* Ernst Krenek, another senior composer, pointed out a further reason for possible complaint, that "most performances of new music take place in an atmosphere dominated by specialists—composers, critics, musicologists, students, musical executive organs, conductors, and the like." These specialists, he added, are mainly interested in the experimental demonstration of new materials and methods, not the "total emotional experience." Nonspecialists— the great majority of us—may therefore have good reason for finding the experience often unsatisfying.

Some, however, appear excited by the latest application of technology, electronic music. Iannis Xenakis, a pioneer in the field, has founded societies to tie music to science. The composer of a work called "Metastaseis," he uses such technical terms as isomorphism and vectorial space structures, and is also exploring the use of probability theory, games theory, and computers in composition. He insists, however, that mathematics is only a tool, and that in the end music must dominate it. Similarly he uses computers to control his music, but emphasizes that he himself remains in control, which is the hallmark of an artist. Having heard little electronic music, I will add only that on lower levels I have a positive distaste for the electric guitar, and so was neither surprised nor distressed when I read that discotheque rock music is literally hard on the ear: a scientist found that it destroyed cells in the cochlea of the ear of guinea pigs, which is much like the human ear.

In any event it is chiefly classical music that has contributed to the recent "culture explosion." While contemporary composers complain that they are not adequately represented on concert programs, to me the important point is the phenomenal increase in the number of such programs. Whereas in 1920 America had about a hundred symphony orchestras, it now has over twelve hundred, and attendance at their concerts has been soaring so steadily that the better known orchestras have been lengthening their season. More millions of people listen to music on FM stations. The market for classical records has boomed correspondingly; statistics showing that sales have doubled or tripled in a short period are outdated so soon that again I see no point in digging up the latest figure. And this startling growth of popular interest has by no means led to the lowering of standards feared by the high priests of culture. As superior recording and high-fidelity equipment have encouraged many superb performances on long-playing records, so concert programs are less conventional than they used to be, and professional musicians report that their audiences have grown more sophisticated and discriminating. The thousands of classical records, many of works once little known, have widely extended popular knowledge of music.

In the other arts contemporary work has figured more prominently in the boom in culture. In literature it has inspired a rather odd complaint, that the avant garde is no more because, as Leslie Fiedler wrote, "anti-fashion becomes fashion" so quickly. "Truly experimental art aims at *insult*," he declared, in the old bohemian tradition of shocking the bourgeois; and apparently they no longer shock so easily. But again the significant point is that Americans who long had no truck with highbrow culture are now turning to it by the millions. The National Book Committee reports that the sales of books have in recent years increased three times more rapidly than the population. Museums, which were once lonely places, now attract crowds totaling many millions. Art galleries, long concentrated in New York, have multiplied and spread all over the country, even in small cities. American artists no longer flock to Paris because New York has replaced it as the world capital of painting and sculpture.

In *The Culture Consumers* (from which I have drawn most of this data) Alvin Toffler adds some significant evidence about this new American breed. Whereas the audience for culture used to be chiefly women, to whom husbands delegated the function of appreciating "the finer things of life," it now contains as many men, including many more businessmen than in the past. In particular it is predominantly youthful—a promising augury for the future. Although it is not representative of the entire nation, since it includes relatively few workers and farmers, it is no longer a cultural elite. In the past, museums, symphony orchestras, and opera used to be subsidized chiefly by wealthy patrons, like as not more interested in social status than in art, but now all kinds of middle-class people contribute to their support. And the new audience includes many amateur artists. The sales of musical instruments have about doubled since 1950, while I suppose no one knows how many hundreds of thousands of Americans have taken to painting as an avocation.[9]

Since culture has become a big business, technology has as usual made some dubious contributions. The organization or bureaucracy that has entered art has brought the usual machinery, ranks of executives, and threats of standardization, or discourage-

9. I do know that among them is my postman, who has exhibited in a local gallery. I also know of an example of the change in popular attitudes. Before the war I wrote a compliment on the theme of a freshman who had written surprisingly well about painting, and he at once came in for a conference to pour out his enthusiasm for painting— an enthusiasm he had kept to himself for fear that his fellow students would think he was queer or effeminate. Today such an interest is quite respectable.

ment of imaginative enterprise. Artists may or may not welcome
the new profession of art management, which already has a technical journal. They are not all prospering either from the manage-
ments that handle the concert business. The fat fees go to a few
conductors and star performers; most recitalists make little money,
as do most composers and symphony orchestra players. The
increasing number of middlemen in the arts, who are neither
creators nor necessarily sensitive judges, bring up the more positive
indignities of commercialization in the marketing or merchan-
dising of the arts. Book publishing perforce remains a business,
which in spite of increasing profits shies away from books, such as
volumes of poetry by the "unacknowledged legislators," likely to
have small sales. The art gallery business in New York has become
more of a commercial racket, especially since the fantastic inflation
in the price of paintings; again the big money goes chiefly to a few,
just as publicity goes to the latest fad. Marketing executives are
talking about the "culture sell," which has attracted Madison
Avenue too.

Even so, to repeat, the "culture consumers" are still a small
minority. Many no doubt take only a casual interest in an art or so,
and with the rest we can never know how fully they understand or
how deeply they appreciate art. But in any case the millions of
Americans who are now taking more interest in the arts represent
a striking change. The change may seem more significant when
one recalls Henry Adams' observation that in 1800 American culture was confined almost entirely to theology, literature, and
oratory, and recalls too that as late as 1900 American theater,
music, and painting were still far behind Europe's. If the new
interest is to some extent merely fashionable, it looks healthier
than the self-conscious aspirations to "culture" nurtured by the
old genteel tradition in America. While an affluent society has
provided a larger audience and more substantial support for the
arts, these have not—Marx notwithstanding—merely reflected the
material conditions of this society. The consciousness of the new
audience, especially the young, is to some unknowable extent
influencing their existence, suggesting that life may imitate art.
However deficient the tastes of this steadily increasing audience
by the standards of the fastidious, it is making high culture more

nearly democratic. I do not anticipate a brilliant flowering, or arts clearly worthy of a Great Society; but neither do I pine for the days when culture was the exclusive possession of an elite, patronized by an aristocracy that had no such passion for the arts as literary people often imply.

17

Religion

As the Christian story goes, "In the beginning was the Word," which with Christ became incarnate in history; but although strange things have been happening to it as to everything else, the subject of religion has been ignored in almost all the studies of a technological society that I am acquainted with. It is not even listed in the index of Jacques Ellul's book. The main reason for this neglect is doubtless simply that religion is no longer a vital concern of most contemporary thinkers, any more than it was a vital inspiration of the Industrial Revolution; never before has there been so little feeling of the "sacred" as a real power, since the powers of science and technology are so much plainer. Another reason is that technology has had a less direct impact on religion than on other major interests, and so has itself been largely ignored by most religious thinkers. Ellul explains parenthetically, "If I have not mentioned religions, it is because they no longer express revolt; they have long since, in their intellectual and spiritual forms, undergone integration"—into the technological society. I should say rather that insofar as Christian sentiment and belief are still a force—and I believe they clearly do count for something—they are another indication that technology does not dominate our society so completely as he insists.

At any rate, the state of religion today seems to me more confused and ambiguous than ever before. Indirectly it has been deeply affected by technology through the rise of a society devoted to business, the most secular society in history. Together with science, this is the plainest reason for the decline of religion as a ruling force, or a guide in living. Martin Buber said that God is "silent" in our age, and recently a "God-is-dead" movement made headlines. At the same time, however, the churches have grown more popular in America, religion has been the theme of best-sellers, and evangelists like Billy Graham get nation-wide publicity. While some churchmen complain that God is being sold like soap, Reinhold Niebuhr, Paul Tillich, and other eminent theologians—a species that a generation or so ago Vernon Parrington pronounced extinct—have also acquired a surprising measure of popularity. Departments of religion in the universities have been attracting many more students, most of whom are not converts but at least are showing a serious interest in the subject. While thoughtful people have been attracted by the "theology of crisis," harking back to Kierkegaard's revolt against the complacent faith of a bourgeois society, churchmen have likewise been responding to the plainer world crisis. In the One World created by modern technology and Western imperialism, there has been an immense effort to make our century the "Christian Century," and now the Christian churches themselves are trying to unite. They have been working actively for world peace, a cause that Christianity was never so devoted to in its heyday. And technology, which vastly increased the importance of the written word, has lately made more of the spoken word too, the use of which the Second Vatican Council stressed more than ever. Now Marshall McLuhan is trumpeting the message that the new medium of television is making a world of difference, and Father Walter Ong —a more reverent student of the Word—is inclined to agree.

This confusion of conflicting tendencies is not hopeless, however. One may find one's way about more easily with the help of an elementary distinction too often ignored in discussion of religion. We can roughly distinguish three kinds of religion— ideal, organized, and popular. The distinction is necessarily rough because they overlap, but it is none the less essential to any consideration of the values and the shortcomings of religion, and of the changes that have come over it in the modern world.

Ideal religion, or religion at its spiritual best, resides in the teachings of the founders of the higher religions and the greater saints, mystics, holy men, and religious thinkers after them. In a long view its influence seems to me no less unmistakable because strictly incalculable, even by the latest polling techniques. In any case it can be respected by those (like me) who are not believers; it introduced universal ideals, such as the brotherhood of man, that have grown more meaningful in the modern world. By contrast, organized religion is the established church or set of churches, which are always worldly institutions, naturally conservative, and have been inclined to alliance with the ruling classes. Commonly addicted to legalism and formalism, they invariably debase to some extent the teachings of the founders, if in part for the humane purpose of adapting them to the needs of frail mortals. The spiritually minded can agree that the influence of organized religion, which is the plainest thing in the historic record, has often been unfortunate. Popular religion, lastly, comprises the thought, feeling, and behavior of the mass of simple believers or worshipers. It of course involves all kinds of behavior, ranging from humility and simple goodness to bigotry and hatred, but it has always been basically materialistic, most conspicuously in prayer or petition for special favors. It is the oldest, most universal element in the higher religions as they came down through the established churches, which accepted much vulgar practice and belief, especially in the efficacy of prayer and sacramental rites that derived from prehistoric magic. The spiritual shortcomings of popular religion may be condoned because of the pathetic needs of all the poor people throughout history; though this excuse is less valid in affluent America.

Inasmuch as the modern Western world has produced few saints or prophets, ideal religion must be looked for chiefly in the realm of high culture. The most obvious influence on thoughtful, educated people has been science and the scientific spirit. In the last century this led to both a decline of faith and the rise of Modernism, more liberal, tolerant varieties of Christianity that came to terms with science by surrendering the traditional belief in the literal, infallible truth of Scripture—a familiar story that I need not go into here, except to add that most unbelievers did not renounce the Christian ethic, and so were not necessarily left homeless or adrift with no guiding light. In our own time science and technology together helped to popularize the theme of *Angst* among intellectuals. More religious thinkers recognized that the truth of Judaeo-Christianity could not be absolutely demonstrated, but required a "leap" of faith that may recall Kierkegaard's desperate wager on the "absurd." Others were deeply troubled by the world crisis. Because of thermonuclear weapons the "end of the world," anticipated by early Christians and periodically by eccentric sects thereafter, became a quite literal possibility for Paul Tillich; and what, then, would be the meaning of Christian history, the whole arduous, painful history of man on earth? Tillich also speculated about questions raised by astronomy and the exploration of outer space, the probability that there are forms of life on millions of other planets, and the possibility that some might be inhabited by beings much "higher" than man. Christians might wonder whether the Word had been made known on these planets too. And so on, through all the questions raised for the thoughtful in One World in which they learned much more about other religions, such as Zen Buddhism, Christianity had no apparent prospect of converting the rest of mankind, and all mankind was in the same storm-tossed boat, while God remained "silent."[1]

Today the prevailing tendency among religious thinkers appears to be an effort to accommodate the interests of a secular society rather than to stress transcendental or ultimate concerns. I gather that they are generally diffident or embarrassed about such subjects as prayer, which has clearly become less habitual as people put their faith in science and technology instead. In

---

1. In *God and Golem, Inc.* Norbert Wiener explored the religious implications of computers. As man-made machines able "to make other machines in their own image," they were the modern counterpart of the Golem of Jewish legend, an embryonic Adam who was a monster; so he asked whether God was to Golem as man is to the machine. But so far as I can see, computers raise no really novel questions about the existence or nature of God. Wiener emphasized the conventional moral, the sin of pride in their creator, and immediately the obvious dangers, the official who will push the button "in the next (and last) atomic war," forgetting that in the new game of atomic warfare "there are no experts." He restated the conclusion that seems to me as obvious, and that I assume God would approve: "Render unto man the things that are man's and unto the computer the things which are the computer's."

warning that religion is not necessarily to be understood as a "resource of an inner life," Krister Stendhal of the Harvard Divinity School writes that contemporary theologians are so uneasy over the traditional distinction between the "spiritual" and the "material" that they are trying to get rid of it. Thus in *The Secular City* (which became a best seller) Harvey Cox argues that secularization is not necessarily fatal to the Christian spirit, but has its roots in the Bible, and that it is the task of theology "to transpose the assertions of the Bible into our contemporary idiom." The Rev. George W. Weber goes further, saying that "it's not a question of restating old teachings in new forms but of totally reformulating them in light of present experiences," and adding that "what is happening today is just as important as what happened 2,000 years ago." The Rev. W. Norris Clarke (a Catholic) even gives a philosophical blessing to "progressive technology" in spite of its admitted dangers: man must "strive to remake or transform, by his own God-given powers, the world that his Father has made for him out of nothing and given him as his workshop." This whole tendency involves some risk of unseemly concessions to the reigning materialism, as well as neglect of the particular pertinence today of Tillich's concern with the ultimate, but it comes out of a long Christian tradition of active interest in the affairs of this world, which Tillich shared.[2]

On organized religion modern technology has had a much plainer influence. For one thing, the revolutionary changes in an industrial society accentuated the long failure of the major churches to keep up with the advances in knowledge and thought; so far from providing intellectual leadership, as the medieval Church had, they kept fighting a losing battle with science, especially over the theory of evolution. Rabbi Abraham Herschel

---

2. In *The Idea of a Christian Society* T. S. Eliot rejected the possibility of compromise with our secular society because it is a commercialized society that has erected Profit into a social ideal. His sound criticism of it was as usual weakened, however, by his traditionalism, which led him to idealize past societies as most literary men have done. Christianity in its medieval heyday scarcely approached his idea of a Christian society, one "in which the natural end of man—virtue and well-being in community—is acknowledged for all." Nor did Eliot's Anglican Church distinguish itself historically by devotion to this ideal.

commented that not secular science but religion itself was to be blamed for its eclipse: "Religion declined not because it was refuted, but because it became irrelevant, dull, oppressive, insipid." Most important was the failure of the churches to provide either moral or spiritual leadership in the new society. In spite of the shocking abuses of early industrialism, they mostly opposed movements toward social and political reform, and then the whole labor movement. In England and western Europe they were adhering to the conservative tradition of established churches, but their policy had an unprecedented consequence—the alienation of the working class. Whereas all through history the poor had looked to religion for solace, many workers now grew indifferent to it, even hostile. The hostility was focused by revolutionary socialism, the more readily because of an apparent hypocrisy that Paul Tillich lamented: while the churches typically condemned all forms of socialism as materialistic and godless, they were consecrating the bourgeois ideal of private property. No other of the higher religions had so sanctified the acquisition of property as Christianity now did.

It was in Protestant America that Christianity was most radically transformed. Here the "Protestant ethic" of individualism was corrupted by the new gospel of economic individualism as churchmen actively supported the ruling business class, whose glorification of competition, the profit motive, and money values was utterly foreign to the Christian gospel. As Lewis Mumford observed, all but one of the seven deadly sins—sloth—became virtues. Greed, avarice, and envy were other words for ambition; gluttony, luxury, and pride were emblems of success. Popular ministers vied in travesties on the teachings of Jesus. Henry Ward Beecher explained in a bad depression year that "God had intended the great to be great and the little to be little," and the little man "who cannot live on bread and water is not fit to live." Russell Conwell went up and down the land giving six thousand times his lecture "Acres of Diamonds," in which he preached the gospel of wealth ("I say, get rich, get rich!") and supported it by declaring that "the number of poor who are to be sympathized with is very small"—the rest of the many poor were being punished by God for their sins. Bishop William Lawrence was only more stately when he maintained that "Godliness is in league with riches."[3] The consequence was again the alienation of workers. A Protestant minister, worried because they were no longer going to church, conducted a nationwide questionnaire in 1900 to find out why, and almost always got the same answer: the church was on the side of the rich, against unions and strikes.

By this time, however, the churches were beginning to face up to the realities of industrial life. In his famous encyclical *Rerum Novarum* Pope Leo XIII declared that while the Church opposed socialism it did not approve of laissez faire capitalism either; he called on Catholic workers to form labor unions. In England the Christian Social Union was formed to combat the evils of slums and work for social justice. In America churchmen began preaching the new "social gospel," which recognized that a great many poor did deserve sympathy and called for economic reforms. The mass media helped to spread mild versions of this gospel, such as the Rev. Charles M. Sheldon's *In His Steps*, a plea that all social problems be confronted with the question "What would Jesus do?"; it sold more than twenty million copies. The churches did not thereupon win back all the workers, since many churchmen neglected to ask this question and popular evangelists like Billy Sunday typically remained conservative. (He crusaded against "sinners, science, and liberals.") But except for Fundamentalists, whose strongholds are in rural America and the South, church leaders have increasingly tended to support social reform since the New Deal. In our time they may even be found on the forefront, as in the civil rights movement and the protests against the war in Vietnam; among their chief critics today are right-wingers. In short, however tardy in adjusting the Christian message to the needs of a new kind of society, the churches are now actively involved in it, less prone to preach an aloof, irrelevant,

---

3. Leaders of the Roman Catholic Church, whose flock were chiefly immigrants, were less inclined to damn the poor as shiftless; but they could chime in by restating the Church's traditional praise of poverty. Thus an archbishop of Baltimore, James Roosevelt Bayley, praised it as "the most efficient means of practicing some of the most necessary Christian virtues, of charity and alms-giving on the part of the rich, and patience and resignation to His holy will on the part of the poor."

or absent-minded kind of spirituality, and no longer so closely allied with the ruling classes. One may better appreciate the striking changes that have come over them by considering the other higher religions, concentrated in the non-Western world, none of which are so actively concerned about the earthly lot of all their poor people.

The organizational revolution that came with modern technology had less effect on the major churches, inasmuch as they were already organized, but in our day there have been some surprising changes on this score too. As the oldest bureaucracy in the world, the Roman Church long continued to exhibit the natural tendencies of bureaucracy that men have grown more aware of; it remained conservative, devoted to its traditional order, hostile to any innovation that might weaken the power of its hierarchy. (Its response to the weakening of its authority was the proclamation of the dogma of papal infallibility.) All the more astounding therefore were the changes in policy and dogma initiated by Pope John, one of the few saints of our time. The enthusiastic response to them revealed that many Catholic laymen, priests, and theologians had grown restive under the rule of the conservative hierarchy. When Pope Paul reverted to tradition by condemning birth control, he was at once obliged to defend his encyclical against widespread opposition, and has since complained piteously of all the rebels against Holy Church. As for the Protestant churches, with their various kinds and degrees of organization, Krister Stendhal emphasizes that institutional religion has become more important in our century. Otherwise I am aware of no significant changes in this respect until the recent efforts at union, notably the World Council of Churches, in which some 170 churches are represented. These efforts clearly require some organization, but as clearly are intended to strengthen the spiritual authority, not the temporal power of organized religion, and so suggest that the interests of bureaucracy may not be sovereign after all. In general, a technological society has created both the means and the felt need of a united Christendom, corresponding to the United Nations, and has helped to make the religious spirit more adventurous.

Popular religion is a rather different matter. Immediately it indicates that the churches still fall considerably short of ideal religion, for their growing popularity in America is not clearly due to their preaching either a loftier spirituality or a more enlightened social gospel. More obviously it is due to the growth of an affluent society, which they have accommodated by a suitably modified gospel of wealth. Too many ministers preach an easy, comfortable kind of faith such as led Kierkegaard to declare in disgust more than a century ago that "Christianity does not exist." In effect they sell religion as a good investment: it pays because it makes you feel better, makes your neighbors behave better, makes America stronger. Will Herberg described popular religion in our day as "a religiousness without religion, a religiousness with almost any kind of content or none, a way of sociability or 'belonging.'" Although religious animosities persist, especially among Fundamentalists, most urban Americans have grown more tolerant chiefly because for their purposes any old faith will do so long as you're a good fellow. Their faith is neither a deep conviction nor a serious commitment, only a "belief in believing."

Hence some leading churchmen have expressed alarm over the popularity of religion. "We churchmen have never had attendance so high," commented the Rev. William Sloane Coffin, "and influence so low." Others complain that the churches have been demanding much too little of Americans. To appreciate how little, one has only to go back to the great prophets of Israel or the teachings of Jesus; the popular theme that God blesses and serves America has been accompanied by no demand that Americans first of all serve God, or make any real sacrifice. And such popular religion is positively dangerous because it confirms the national complacence and self-righteousness at a time of world crisis. Thus Congress was pleased to pass a law putting the nation "under God," as if a supposedly Christian society could be anywhere else, and the American Legion led a "Back to God" movement that made it no less chauvinistic. Assured that God was on our side, especially because the Communists are godless "materialists," Americans could go on wallowing in their material goods, applauding while their leaders laid down the law to the rest of the world.

The mass media have on the whole tended to propagate the

most vulgar tendencies of popular religion. Henry Luce's magazines plugged moral and spiritual crusades that called for no
sacrifice of the profit system, the heart of the American Way. Tin
Pan Alley contributed popular tune hits, such as one some years
back entitled "The Man Upstairs"—all about how He would take
care of all your troubles. More lately an enterprising adman sold
one of the big churches the idea of broadcasting on radio, as paid
commercials, religious jingles he composed:

> Doesn't it get a little lonesome sometimes
> Out on that limb without Him?

Television stations, which like their audience largely confine their
service of religion to the unpopular Sunday morning hours,
offered more solace in the programs of the Rev. Norman Vincent
Peale, who taught millions easy guaranteed ways of acquiring
"the power of positive thinking." Or perhaps most characteristic
of our day is the Billy Graham Evangelistic Association Inc., a
thriving enterprise that sends out the founder with a team of
carefully drilled religious salesmen on Crusades, well-organized,
well-heeled, and well-publicized, with hospitality cards at all
Hilton Hotels, and with press-handouts suggesting that souls are
saved for as little as $2 each. Graham does not make religion
simply easy, inasmuch as he requires some spiritual effort of the
people who come to get their "batteries recharged" or step up to
make the Act of Decision; yet he preaches a crude, primitive kind
of Christianity, which is likely to make people more concerned
with personal salvation than with love of God and neighbor, and
which in any event suggests a depressingly low level of religious
literacy in the crowds he harangues, or the national leaders who
invite his blessing.[4]

---

4. I have had a mournful letter from a Protestant minister in the
Mid-West about the difficulties of trying to preach honestly and
intelligently to a congregation most of whom have never advanced
"beyond a concept of God understandable to and by the average five-
year-old." He added that what is called "Christian education" has been
neither Christian nor education. It is perhaps for this reason that there
has been a sharp decline in the number of graduates of theological
seminaries who enter the pastorate.

Recently technology has made a different kind of contribution to religion by another wonder drug, LSD, which has also got much publicity. Popular especially among young people repelled by American society today, this nevertheless strikes me as typically American. It offers an easy way to spiritual bliss, aptly called a "trip," which requires no spiritual discipline or even effort, as yoga does, and raises some question about what "God" means to the young. One who respects the great religious teachers may doubt that the experience of LSD is any more truly religious than that provided by opium or "pot," another apt word for where this kind of religion has gone to.

In a historical perspective, however, one may also doubt that popular religion in America has become more uniform, or more uniformly materialistic. It embraces a wider spectrum of sentiment, aspiration, and belief than the religion of the illiterate societies of the past. Some awareness of its shortcomings was intimated even by a poll that revealed the ineffable complacence of Americans, most of whom said they thought they lived like good Christians, but also said they thought most other Americans didn't; at least they recognized strong un-Christian tendencies in the land. A clearer illustration of mixed or confused aspirations was a poll conducted some years ago by *This Week* on "Pick the Sermon You'd Like to Hear." The top choice, How Can I Make Prayer More Effective?, pointed to the pathetic, age-old need served by religion, since even affluent Americans still have plenty of problems. Another favorite, How Can Religion Eliminate Worry and Tension?, suggested more plainly that Americans are not so complacent as they appear on the surface; hence millions of them are seeking "peace of mind" (the theme of a best-seller), which evidently they do not get from the tranquilizers they have grown addicted to. And two of the six leading choices in sermons, How Can I Make the Greatest Contribution to Life? and What Can the Individual Do for World Peace?, indicated that there are still many earnest Americans who have unselfish motives, because of which popular religion cannot be distinguished absolutely from ideal religion.

Inasmuch as I am approaching a chapter on my ultimate concern, "People, Just People," I repeat that like all polls this one does not indicate how deeply Americans felt these needs, or how earnest the more idealistic ones were, how much effort they were prepared to exert on behalf of unselfish causes. Likewise the significance of still another favorite choice in sermons, How to Increase Religious Faith, depends on what they mean by faith and how they conceive its values. For such reasons I am not sure that religion has deteriorated in modern America, or at least as much as some thinkers believe. Government and business, the controllers of our technology, are not governed by Christian principles, of course, but neither were they a century ago, in the corrupt Gilded Age, nor for that matter have they ever been in any allegedly Christian nation; it is a commonplace that the author of the Sermon on the Mount would never recognize as Christian the societies that call themselves by this name. The rampant commercialism in America should have weakened the Christian spirit, but again such commercialism was conspicuous enough a century ago, and Yankees have always tended to this kind of smartness. The Peales, Grahams, and other popular preachers today are at least a cut above the vulgar materialism of the Rev. Russell Conwell. That he edified large audiences still given to reading the Bible, saying grace at meals, and saying family prayers may make one doubt that the Christian ethic in private or domestic life has necessarily suffered from the decay of Christian forms. And though I do feel sure that popular religion has not been growing purer or loftier in our affluent society, I would add a qualification here too. Freedom from the want of basic material necessities that preoccupies the poor can enable people to devote more time and thought to social causes, as it has with many Americans. The Negroes are now testifying that poverty is not necessarily good for the soul.

Since I am nevertheless inclined to take a jaundiced view of popular religion in America, and as an unbeliever may be prejudiced, I shall again give the last word to the more cheerful views of Father Walter Ong, a much closer student of what is happening to the Word in the shifting sensorium of modern man. He believes that we may well be "moving toward some new realization of the personal and thus new opportunities for Christian living and giving." The mass media, in particular the new electronic media, have helped to give modern man an entirely novel sense of "a peopled world." With this he has developed an "open-end style of activity," which in recent years has become conspicuous in Christian efforts at unity and at better relations with other religions.[5] The catechism, a product of typography, has given way to dialogue, better suited to an oral age. And though the increased sense of human presence is often an overwhelming psychological burden, leading to the belief that God is dead, Ong sees in it a promising means to the sense of God's presence, which had originally grown out of human community.

His optimism carries him pretty far, I think, from the present realities of popular religion, a shallow community of consensus on the values of an affluent society, in which personal relations are too often reduced to "roles," as of buyers and sellers, that have nothing to do with a Christian spirit. He is nevertheless pointing to real changes in the thought and feeling of many good people. Most important to me is that the major churches have taken the lead in the movement for world peace—a movement that has grown directly out of modern technology rather than Christian tradition, but is surely in keeping with the spirit of Jesus, or the Word incarnate.

---

5. Such openness might also be illustrated by the belief of Harvey Cox that "today's emerging religious sensibility is a comic one." It trusts and delights in personal experience, "encourages fantasy instead of repressing it." Among the examples of this "comic elan" cited by Dr. Cox were the joyous "jazz liturgies" and the return of the dance to the sanctuary. All this might seem remote from the idea of the "sacred," but those for whom religion is no laughing matter might note that it was associated with gaiety as well as sanctity by many peoples, including the ancient Greeks and medieval Catholics. Protestantism was a preternaturally solemn religion.

18

"People, Just People"

This chapter, which to me is the most important one, will also be the most inconclusive. In all the others I indicated plenty of uncertainties, but since their subject was more circumscribed it was easier to specify the major effects of technology, define the kind of problems they raised, point out the obvious abuses; and if I usually suggested that the prospects of eliminating these abuses were not bright, at least the problems seemed relatively clear to me, as did some possible ways of doing something about them. But a consideration of what technology has done to people involves all the changes surveyed in the other chapters, and much more that is difficult to specify. As I have periodically repeated, we do not know for certain how deeply people have been affected or how much they have changed. Though a principle of ambiguity is a helpful guide in the welter of conflicting tendencies, we cannot measure the effects or by any process of addition and subtraction arrive at a total. Nor can there be any simple programs for eliminating the bad effects on people. "If you are looking for a Moses to lead you out of this capitalist wilderness," Eugene Debs once said to his followers, "you will stay right where you are. I would not lead you into the promised land if I could, because if I could lead you in someone else could lead you out." Now we are all wandering in the much vaster wilderness of a technological society, no leader can possibly lead us into an essentially different kind of society, and I assume that few thoughtful Americans can believe any more that either America or the Soviet Union is the promised land.

We still have a choice of prophets, however, and of visions of a promised land; so I begin with some cautions. "The light of human consciousness," writes Lewis Mumford in the spirit of Hegel, "is, so far, the ultimate wonder of life, and the main justification for all the suffering and misery that have accompanied human development." In his generous vision of a full, harmonious realization of human potentialities, Mumford passionately denounces all the technological developments that have narrowed, warped, blunted, or darkened consciousness. But because I agree with his basic criticisms of our society I would note that he obscures the inevitable tragic aspect of the human condition that Hegel stressed. Everything he denounces was also a product of human consciousness; all "inhuman" behavior is always human. The enemy is not a devil out there called technology—he is Man, the creature we are trying to save. Only because he has become more conscious of his powers is he capable of so much folly and evil.

To me, as to Mumford, the most conspicuous tendencies of modern technology can be pretty alarming when spelled out in detail. So again I would first caution against a tendency to wholesale, unqualified indictment, best exemplified by Jacques Ellul. "It blots out all personal opinions," he writes characteristically of technology. "It effaces all individual, and even all collective, modes of expression." It enslaves modern men completely because they are so enthusiastic about it, so immersed in its milieu, that "without exception they are oriented toward technical progress." If he is right, there is nothing we can do about our plight; but my point is simply that such statements are not true. Ellul grows ludicrous when he maintains that "no one will publish a book attacking the real religion of our times, by which I mean the dominant social forces of the technological society"—this in a book that was not only published but translated, widely read, and highly praised. Given the many real goods that have come with the evils, and the visions both of a much better life and the end of man's life, we are dealing like Hamlet with "mighty opposites."

A more common kind of confusion is exemplified by Jacques Barzun, another extravagant critic of our technological society. In *Science: The Glorious Entertainment* he wrote at length about

"the burden of modernity," how we are all suffering from alienation, self-hatred, *Angst*, despair—"all discontents and no civilization." By "we" he sometimes seems to mean, like Ellul, literally
all of us. At other times he plainly means all ordinary people, who
are helpless victims of technology but still have a naive faith in it:
"Our strongest faith is that there is no limit to the good and the
useful." Most often he means we literary people or intellectuals
(excluding most scientists); thus he writes that although "we" are
very lonely, "we reject togetherness as a worse horror." As one of
this group who still enjoys some contents, I think that like many
writers Barzun exaggerates our plight; but my main point is that
he woefully simplifies an extraordinarily complicated situation.
While modern technology has had some common effects on all
people in America, and more on the majority of people, it has also
had different effects on different kinds of people—not only workers, businessmen, professionals, and intellectuals but men and
women, young and old, poor and rich, white and black.

For this lengthy chapter I have therefore laid out a simple A,
B, C outline. First I shall review the deplorable effects of technology on many or most people, the familiar complaints about the
human condition today. Then I shall consider some reasons for
discounting these complaints, in terms of both the compensations
that people enjoy and the powers of resilience and resistance they
exhibit. Finally I shall consider the distinctive effects on some of
the different kinds of people. Needless to add, no study could
possibly be exhaustive, even if it were undertaken by a team of
specialists; not to say that none could comprehend the diverse
experience of every Tom, Dick, and Harry (or of H. J. Muller,
who wrote this book all by himself and could write another just on
his own experience in our age). Again I urge readers to check all
my generalizations against their own personal experience.

The most familiar cluster of complaints is centered on the
pressures to conformity in a mass society. This story begins with
the fact of increasing uniformity. People live in cities and suburbs
that are growing more alike, in factory and office they go through
much the same kind of mechanical routines, they buy the same
standard brands, they drive the same kind of automobiles, they
watch the same television shows, they share the same mass culture.

Now most people—the great middle class—apparently want to be just like other people, are afraid of being different from the Joneses. The independent, genuine individual, prized in both democratic and business theory, is not in fact esteemed in middle-class America. A bizarre example of the reigning mentality was a suit filed by neighbors of a man in a housing development because his mailbox was white with black lettering: they went to law to force him to paint it black with white lettering like all the others.[1]

In his classic study *The Lonely Crowd* (written in collaboration with Nathan Glazer and Reuel Denney), David Riesman described the change in Americans as the beginning of a social revolution. Whereas not many years ago critics worried over the fierce competitiveness or rugged individualism in American life, now they worry over all the togetherness, groupism, and conformism. In Riesman's terms, the "inner-directed" man who came out of the Protestant ethic is giving way to the "outer-directed" man, who gets his goals and his signals from all the people around him, a crowd likewise lacking personal autonomy. He wants above all to be secure and popular, not to achieve power or fame. Aspiring to be liked by all the people all the time, he is "at home everywhere and nowhere." He feels at home only in groups because he is afraid of aloneness. He thinks he knows what he likes, but he

---

1. A more frightening example of the slavish obedience of Americans was an experiment conducted by a Yale psychologist, ostensibly to discover the effects of punishment on learning. He seated a mixed group of forty subjects at a console with a bank of thirty switches supposed to give increasingly painful electrical shocks, ranging from "intense" to "Danger: Severe Shock." The forty were "teachers" in touch with as many "learners" in another room, and whenever a learner made a mistake his teacher was instructed to punish him with a worse shock. Although the learners were not actually getting any shocks, they were instructed to cry out as if they were suffering intense pain, but to keep dutifully calling for more punishment. It was thought that most of the teachers would soon refuse to obey their instructions to inflict more pain, but in fact only fourteen refused; the rest flipped all thirty switches, past the danger signals, even though they were much distressed themselves. I would not make much of this gruesome experiment because of its artificial conditions, and the possibly mistaken respect of the subjects for a psychologist. But no doubt he was right when he explained that it threw light on the many Nazi underlings in concentration camps who dutifully tortured and slaughtered millions of people.

does not know or believe in himself enough to know what he wants. When most pleased with himself he still lacks genuine self-respect. Other social scientists describe him as a mere product of mass society and culture, lacking personal identity. He appears to be strictly the nonentity pictured by Ellul—a creature of, by, and for technique. In the economy he might help as a nobody to fulfill Samuel Butler's prediction in *Erewhon* that man would finally become merely a machine's device for creating more machines.

The loss of "selfhood" brings up another familiar cluster of complaints, centered on the theme of alienation. Harking back to Karl Marx, this theme begins with the insistence that modern technology has alienated man from nature, from his work, from his fellows, and from himself, so much so that Erich Fromm has asserted that alienation is "almost total" in modern society. Among the many symptoms of the disorder noted by psychologists and sociologists are feelings of isolation, frustration, insecurity, anxiety, impotence, emptiness, meaninglessness, and anomie, and in behavior increasing drug addiction, juvenile delinquency, destructiveness, violence, and crime. Harking back to the Romantics, literary people dwell in particular on the reasons why they feel estranged from our society. I should say at once that much of the talk about how alienated "we" all are seems to me not only loose and somewhat mawkish but grossly exaggerated; one might wonder how in a state of almost total alienation so many people were free to diagnose and attack it, and so many more to get some fun out of life. Yet all the talk expresses at least a partial truth, perhaps the most important truth about the human condition in a technological society. I think it certainly important enough to warrant a review at some length of the reputed reasons why people have suffered from some kind or degree of alienation.

Hegel introduced the term to denote the kind of detachment of man from nature simply by virtue of his self-consciousness, but the more specific alienation of modern man from nature was announced by Wordsworth:

> The world is too much with us; late and soon,
> Getting and spending, we lay waste our powers:
> Little we see in Nature that is ours.

In the early industrial towns men saw still less in it; they were cut off from nature as men had never been before. Industry saw in it only something to be exploited, with a ruthless disregard for scenery. With the growth of an urban civilization ever fewer men lived close to the soil or were rooted in any natural locale; the average man became "the man on the street." In modern America people have in some ways got closer to nature again, since many live in green suburbs and more can escape to the country in their automobiles; yet few have a deep sense of its permanent order, elemental rhythms, or the deep connections with man's life that made Wordsworth sense in it "the still, sad music of humanity." And now all the bulldozing and the "ugly-making" along the highways suggest something worse than alienation. I repeat that men have never shown less respect for the natural environment than have modern Americans.

About the alienation of workers from their work, what Marx said is still substantially valid. Most work in factory and office remains "*external* to the worker, i.e., it does not belong to his essential being." It is only a routine "piece" of work, not meaningful in itself; industrialism has reduced workmanship to "labor." The increasing attention now given to "human relations" has eased the mechanical routines but failed to humanize them, for the relations are too superficial or artificial, neither truly personal nor communal. Surveys continue to reveal that most industrial workers still dislike their work, considering it not only boring but degrading. At the same time their unions have shown no interest in the problem of making work more meaningful and satisfying, nor have the workers themselves. Although they may take out their dissatisfaction in wildcat strikes, they do not rebel against the factory system; it appears that they can no longer imagine anything different. The basic problem of making industrial work more humanly satisfying may well be insoluble, since it remains as mechanical in the socialist countries; enlightened Marxists now recognize that it cannot be explained away as simple "exploitation" of workers by capitalists. In any case little is being done about it except to sugar-coat the daily pill.

One possible compensation for the boring jobs workers are condemned to, the shorter work week and the prospect of increasing leisure, is regarded rather as another source of alienation. As

Sebastian de Grazia points out, the villager of the past had no problem of free time; he always knew what to do with it and had traditional, communal ways of enjoying it. The problem arose with industrial man, who had no such tradition. Uncreative or unmanly work makes it harder for people to use their leisure creatively, for they may see in it chiefly a means of distraction or escape from work; unable to find themselves on the job, they may be as unable during their free time. Sometimes they may work as hard at enjoying themselves or feel as harried by the clock— they "don't have time" for a lot of things. More often they are passive, merely taking in. Since they have not been educated for leisure, they are largely dependent on the mass media and the admen for ways of passing time, and their consumption of entertainment may become as compulsive as their consumption of other goods.

Whether all this constitutes an "alienated form of pleasure," as Erich Fromm calls it, or makes people "alienated consumers," seems to me questionable. The term is being used so loosely these days to describe almost any kind of behavior writers deplore that I would prefer to say only that the enjoyment of leisure is too often not spontaneous, active, creative, or deeply satisfying. Still more questionable is the common assertion that modern man is alienated from his fellows or his society. Apart from all the friendly people I have known, most Americans seem too pleased with their affluence to be deeply estranged. If one adds that they are "unconsciously" alienated, one is talking about something different from the alienation of young radicals, Negroes, and other overtly discontented types. And to me the assertion that modern man is now totally alienated "from himself" does not make clear sense—unless one has a much clearer idea of man's "real" or "true" self than I have been able to get from studying history or his life today.

But I bring up this question of fuzzy naming chiefly in order to get at the disorders in American society, which are plain enough. By and large people are not so self-satisfied as they may often seem. Millions of Americans crack up every year, more millions feel the need of tranquilizers to get by. The rates of alcoholism and suicide are among the highest in the world.

Although crime statistics are not so reliable as J. Edgar Hoover would have it, crime has unquestionably been increasing in the cities. So has juvenile delinquency, a clearer sign of revolt against our society. And these are only the more lurid symptoms of a widespread anxiety: not a healthy anxiety over such dangers as thermonuclear war, but a neurotic kind sprung from nameless fears or frustrations. Even the popular press often refers to the "sickness" of America.[2]

Again some historical perspective on the underlying disorders can be helpful. From the beginning the dislocations of the Industrial Revolution meant in human terms many uprooted people, such as the workers in the new industrial towns. The spread of industrialism meant that ever fewer people lived a rooted way of life that might steady them. By now rootlessness is the normal condition of urban Americans, as is their mobility. Although most of them welcome this way of life, they pay some price for it in strains. Immediately their mobility, both geographical and social, forces them to keep on adapting and proving themselves in new communities and jobs, making it harder for them to feel that they "belong"; they have to go on *making* new friends, establishing human relationships that are seldom as satisfying or intimate as they were in the "good old college days" that they accordingly get sentimental about. They are more liable to psychological insecurity because they have little sense of permanence, in which people can form deep attachments. They have no ancestral homestead, most of them no permanent residence; on the average they change their homes every five years. Like their automobiles, many of their possessions will last for only a few years. Under the constant goading of the admen they will keep on buying new things, but despite their pride in their possessions most people are no more deeply attached to most of them than they are to their work.

---

2. When *Life* magazine complained that our novels do not portray average American families that have no alcoholics, addicts, perverts, jailbirds, and other deviants, James Farrell answered that the total of these deviants was greater than the number of families. I do not know where he got his statistics, if any, and suspect a degree of exaggeration; but I also suspect that those "average" families include a lot of people who are not in good mental health.

As Max Weber pointed out long ago, they also live in a *"disenchanted"* world. The process of disenchantment began with the rise of civilization, but it too has been greatly accelerated by the rise of modern science and technology. While science gives us all kinds of knowledge and power, it is "meaningless" in that it does not answer the most important question: "What shall we do and how shall we live?" Religion had answered this question, but it has lost its authority; popular religion gives most people no real certainty for their living purposes. Technology of course gives no answer to the question—it provides only an abundance of means that too often confuse or obscure it. Hence, Weber observed, both life and death may become meaningless. Like Abraham, the peasant of the past could die "old and satiated with life" because "his life, in terms of its meaning and on the eve of his days, had given to him what life had to offer." Modern man, by contrast, may become "tired of life" but not "satiated with life" because he has no such clear idea of its meaning. He knows only that it offers all kinds of goods, many more than he can hope to acquire, and "what he seizes is always something provisional and not definitive."

Or one might say that the affluent American can be satiated with life in the sense of being simply fed up with it. His "happiness" in possessing and consuming too often does not make him feel more alive. Hence the common feelings of hollowness, pointlessness, or what Robert MacIver called the "great emptiness." The most resolute go-getters may get nowhere because they have nowhere to go, get only means without end. On their "success" MacIver quoted the poet:

> "But what good came of it at last?"
> Quoth little Peterkin.
> "Why that I cannot tell," said he,
> "But 'twas a famous victory."

So Americans remain conventionally optimistic, because that is the way they are supposed to feel, but as Erich Fromm suggests, the feeling of many is "a peculiar kind of resigned optimism" that is better described as "unconscious hopelessness." Although they agree offhand that life is real and earnest, for many it is not really earnest. A widespread indifference is typified by the popular saying "I couldn't care less." (Thus thirty-odd people in New York who heard the screams of a woman being attacked in the night on the street, and after a struggle finally stabbed to death, didn't care enough to come to her aid or even to call the police.) "What matters today," said Abbé Pire, "is not the difference between those who believe and those who do not believe, but the difference between those who care and those who don't."

One excuse for such apathy is that the ordinary person is likely to feel insignificant, more impotent than ever before, again just because of the power and massiveness of his society. The powers that make the critical decisions—government, the Pentagon, big business—are all remote and gigantic, beyond his power to influence or even to understand. Too often he has a limited capacity anyway for strong interests, devotions, or enthusiasms, but he may feel that his opinions don't count for anything, they won't affect the major happenings like war or all the bewildering changes going on. And the group pressures against him that weaken his self-respect may make him feel more helpless than he is, afraid to assert himself even when he safely could. He cannot fall back on his work either, or like Candide cultivate his own garden, since he commonly has no garden of his own.

For all such reasons he contributes to the national mood of frustration. This is something new in the history of a people long known for their self-confidence, even their boastfulness. The plainest reason for it is the war in Vietnam, but the country is faced with baffling failures on other fronts too. It was unable to prevent the riots of the Negroes in the cities, the insurrections of students on the campus, the shocking assassination of national leaders. And in a deeper sense it appears to have lost its way. In spite of all the patriotic bluster, it lacks a clear faith in its destiny because it lacks a high sense of national purpose. With most people patriotism is not typically a clear or strong devotion to a cause, only a feeling that America ought always to win or get what it wants, hence a blind resentment when it doesn't. Their feelings of impotence confirm their "unconscious hopelessness." They know only that things have somehow changed for the worse, and that there is little or nothing they can do about it as individuals.

They are still capable of outbursts of passion, however, and because of their frustrations are prone to violent irrationality. A frightening number of them supported George Wallace, a blatant demagogue who preached violence in the name of law and order. Although an old American custom, violence seems more neurotic in our affluent society, as evidenced by the popularity of sadism. Similarly destructiveness has grown blinder and more pointless, sometimes in rage, sometimes just for the hell of it. The relentless kind of impersonal rationality in technology itself produces outbursts of irrationality, nasty kinds of Happenings. And worst of all is the common inhumanity. Although this is as old as history, Erich Kahler suggests that we are now faced with a new kind of "a-humanity," a culmination of the dehumanizing tendencies of modern technology. Thus the "strange, systematic bestiality" of the Nazi concentration camps, which most Americans already appear to have forgotten, just as most have not been horrified either by our own barbarous tactics in war.

So the story goes about the "alienation" of modern man. It goes, I have said, a little too easily, and it may carry one too far from the realities of life in America, or from the thought and feeling of the many people who manage to be decently contented, do not lead "lives of quiet desperation." (Here readers might ask themselves to what extent they recognize all these symptoms in their friends, their neighbors, or themselves.) But before going into the other reasons for qualifying the theme, let us take a closer look at people at home, specifically at what technology has done to the basic institution of the family—an institution that all known societies have preserved in some form all through history.

Again the changes most often discussed have been troublesome. The family is notoriously a much less stable institution than it used to be, and less central in the lives of its members. The decline in its importance began early in the Industrial Revolution when more fathers went off to work, more and more young people left home permanently to seek their fortunes elsewhere, and the whole kinship group was more widely dispersed. In our day family ties have been weakened much more by not only the high divorce rate but the constant shifts in homes, and the young leaving their parents as a matter of course; like so much else in our society, the family is here today and gone tomorrow. The large kinship group exists in little but name; family reunions are rare because their members are scattered all over the country, and few families have a permanent gathering place anyway. Lacking cohesion, the family has less sense of tradition, continuity, and therefore purposefulness, guides especially needed by children. There are no longer close bonds between young and old. In a rapidly changing society the grandparents have no traditional lore to pass on to the young, and in most homes there is no room for them to live. Parents and children gather only at nights and on weekends, with less regularity. While teenagers may join their elders watching TV, they often go out for their recreation, just as they make love in parked cars instead of parlors. Similarly most of the domestic arts and skills that once were handed down through the generations—sewing, canning, making preserves, baking bread, curing meat, etc.—are no more because of mass production. Housewives once proud of their biscuits, cakes, and pies now buy mixes. Family life remains important because children still have to be brought up, but otherwise it no longer seems to many people enough for a full life.

At the same time, bringing up children has become much more of a problem than it seemed in the past. One reason is again a changing world, in which parents may not keep up with the young, or may work too hard trying to keep up. Another reason is the many modern techniques that have replaced traditional ways. Babies once fed at the breast now have "formulas"; their diet is a problem. Mothers are prone to more worry over the techniques suggested by child specialists, who have different theories about how children should be brought up. And though all such problems also mean opportunities for rearing children better, the very wealth of opportunity for young people is a source of anxiety. Parents may fret more over their tastes or their choices in recreation. In the suburbs mothers are likely to worry about their schooling, how well they are doing, because they must go on to college, preferably a good one that is hard to get into. One study of a suburb famous for its excellent school system indicated a high rate of neuroses in mothers and allergies in children.

Another domestic problem that has recently been aggravated is sex. The rapid breakdown of the traditional taboos has led to a degree of frankness and freedom in the young that may disturb older people, but parents too may suffer from a kind of compulsive attention to sex that makes it less healthy, normal an interest. The mass media and the advertisements are constantly making sexuality alluring, reducing it to sexiness, and linking it with other forms of consumption. It has become a way of combatting the "great emptiness," proving one's selfhood and vitality. Once a common theme for humor, it is now for many emancipated people a very solemn matter, especially because of the recent stress on efficient techniques. The ads feature books on the art of love and marriage, methods of improving "sexual performance." These may well be more satisfying than the old untutored ways, and may make for a happier married life; but again many people worry over whether they are giving their partner enough sexual satisfaction, or getting enough themselves. Higher expectations have contributed to the higher divorce rate. One may doubt that the American male has become a great lover, or that technique can guarantee happiness.

Domestic life has not been simply eased either by the commercialism of the American way. This has exploited decent sentiment by making a racket of the ritual of Christmas, inventing a Mother's Day and a Father's Day, creating other occasions for making the purchase of things a necessary way of expressing sentiment. The family may be most closely united as a bulwark of consumption. With affluence it enjoys still less of what Herbert Marcuse suggested economic freedom should mean—"freedom *from* the economy." Although housewives apparently do enjoy the slick service provided by shopping centers, often "personalized" by a smile or the glad hand, the relationships are still not so easy, friendly, and truly personal as they were with the old-time shopkeepers and craftsmen. Many of the surviving craftsmen have been corrupted by the commercialism, if with some excuse because of the obsolescence built into the economy. Garage mechanics and radio and television repairmen have become something of a scandal because more often than not they are dishonest as well as incompetent. Their customers may not feel compensated by the thought that they themselves suffer by missing the satisfactions of good honest craftsmanship, and so become eligible for alienation.

Altogether, domestic life has lost much of its old savor.[3] Homes gleaming with appliances may lack the warm glow of a gracious family life, give little suggestion of a steadying influence. In the city the family is commonly cramped for space, living in apartments designed chiefly for efficiency. In the suburbs it often has little of the privacy that people used to cherish. This is usually its own choice, in the interests of togetherness or organized social life; as a group it too is afraid of aloneness. But the loss of privacy is another sign that family life is no longer so central or sufficient as it used to be. Both parents and children are engrossed in other concerns, and may have in common chiefly colds.

Or so again I often read. Once more, however, it is too easy to exaggerate the decline of the family and the loss that people have suffered thereby. Obviously there are still many devoted families. Husbands are if anything more given to boasting about their wife and kids, displaying pictures, showing off their homes. Children get more attention from their parents if only because bringing them up is regarded as more of a problem. The old rituals of the family circle have been replaced by some new ones, such as barbecues in the backyard. People enjoy other compensations for the losses even apart from all the new comforts and conveniences. That few people have a permanent home means that they are freer to rent, buy, or build better houses, in more attractive neighborhoods. There is more flexibility in domestic arrangements, more give-and-take among the members of the family. Above all, it is too easy to idealize the closely knit family of the past. The close relations were of course often unhappy, in homes that could be like prisons, under the rule of fathers who could be tyrants.[4] In particular many daughters would have

---

3. It may recall the notoriety of American bread as the most savorless in the world. So purified that it lacks any taste or vitamins, bread is then sold under fancy names in wrappers featuring the few vitamins that have been added. I remember one brand years ago that was sold under the name of Butternut Bread until the Federal Trade Commission charged its producers with misleading labeling, pointing out that it contained not only no butternuts but no butter or nuts of any kind.

4. As an example of the patriarchal tyranny he abhorred, Charles Fourier cited Abraham, who sent his wife Agar and her son Ishmael out into the desert to starve just because he no longer desired her.

envied the freedom of young girls today; they pined for marriage as an escape from the home. The domestic skills involved considerable drudgery, routine tasks that the young did not perform with loving care. Housewives were no more given to singing over their sinks and washtubs.

In a historical perspective all the alarm over the condition of modern man needs to be discounted somewhat. Conformity is an old story, inasmuch as people have always tended to an unquestioning acceptance of the ways of the group. In America de Tocqueville and Emerson were worrying over it more than a century ago, before the rise of a mass society. Similarly with the "alienation" of modern man. His alienation from nature may seem less lamentable when one considers the peasant masses in history: working and living close to the soil, they manifested no such love of nature as the Romantic poets expressed, and had a more vivid idea of its elemental menaces to man, as often appears in mythology and folklore. As for the alienation of men from their work, this seems worse because of a novel idea, that work *ought* to be humanly satisfying. Although we can never know how much peasants and artisans enjoyed their work, on the record it was called toiling for their bread, and their betters never maintained that toil ought to be interesting and pleasurable. Like Aristotle, St. Thomas Aquinas declared that "only the necessity to keep alive compels to do manual labor." At least the word "work" has more agreeable connotations than did "toil" and "labor," the more common terms until our century.

The alienation of people from society is distinctly less pronounced than in the past, or at least much subtler. In times of trouble, which were common enough, men were wont to talk like the Egyptian chronicler of four thousand years ago: "Great men and small agree in saying: 'Would that I had never been born.'" Again we cannot know how all the small men felt, but we do know that peasants were often oppressed, city workers were often sullen when not apathetic. However they felt, they had much more reason to feel sorry for themselves than most Americans do. And if life now seems "meaningless" to many people, we should remember that the meaning religion gave it was not always inspiring. Erich Fromm himself declares that "the concept of alienation is the same as the Biblical concept of idolatry," in that "it is man's submission to the things of his creation"; but he neglects to add that popular religion through the ages has always been idolatry. Likewise the laments over the anxiety of modern man can be absent-minded. Although it differs from the brute anxieties of the great majority of poor people through the ages, it is surely no more intolerable than the miseries and the terrors they often had to endure. In the literary talk about our Age of Anxiety it is commonly exaggerated by the implication that it is unique, whereas it is written all over the historic record, most plainly in great creative ages like the Renaissance. In Western history especially, from ancient Greece to our times, no great period has been conspicuous for peace of mind.[5]

Above all, some historical sense can give a fuller realization of not only the extraordinary complexity but the fluidity of the human condition today. David Riesman—whom I am making a point of citing because he is both humbler and more flexible than most of our professional social analysts—has said that he could not emphasize enough "how rapidly our country is changing, and how hard it is even for the best informed among us to grasp what is going on." In *The Lonely Crowd* he admitted that it was hard to be fair to other-directed men, but took pains to point out that they often had the virtues of their defects. He has since found more to say on behalf of "groupism," first of all that its durable achievements made individualism possible. And he has also dwelt more on the remarkable capacity many Americans are displaying in adapting themselves to a fluid society.

In taking up these ambiguities, I would first note an apparent inconsistency in the familiar complaints. "A comfortable, smooth, reasonable, democratic unfreedom prevails in advanced industrial civilization," Herbert Marcuse begins *One-Dimensional Man*. This statement suggests the popular themes of complacence, conformism, togetherness, the "escape from freedom," the utter subjection of people to the needs of technology. But the no less popular

---

5. The Victorian Age was once described—and ridiculed—as ineffably complacent, but recent students of it have dwelt on the many signs of anxiety beneath its surface. Among literary people it has thereby become more respectable today.

themes of alienation and anxiety give a rather different idea of people: it appears that the "democratic unfreedom" is not really comfortable and smooth after all. One might reconcile the two major themes by saying that the unfreedom is only superficially comfortable, or that people have been so enslaved by technology that they think they are happy. I prefer to think that to some extent they resent the pressures against them, have some powers of resistance, and still want to live their own lives, but also that they conform both to convention and to the requirements of technology in part because they have some good reason for doing so, enjoy some clear compensations. The disorders beneath a technically rational order imply that people are not completely dominated by technology, but except in the minds of some Negroes and young radicals, the disorders stop short of revolution because people are not completely alienated by it either. Marcuse asked some rhetorical questions:

> Can one really distinguish between the mass media as instruments of information and entertainment, and as agents of manipulation and indoctrination? Between the automobile as nuisance and as convenience? Between the horrors and the comforts of functional architecture?

My answer is Yes—we can and must distinguish.

The effort is necessary first of all in order to appreciate the compensations, or hang on to the goods that have come with the evils. The impersonal organization that dwarfs the individual makes human relations smoother and conserves energy for potentially higher purposes; an anonymous underling in General Motors is freer to live his own life because he has a limited loyalty to the corporation and would never think of dying for it. The affluence that has created so many new needs has enabled many people to satisfy wholesome desires, as for travel, and can still be an inspiring dream for the many millions who do not yet enjoy it. The goals of other-directed men who seek security and happiness are not necessarily more unreasonable, unsatisfying, or inhuman than the goals of wealth, power, or fame sought by inner-directed men, who could easily be ruthless, or as aggressive as some popular books say man is by nature. The common indifference and political apathy are at least a possible protection against the charisma of demagogues, temptations to hysteria and violence, and perhaps unreasonable demands on people trying to live their own lives. (Paul Valéry observed that politics, once "the art of preventing people from minding their own business," was now "the art of forcing people to decide things they do not understand.") The mobility that keeps people rootless also enables them to feel as free to move on as the early pioneers did, and to escape uncongenial communities or rooted ways of life that would not satisfy most devotees of "organic society" either; it helps to explain why in spite of all the conformism Americans have grown more tolerant and enjoy more freedom in some matters, such as dress, sex, religion, and the upbringing of children. In general, they may be forgiven some complacence because in various ways most people have indeed never had it so good before.

In particular the groupism and conformism are not simply mindless. In suburbia they are a problem just because they involve so much real friendliness and good will, and are in many ways helpful. (Writers who lament the "loss of community" in our mass society might consult the many other writers who complain of too much community in suburbia.) Everywhere they rest upon gregariousness, habits of cooperation and team play, an active pleasure in working with people, and powers of adaptation that have been developed by modern technology. Even Jacques Ellul grants that these "human techniques" have a real value:

> It cannot be denied that this kind of conscious psychological adaptation, which gives the individual a chance to survive and even be happy, can produce beneficial effects. Though he loses much personal responsibility, he gains as compensation a spirit of cooperation and a certain self-respect in his relations with other members of the group. These are eminently collectivist virtues, but they are not negligible, and they assure the individual a certain human dignity in the collectivity of mass men.

Similarly the emphasis on "human relations" in modern industry assures a measure of human dignity. However superficial or artificial, as a means of manipulating people, these relations are surely preferable to the often brutal impersonality of industry in the last century. They can afford to be superficial because no deep understanding of people is necessary to get along with them decently. Neither is their artificiality merely deplorable in a society that in some respects appears to be setting a premium on the virtue of being "sincere." Sincerity can be trying enough on other people to warrant a defense of the right to be insincere. The "glory" of our kind of society, concludes Riesman, "is that it has developed the social inventions . . . which allow us to put forward in a given situation only part of ourselves—which allow us to get along and, usually, not to kill each other while retaining the privilege of private conscience and of veto over many requests made of us by our fellow-men."

Like him, I would also emphasize the flexibility and resourcefulness that people are developing. However intent on being "well adjusted," they have to make many more choices than people used to, in vocation, habitat, consumption, and culture. In a changing society they also have to choose between conflicting traditions and to adjust themselves to the fact of change, become always *adjustable*. The surprising thing is not that many people are afflicted by anxiety or apathy, but that so many manage to feel decently at home. Many, too, are taking advantage of the variety and change by broadening their acquaintance with different possibilities of life and developing new styles of living. The constant change has helped to stimulate a fuller awareness of their condition, or more self-understanding, than most people had in traditional societies. From best-sellers they have learned about the lonely crowd, the organization man, the status-seekers, the hidden persuaders, etc. They can still laugh at themselves too, enjoy satires on the American way of life.

Such awareness may strengthen the individual's powers of resistance to the pressures against him. When Jerome Bruner writes that the last half century of psychology has taught us that "man has powerful and exquisite capacities for defending himself against violations of his cherished self-image," a skeptic might say that this image is not really his own but got from his group, with the help of Madison Avenue. Nevertheless my impression is that many if not most people have some real desire to be independent, make their own choices, and be themselves. The most striking evidence of their powers of resistance is the restiveness in the Communist world, where people are asserting themselves even though for a generation they have been subjected to incessant, massive propaganda. In America such independence of spirit is less apparent because the propaganda of politicians and admen has been much more flattering and the rewards of conformity much greater, but I have remarked signs that people still have some immunity to propaganda, while they can also draw on a much older, deeper tradition of personal freedom. And they may go their own way because willy-nilly they have to make some key choices. In another qualification of his study of the other-directed man, Riesman observed, "We are only beginning to understand the power of individuals to shape their own character by their selection among models or experiences." Again our society offers an uncommonly wide variety of models or experiences.

Hence I would also qualify the complaints about how Americans spend their leisure. Of course they are not all compulsively consuming or passively taking in entertainment all the time. Many people still enjoy gardening, carpentering, or puttering about the house; many develop skills in sports, games, and other forms of play. Many too still do read, just as many are learning to appreciate other forms of traditional culture. And most have hobbies of one sort or another. The commercialized "do-it-yourself" hobbies had a pathetic aspect in their reminder of how many people cannot express their craftsman's impulse in their work, just as the arty vogue of folk ballads and folk dances recalls that these can no longer be spontaneous expressions of a society that has no folk except in a few surviving subcultures; yet the important point is that people are not simply passive or content with mechanical entertainment, there is a hunger in the land for meaningful, creative uses of leisure, many people are managing to express themselves, and a technological society provides an unprecedented wealth of means for doing so.

For such reasons family life too has been enriched in some ways, especially among the younger people. Americans have been developing more sophisticated tastes in house furnishings, food, wine, and other long-neglected amenities. We might all be more grateful to technology simply for our better food, beginning with the much greater variety of fruit, vegetables, and meats available the year round, even if too much of it comes wrapped in cellophane. Supermarkets and delicatessens offer a rich variety of imported delicacies, publishers an abundance of new cookbooks, newspapers all kind of recipes. Housewives now pride themselves on their hors d'ouevres, salads, and casseroles. To be sure, they are not developing a specifically American cuisine as they scour the world's cuisines, and they are likely to be too arty and self-conscious about their dinners, served in candlelight. Riesman believes that the anxieties in suburbia are due in part to the exploration of new frontiers in leisure. But it is something that people are exploring, as in culture too.

Well, readers must judge for themselves what all this adds up to—how much people have changed, and whether the changes are on the whole for the better or for the worse. Simply because I believe that people have not been hopelessly enslaved by technology, and the prosperous should still have the capacity to be full-fledged individuals, make their own choices, and live like free men, I am inclined chiefly to deplore the choices that most of them have made or failed to make. Certainly Americans have not by and large taken full advantage of the opportunities opened up by technology. As certainly they are not yet rising as a people to its challenges; at a time when a new national administration is seeking only to consolidate the small gains made on the domestic front, offering no bold programs or demanding no real sacrifices of the prosperous, I would repeat that they could do with more anxiety. But I say "as a people" to bring up the final complication —the different effects of technology on different kinds of people.

First of all, the charges of conformism, groupism, etc. do not apply to the many millions of poor people. (How many millions there are depends on how one defines poverty, but up to thirty millions have been officially classified as poor.) The "unfreedom" of those in the rural and urban slums cannot be called democratic. Although middle-class Americans still look down on them as lazy, improvident, addicted to drunkenness, crime, and sex offenses, as no doubt many of them are, most have never had a fair chance.

Industrialism, which slowly came to the aid of the poor in the last century, has been harder on them in our day, denying many men the dignity of being breadwinners because there is less room for poorly educated people in modern industry, not to mention the many employers who discriminate against Negroes, Puerto Ricans, and Mexicans in any case. As the New Deal did relatively little for the poor, so the misnamed "war" on poverty is doing little more. Welfare programs, administered by middle-class people who have prospered on them, have notoriously failed to relieve the problems of poverty, while never allowing the poor to forget that they are receiving charity.

The recent revolt of the blacks has obscured a significant change in the poor. The millions of immigrants who came to America to make a better living generally had a hard time of it, but they retained a hopeful spirit; most of them felt that at least their children would come up in the world. Today most people in the rural and urban slums no longer feel such hope. If they are physically better off than the immigrants, they are worse off spiritually, and not merely because of all the signs of affluence around them. They know they are on the bottom and feel they are doomed to stay there. They are not so much resigned as apathetic, feeling they are simply out of it. Gunnar Myrdal has therefore called them "the world's least revolutionary proletariat." I think Mexican Indians, among others, might compete for that distinction, but the American poor have earned it because so many now lack the celebrated American spirit. The embattled Negroes who do have this spirit and are standing up for their rights represent another significant change, although not one relished by most middle-class Americans. These can remain indifferent to the whole problem of poverty more easily because the millions of poor whites remain apathetic.

The wealthy people at the opposite extreme represent a unique distinction of modern America, that unlike virtually all the civilized societies of the past it has no leisure class. The aristocrats in traditional societies felt no such obligation to work as do most of the wealthy and their sons in America; pure playboys are as rare as pure gentlemen of leisure. But the gospel of work has obscured still another distinctive development in America, the rise of what Kenneth Galbraith has called the "New Class." This is made up of the many kinds of professional people who enjoy their work, and identify themselves primarily with their job instead of their income or their social status. Although there have of course always been such people, the class is "new" in that it is so large and is growing so fast, now numbering in the millions instead of the thousands of a century ago. These people have their complaints about some routines, as everybody does, but they are by no means alienated from their work. Despite the prospects of increasing abundance and leisure they can be expected to go on working hard because they find their work satisfying. They almost never leave the class voluntarily. "No aristocrat," writes Galbraith, "ever contemplated the loss of feudal privileges with more sorrow than a member of this class would regard his descent into ordinary labor where the reward was only the pay"—what most businessmen still think is the chief incentive.

As one who is pleased to be a member, I proceed dutifully to introduce the usual complications. In an issue of *Daedalus* devoted to the professions, the editor remarked that it was extraordinary how little we know about them, considering the extent to which our society is now professionally run, and that otherwise we might be more conscious of the "crisis" that now confronts them. This again seems to me too strong a word for the problems confronting professional people as such, but at least they too are subject to frustration. Although traditionally they were "retained" or "consulted," not hired, in fact most of them are hired by business and government, in the service of which they may be harried by other professionals, such as politicians and business executives. I have noted some of the problems confronting scientists, engineers, industrial designers, architects, city planners, and others. Their different complaints also emphasize the motley character of the "new class," which lacks a common tradition and *esprit de corps*. Ideally, as Galbraith suggests, it might help to set the educational goals for a society in which so many people do not deeply enjoy their work, but actually it is not qualified for such purposes. For professional people are specialists, often narrow ones. Typically they support the tyranny of intensive specialization in the universities, while the many technicians among them remain too often indifferent to the values of traditional culture.

Another problem arises simply from their prosperity in a technological society that has produced so many of them. The older professions have had some tradition of devotion to the public interest, and in so highly commercialized a society this interest may suffer. Doctors in particular have stirred considerable complaint. Their AMA, the tightest union in the country, has been about the most conservative, but in defense of the professional interest, not the Hippocratic oath. Its long, bitter opposition to anything like medicare could look selfish when so many millions of Americans were unable to afford proper medical care, which was getting so much more expensive, and American health standards were falling well below those of European democracies. Its criticism of the government was more questionable because it remained uncritical of the drug industry, whose advertising provided more than half its income. And its standard argument of the supreme importance of personal relations between doctor and patient made less sense because of another typical development: the old family doctor has been giving way to specialists, banded in clinics. Whereas he responded to calls in the middle of the night, his successors are much less willing to pay house calls at any time. It is perhaps unreasonable of pampered Americans to complain because doctors too now have an easier life, but in any case the old tradition seems doomed. Medicine too is growing more impersonal in a technological society.[6]

Since I could not cover here all the many professions, even if I knew enough about them, I shall add only a brief comment on a characteristic modern one that is a possible exception to the privileges of professional people—the people who produce the programs and the commercials for radio and television, or perform in them. As talented people they constitute a creative elite of sorts, but a subservient one distinctly unlike the elites of the past. Many of them no doubt share the tastes of the public they cater to, or

6. In their conservatism medical men were likewise inclined to hostility to a quite different modern development, the entrance of psychoanalysis. Even so it caught on much more quickly in America than in Europe, for reasons not altogether clear; but I suppose its popularity is now symptomatic of the underlying anxieties of an affluent society. Psychiatrists have acquired some reputation for often being neurotic themselves.

are at least content to sell their skills for good pay. (Liberace—or his press agent—once said he was so hurt by the unfavorable reviews of critics that he cried all the way to the bank.) Some are simply cynical, contemptuous of the boobs they address; they belong with the too many successful Americans who cannot really believe in their life's work. Others, like the TV playwrights of some years ago, quit in disgust. But whatever their attitude, few can work as individuals, speaking to other individuals. They have to turn out standard products, agreeable to their business sponsors. Insofar as they are creative, they are rarely allowed to develop and express their talents freely, as artists can.[7]

Intellectuals, who cut across the professions, have become a more amorphous class. Those who have prospered most are the highly professionalized types, the technical specialists. Zbigniew Brzezinski notes with approval a "profound change" in the community, in that "the largely humanist-oriented, occasionally ideologically-minded intellectual-dissenter, who sees his role largely in terms of proffering social critiques," is rapidly being displaced by "organization-oriented, application-minded intellectuals," a new elite under whose leadership America has become "*the* creative society." The "displaced" types naturally take a dimmer view of this change. As Conor Cruise O'Brien observed, "Power in our time has more intelligence in its service, and allows that intelligence more discretion as to its methods, than ever before in history"; but he fears that the outcome may be "a society maimed through the systematic corruption of its intelligence." If the new elite is not actually corrupted by its eminence, at least it is not inclined to be highly critical of the powers that employ it. And its professional devotion to method or technique raises the usual questions about its ruling values, or the measure of its wisdom.

---

7. Because the "organization man" has become so familiar a type, I shall not go into the price he commonly pays for success in the "rat-race." But a writer in the *Harvard Business Review* deserves mention as a spokesman of the values of this semi-professional community. He warned wives that they should not demand too much of their husband's time and interest: "Because of his single-minded devotion to his job, even his sexual activity must be relegated to a secondary place."

Literary or humanistic intellectuals are harder to classify, but exhibit some common tendencies in their still active role as critics of our society and culture. Of all professional people they are the most liable to some degree of self-conscious alienation, the term they have popularized. For one thing they are acutely aware of the various crises. Paul Valéry's remark after the shock of World War I still describes their feelings: "We hope vaguely, but dread precisely; our fears are infinitely clearer than our hopes." Their fears are more acute because both our experience in this century and our increasing knowledge of man and society have tended to weaken faith in the power of reason, their stock in trade. But the plainest reason for the alienation that many profess is their feeling that they are a lonely, beleaguered minority, in a society hostile to them; and in my opinion this feeling is not clearly justified. They are not really being displaced by the technical intelligentsia. They arouse hostility, especially in business and political circles, just because they have grown more numerous, have a larger audience, and have more apparent influence. And they themselves still tend to profoundly ambiguous attitudes because of their unhappiness over the culture of the great majority who do not share their interests and tastes, or their anxieties. As a self-conscious elite they may manage at best to be patronizing.

So I repeat that the popular concepts of mass society and mass culture are too often crude. It was not in any case the "mass-men" who created the crises that worry intellectuals, but they include all kinds of people. I am still generalizing very roughly as I now consider, lastly, some much broader classes of people.

Perhaps most pathetic is the state of the old. Relatively more numerous because of the longer life-span of modern man, they are among the chief victims of a technological society. True, many of them remain alert and independent, lead a full life to the end. All but the most wretched are at least better off than the laborers of the last century, who went to work as boys and often were worn out by the time they were forty, long before the present retirement age of sixty-five. Now government has begun to provide for the care of their frequent ailments. Yet most of them are poor. Only a small proportion have private pensions to supplement their social security benefits; their average income from all sources is well below the minimum that the Labor Department has set for a "modest but adequate" living. Many do not welcome the retirement that is forced on them, for they feel lost without a job and are unable to find other work. Commonly they lack the intellectual resources that might sustain them. Worst of all, they are likely to feel like outcasts, discarded by a society that has no further use for them. As one old lady told Adlai Stevenson, "We just want to be wanted." Because of the rapid change they lack both authority and responsibility, have no wisdom to pass on to the young. Usually they live alone; on a recent poll an overwhelming majority of Americans said they opposed the idea of old people living with their children. At best treated with a patronizing affection, the old are too often made to feel they are simply a burden. They suffer from a distinctly higher rate of mental disorder than do the rest of Americans. Except for the poor in the slums, no class of Americans is more "alienated."

The status of women, who have also been described as alienated, seems to me much more ambiguous. Most obviously technology has eased their domestic duties by providing them with a wealth of canned and prepared foods, electrical appliances, laundries, dry-cleaners, diaper services, all manner of services. In suburbia they preside over a new kind of matriarchal society, while also taking charge of the consumership. This "role" keeps them pretty busy, however, and its mixed blessings involve the usual costs. Housewives no longer have servants, at most only a cleaning woman to come in once a week; they might be glad to exchange all their electrical appliances for one good family servant of the old school. (From my experience in Mexico and Turkey I know that American women work much harder than middle-class women in these countries, who have servants as a matter of course.) As for their "free" time, they may fill it with not only bridge but club work, community service, meetings of parent-teacher associations, etc.; and they may or may not find such activities rewarding, just as continual social life may become a burden. They may find that they "don't have time" for things they might enjoy more. The reports I have read of their life in suburbia seldom suggest that it is exhilarating.

And so with a more fundamental matter, the freedom that

women enjoy, now including control over how many children they choose to have. (Most have their last child at a distinctly younger age than women used to.) How much happier is the emancipated woman? Although I don't think "alienation" is the word for most of them, I assume as usual that it is an open question, no less because I approve of the emancipation. With the breakdown of the traditional inhibitions women in particular can more fully enjoy sexual experience, but they may not enjoy the promiscuity many indulge in, sometimes as a sort of obligation of modernity; and it is not clear that marriages are happier. The high rate of divorce indicates that women are freer to escape loveless marriages, or it may indicate a too restless pursuit of happiness, misguided by the assumption that sex is a woman's main business. In any case, the new freedoms are still too self-conscious to be easy and natural, and they are still confused by old ways of thinking and feeling. Women who want careers soon discover that they have nothing like professional equality, while running into further problems if they want children too. It appears that most of the millions of working wives want only jobs, not careers, and are not feminists seeking more freedom; but their jobs complicate the problem of running their home too, managing as wives and mothers.[8]

As young mothers women work harder because technology has freed them from many household chores, which used to keep mothers so busy that they left children on their own. Even those who are well off have no nurses or governesses, of course, but most would not want them anyway. David Riesman writes that they have gained "the right to enjoy their own children." I would repeat

---

8. They might not be helped much by the findings of a professional researcher: "Sociologists studying the family . . . have pronounced sex to be a universally necessary basis for role differentiation in the family. By extension, in the larger society women are seen as predominantly fulfilling nurturant, expressive functions and men the instrumental, active functions." It followed that "intellectually aggressive women" showed signs of "role conflict" and "role confusion," but many others may be confused in a changing society that has not actually agreed on any one predominant role. At least, however, they may rejoice in their sociological respectability as creatures having "functions" and "roles."

that often they have also acquired enough anxiety to mar this enjoyment, and perhaps to infect their children with it. They have learned from child specialists about baby feeding, toilet training, thumb sucking, and other worrisome matters having to do with early character formation. They may be too fearful of leaving their children on their own, or on the other hand of repressing them too much. (I write as a parent who belonged to the George Bernard Shaw school: Never strike a child except in anger.) If working outside the home, they may worry too because they have heard that leaving children with other adults may make them insecure. All such worries are aggravated by the common assumption that when anything goes wrong, it is always the parents' fault.

Hence I am brought to my final concern. What are these children like who have been getting so much more attention than they used to? What are they like as teenagers, when technology does so much more for them directly through the service economy, industry, and the mass media? What in general is the state of our youth, who are inheriting the opportunities and the messes created by their elders? The question is more important because they have grown so self-conscious and self-assertive as a class.

Now, I think there has been too much stress on youth as a problem, and too much generalization based on the minority who clearly are a problem. I also find it hard as always to size up their state. We indeed know far more about the subculture of youth than people once did, if only because it is the most advanced and distinctive in the world, but it is not so homogeneous as the familiar descriptions often make out. "Isn't it easier," asked Reuel Denney, "to call to mind a typical young person in the United States today than it would have been when I was between twelve and nineteen, from 1925 to 1930?" Up to a point it is easier: we all talk about a new type, the teenager. From there on, however, it is not at all easy. It makes a great deal of difference whether one is talking about the children of the many poor or the children of suburbia. With teenagers I am struck by the usual ambiguities. They are commonly described as conformists in their own culture, potentially like their parents, in particular as ardent consumers; but there is also talk about their lack of deep commitments to adult values, sometimes their rebelliousness. Once they are in college I do not know what a "typical" student is. And the youth too have been changing under our eyes. In 1962 Kenneth Keniston, a sensitive observer of college youth, stressed "the decline in political involvement" among them. Today such involvement is more common than I have ever known it to be.

At any rate, I begin with the children of the poor, who are easiest to generalize about because they get no more attention than they did in the past. As teenagers in the city ghettos they suffer from unemployment; while their number has considerably increased, the number of jobs available to them has gone down. They have a harder time because the handicaps of their poor background are compounded by poor schooling, another example of how much more aid our government gives to the prosperous or even the rich than to the poor. Many of them drop out of school, which apparently serves chiefly to keep them penned up in a society that has no use for them, and so find it still harder to get jobs. Since the high-school diploma has become a requisite for jobs with a future, what is available to them is chiefly menial, poorly paid ones. Their resentment of these makes them more prone to juvenile delinquency; only as "tough guys" or members of a gang can they maintain self-respect. Then they become conformists to the obligations of nonconformity. No doubt most of the slum teenagers settle down to steady jobs as they grow up, but they have a poor chance of realizing the middle-class myth that all Americans can hope to come up in the world if they only work hard. As a class they are clearly liable to alienation. To add that they are statistically a small minority is not a high tribute to America.

About the children of middle-class Americans the first thing to say is that they owe to modern technology their childhood of play and school, goods enjoyed by relatively few youngsters in pre-industrial societies. They have many more games and toys, including mechanical ones that help boys to develop a suitable fondness for machines, their schooling is supplemented by a vast children's literature, and they are provided with much more entertainment by the mass media. At an early age they begin to grow sophisticated as consumers; appropriately the boys wear

trousers long before they grow up into young men. If by traditional standards the children are spoiled, they are at least being
prepared for life in a society of abundance, where the traditional
wisdom based on scarcity is not clearly appropriate. For the same
reason it may not be simply regrettable that the chief influences
on them tend to make them other-directed persons, or to stress
more the acquisition of "social skills" than the development of
personal character; as early as nursing school the stress is on their
"socialization," getting and keeping them "involved" with one
another. But if, according to Marshall McLuhan again, the
"electronically processed data and experience" provided by television make for deep or total involvement, others remain concerned over the more apparent influence of the content of the mass
media, which may give children their ideas about the world, and
their styles of both speaking and living. The youngsters might
even enjoy more the stories that grandparents used to tell them or
parents read them. They do not seem as happy as one might
expect in view of all their advantages, and sometimes seem as
restless or bored as grown-ups.[9]

In spite of the modern emphasis on the supreme importance
of the earliest years in character formation, I think it more helpful
for my purposes to look at the youngsters as teenagers. At this
stage their basic character (however formed) becomes more
apparent and the social influences on them more marked. As a
historically novel type they more consciously exploit the resources
of a technological society, more fully reveal what it has done to
them. And they begin to decide about their future. Edgar Fried-
enberg has defined adolescence as "the period during which a
young person learns who he is, and what he really feels." Erik
Erikson sees it more specifically as "the search for something

---

9. While I was finishing this book in Taxco, Mexico, I had as
neighbors lower-middle-class families who had half a dozen small children I enjoyed watching. The Mexican youngsters had almost no toys,
but were almost always smiling or laughing, amusing themselves all
day by forms of play or make-believe of their own devising; rarely did
one sulk or cry. My impression was that they were distinctly happier
than most American youngsters. My occasional presents of candy were
a real occasion for them, in which they were also happy to trot out
their two words of English: "Mucho thank you."

and somebody to be true to." In the city ghettos, once more, the adolescent may be discouraged in his search and find it hard to be true to the avowed ideals of a society that denies him opportunity; but most teenagers of the great middle class have real choices and abundant means of achieving their heart's desire. To aid them in their search they have high school and the prospects of college, the resources of a popular culture that idealizes youth, freedom from the need of earning their living or helping the family out, and big industries to gratify their wants.

They are least engaging as consumers, with new wants that have become needs. At last report teenagers were spending up to $20 billion a year on things they "had" to have or were "dying" to own, in addition to the billions spent on them by their parents. One sign of the new times, shortly after the World War, was the shift of publishers of pulp magazines to slicks, *Mademoiselle*, *Mademoiselle's Living*, and *Charm*. Another sign was the shopping list of a twelve-year-old girl: "water-pistol, brassiere, and permanent." Among boys Erik Erikson sees the key to their restless search as "the craving for locomotion," indicated by such expressions as "being on the go," "tearing after something," and "running around," and also by locomotion proper, as in sports and dancing. Above all most crave an automobile. They look forward less to becoming twenty-one than to reaching the age when they can get a driver's license; this marks their passage into manhood. For many it is no longer an exciting passage, but only the satisfaction of an ordinary need.

While teenagers have grown much more sophisticated as consumers, their goals in this capacity are seldom richly rewarding, and tend to be to some extent self-defeating. They want to have fun, of course, but they now need more things to have it and may be discontented because they can't have everything they are dying to have. The girls also want to be popular and glamorous, the boys to be popular and virile, but many are naturally doomed to disappointment. Intent on keeping on the go, they may not ask where they are going and why, but just keep running around like everybody else. In this respect most of them have their own way enough so that I think they cannot be called "alienated." But I also think they too are not so happy as they ought to be in view of all their advantages. A particular reason is again the pressures to conformity. These make it harder for them to learn who they really are as individual persons, what they really feel, and what they need to realize and express themselves.

Although young people have always tended to conform to the ways of their group, the pressures on them are now stronger because of a premature, shallow sophistication. Organized social life, in imitation of older people, begins at a much earlier age. High schools have fraternities, sororities, beauty queen contests, formal dances, and the like, which leave many unhappy outsiders or losers in the social game. Even in their early teens youngsters are prone to anxiety over their popularity with the whole group. Girls at this age may feel obliged to "go steady" with a boy friend, else they will be thought unpopular or queer, and young boys must also prove themselves in the same fashion. All must have the right tastes and interests, wear the right kind of clothes, use the right vocabulary. And all such pressures are strengthened by most parents, who are likely to worry if their children have unusual interests unless they have the right ones too.

Nevertheless the teenagers are not simply aping their elders or learning from them who they are. They pride themselves on doing their own thing or having their own culture, which is to some extent shaped by the creative or sensitive spirits among them. For various reasons many refuse to commit themselves fully to the values of their parents, preferring to stay "cool." They want to be independent, "different," even if they then conform to the latest fashion in independence and are all different in the same way. In a pluralistic, changing society they cannot in any case simply take over the wisdom of the fathers, but have to learn some things for themselves; among their favorite expressions years ago was "That's what *you* think." Although as a parent I did not always relish this, I did appreciate their sense that the conventional wisdom they hear may include antiquated or irrelevant ideas, as about the virtues of rugged individualism—virtues they may not recognize in the fathers. The older ones are often in some respects better informed and more realistic than their parents, more aware that their society is rapidly changing. For all their apparent unwillingness to assume adult responsibilities, the

teenagers often strike me as surprisingly thoughtful and level-headed, less prone to a commitment to simple solutions. (I have known a number of spoiled brats who grew up very well.) "In certain ways," Reuel Denney concluded, "modern American young people seem to walk on eggs more than any generation in the twentieth century. Their talent for the 'delayed reflex' may prove to be one of our main resources in the coming culture and politics of the nuclear age."

Because of their "coolness" most of the young people do not appear to be overtly rebellious against the older generation, and indeed might feel more exhilarated if there were a distinct conflict. Most accept the basic values of their parents in their own version, and find it easier to do their own thing because the parents tend to be compliant. They can tyrannize over the parents because so many Americans pride themselves more on being youthful in both looks and spirit than on being mature. Yet the middle-class teenagers have also been undergoing a significant change in recent years. Today many do appear to be rebelling, in various ways that distress their parents. They are dropping out of school and college. Juvenile delinquency is on the rise in suburbia too, with much pointless destructiveness. Many young people from the best families have taken to smoking pot, experimenting with other drugs. They have rejected parental authority by sexual adventures as well. Some have been leaving home to join the beats, hippies, or yippies. The symptoms of revolt are so widespread that we now hear of an "alienated generation."

As I see it, much of the apparent revolt is due to the self-indulgence of pampered teenagers, who have searched for little but new sensations or thrills. Many who kept running around are now running away from themselves, or are in search of a self by the latest fashion in "turning on." Yet the change is significant because many are rebelling for serious reasons. In spite of all their freedom they feel trapped, thwarted by the social system. They are dropping out of school in something like the feeling suggested by Paul Goodman, that they were "growing up absurd." They are repelled by the too frequent hollowness and hypocrisy of the fathers' way of life, the lack of high goals or ideals of personal honor. They are seeking something more meaningful and challenging, worthy of their loyalty.[10] Erikson noted that their craving for "locomotion" also found expression by participation in the movements of the day, which now include radical movements. They want something or somebody really good to be true to. Most of the rebels are vague about their ends, and may be deluded about their means, but this is only to say that they are still very young. The important thing is that they really care, about something more than the automobiles to which many others give their most loving attention.

To repeat, most middle-class teenagers appear to be basically satisfied with themselves and their prospects, by no means alienated from their society. They are prepared for a technological society by their acceptance of the common belief that what is good is what works, with people as with machines. Most of the millions who go on to college remain complacent and not too much concerned about ethics or ideals. (When the hero of the rigged quiz shows on television was exposed, the overwhelming majority of students polled at his university sympathized with him, and otherwise couldn't care less about the scandal.) Having already considered the various types of college students in my chapter on higher education, I now add chiefly that the revolt of the serious teenagers has become much plainer and still more significant on the campus, in part simply because of the greater maturity of the students, but also because of restlessness caused by the many years they have to spend in preparing themselves for life in our kind of society when many feel like taking an active part in it.

With these older students who want to be involved and committed, not merely "cool," the common vagueness or uncertainty about ends and confusion over means is perhaps less forgivable. "Higher education," said Tom Hayden, "is confusing the teaching of subject matter with the learning of subjects that matter"; but

---

10. The kind of values they may resent was expressed by a suburbanite who was angered by the bad publicity his exclusively white community got because of the arrest of some teenagers on narcotic charges. "This is a damned fine town," he told a reporter. "We've got Kiwanis, Rotary, and Lions here. . . . Our football team was tops in eastern Connecticut—right up there with the leaders. These kids were just kooks."

those of us oldsters who agree with him might therefore wish that the ardor and energy of the rebels were focused on the critical national and international problems created by modern technology, with clearer ideas about a new order, and less stress on the need of wearing beards and using the four-letter words. Still, the radicals won the sympathy of many more students after the brutal violence of the police, the guardians of the old order, who at the 1968 Democratic National Convention in Chicago further alienated young people. And the stir on the American campus may seem more heartening in the light of the world-wide student revolt. The rioters in other countries (especially France) had plain academic or economic grievances. White American students had no such serious grievances, but enjoyed more advantages and better personal prospects than they ever had in the past. Again the important point to me is that so many of them have come to care for more than their immediate selfish interests.

It is just possible, I think, that the future may belong to them. But since again I cannot say confidently what all these changes in the youth add up to, or point to, I conclude this chapter with one simple implication of the changes in people generally. Human nature remains as plastic as it has been all through history. Even ordinary "simple" people are much more complex than most politicians, advertisers, and social scientists appear to assume. As Montaigne said, "There is as much difference between us and ourselves as between us and others." What "people, just people" will be like in another generation is anybody's guess; but mine is that they will still be all too human.

IV

Toward the Year 2000

19

A Note on Utopia

The dream of a just society, wrote George Orwell, "seems to haunt the human imagination ineradicably and in all ages, whether it is called the Kingdom of Heaven or the classless society, or whether it is thought of as a Golden Age which once existed in the past and from which we have degenerated." In our civilization the idea of progress led to a novel utopianism, the conviction that the ideal society was positively going to be established on earth. Among the reasons for the spread of this popular conviction were the hopes stirred by America, a New World in which the common people could at last realize their dreams. And thereby hangs still another paradoxical tale. Today our fabulous technology gives us the means of doing almost anything we want to do, which in popular journals inspire many a vision of a future gleaming with new wonders. Yet in our time there has been a profound revulsion against utopianism, in which Orwell himself shared. Aside from scientists, most intellectuals in the democracies—the natural habitat of dreams of a just society—are not only disillusioned but often hostile to utopian dreams. They do not deny that modern technology can perform all the feats men predict—they fear it will. Instead of utopias we get Huxley's *Brave New World* and Orwell's *1984*, "the future as nightmare."

This whole development is only superficially paradoxical. Simple common sense makes it impossible for most of us to believe in anything like a perfect state for man on this earth, but in view of our experience—all that I have surveyed in these pages—the revulsion against utopianism is not at all startling. It is still easier to understand when one takes a hard look at the modern utopias, down to B. F. Skinner's *Walden Two*. Still, it is trouble-some. As George Katep points out in his admirably balanced study of it, *Utopia and Its Enemies,* it has involved dangerous antisocial tendencies. It is occasionally so extreme that it threatens to kill even reasonable hopes for a simply better future, and to discourage the always necessary effort to make it better. We cannot afford to lose all dreams of a just society, lest the injustice that has always been with us have a still easier time. Or if we are going to "think about the unthinkable" we may need to dream simply to preserve our sanity. The real challenge remains that we do possess the technical means of doing almost anything we have a mind to, short of making angels of men. Today wide visions of ideal ends are more necessary than ever before.

Now, utopianism is by no means dead. It is most clearly alive in the Communist world, which we should remember still claims the future. Although European Communists seem considerably less fervent than they used to be, I assume that most of their serious thinkers still cling to something like Marx's dream of a classless society. In America the indignation of many intellectuals over all the poverty and social injustice in the land, the treatment of the Negro, the war in Vietnam, and other contemporary forms of ancient evils implies that the evils are unnecessary. Some have a more explicitly utopian mentality, notably Lewis Mumford. The American Dream is still a living inspiration, as in the call for a Great Society. Yet utopianism is clearly out of fashion—except for the visions of more technological wonders, which are not dreams of a just society. In the world of affairs it is countered by the inveterate tendencies to a narrow, short-sighted, usually selfish realism, a practical spirit that for civilized or simply humane purposes can be profoundly impractical; so the Great Society has become a theme for scorn. In the world of thought pessimism is grounded on a much more clear-sighted realism, but it also springs from a possibly excessive fear of being gullible or simple-minded. The widespread indignation is mostly bitter, accompanied by little of the generous enthusiasm of the utopians of the last century.

Hence we might do well to take a closer look at these

utopians, who led Karl Mannheim to assert that without the utopian mentality "man would lose his will to shape history and therewith his ability to understand it." They were not so utterly foolish as they may now appear. Let us go back to Saint-Simon, who in 1814—the year before Waterloo—hailed the new era:

> The imagination of poets has placed the golden age in the cradle of the human race. It was the age of iron they should have banished there. The golden age is not behind us, but in front of us. It is the perfection of social order. Our fathers have not seen it; our children will arrive there one day, and it is for us to clear the way for them.

This vision was inspired by a prophetic insight into the possibilities of industrialism, still in its infancy. Saint-Simon heralded a radical change in the familiar visions of an ideal society, which from Plato on had usually been of a more or less austere way of life, reflecting the actual scarcity in all past societies that bred the traditional wisdom of learning to do without. He rightly foresaw an era of increasing abundance, in which all human needs could be amply provided for. Whereas the utopias of the past had been a-historical, typically located on an island, he set his squarely in the mainstream of history. No idle dreamers, he and his followers set to work to "clear the way" for their children. They had their eye too on the whole society and thus avoided the futility of the utopian communities in America in the nineteenth century, the quixotic effort to realize ideals of perfection in islands within a very different society. To Karl Marx Saint-Simon belonged with the various "utopian" socialists he attacked only because this golden age was to be entered by peaceful means, with the help of the bourgeois.

As a realist, Marx too nevertheless banked on the productivity of bourgeois industrialism for the sake of his classless society. While uniting utopianism with the demands of the poor and oppressed, and giving a philosophical authority to the dreams of the millennium harking back to the Middle Ages, he more emphatically rejected austere conceptions of the good society. Even the radical proletarian movement of Babeuf in the French

Revolution he described as "reactionary" because it "inculcated universal asceticism." His classless society would assure perfect equality and freedom more easily because of its material abundance, which would preclude the social injustice that had always resulted from scarcity in civilized societies. In his own words, capitalists had at least developed the productive forces of society and created "those very material conditions which alone can form the real basis of a higher form of society, a society in which the full and free development of every individual forms the ruling principle." Believers in democracy might consider whether there is a better guiding principle for affluent America. Marx also backed his revolutionary program by a comprehensive analysis that remains useful as a "model" even to those who do not accept his basic doctrine of historical materialism or economic determinism. Unquestionably he threw much light on Western history, in particular on major developments not consciously planned or willed by men; so he made it possible to deal more intelligently with economic forces that had been operating blindly.

Engels himself admired Charles Fourier, a utopian long neglected, especially in America. He was neglected because he was a primitivist, who in the dawn of industrialism proposed as an ideal an agricultural society centered in villages of just 1,620 people, but in other respects he was a surprisingly advanced thinker. He sought to give "industry" its best meaning by making work gratifying, avoiding the monotony as well as the misery in store for industrial workers. To prevent the curse of boredom, he likewise stressed the primary need of variety, as would John Stuart Mill for the sake of liberty. In his fear of trends to uniformity and conformity he attacked the authoritarianism that would appear in most utopias, and in the states set up in the name of Marx. Hating especially the tyranny of patriarchs, he attacked as well the institution of marriage, even to maintaining the right to commit adultery. In this respect still an advanced thinker for our age of sexual freedom, he more clearly anticipated the wave of the future by arguing for the complete equality of women, concluding—like Engels in his study of the family—that "the extension of the privileges of women is the general principle of all social progress." He was as "modern" in his hatred of the hypocrisy resulting from a denial of the facts of sexual passion.

Recognizing that man is a creature of passions, he was utopian in his belief that none of them were naturally evil, there were "only vicious developments" when they were choked and turned into "counter-passions," such as jealousy; but he anticipated Freud's concepts of substitute gratification and sublimation in his efforts to assure both a release and a harmony of the passions. Although the future would belong to the industrialism Fourier spurned, it would also vindicate the judgment of Engels that there was more of value to be found in his thought than in the school of Saint-Simon.

In our own century the most indefatigable prophet was H. G. Wells, who George Orwell said had more influence on young men of his generation than any other writer between 1900 and 1920. His utopian visions were based on science: "the shape of things to come" was not only a host of scientific wonders but a society run by a scientific elite. In *A Modern Utopia* Wells met more squarely than most dreamers before him the basic issue of the nature of man by explicitly rejecting the doctrine of Original Sin. Most of us might agree that they all had an excessive faith in human nature; but now that Original Sin has become popular in some sophisticated circles, while realists insist on the ineradicable evil at the heart of a selfish, aggressive creature, they may obscure the simple truth that many social evils are to some extent remediable, and many social reforms now accepted as a matter of course were once considered wildly utopian. Like Marx and other utopians, Wells showed more concern for the common people than have most of the greater modern writers.

He was not the naive optimist of popular repute, however, but like Orwell an ambiguous figure. Because of his high hopes for man, Wells had darker fears for the future. He attacked the popular notions of an automatic progress, foreseeing that the new scientific wonders could be horrors. In such novels as *The Time Machine* he anticipated the horrors of much science fiction today, and as early as 1914 he even wrote about "atomic bombs" that would be used in a war in the 1950's. Long before he recorded his absolute despair in his last work, *The End of the Tether*, Wells was a major source of the revulsion against utopianism. And this too calls for a closer look.

In our time the plainest and soundest reasons for the revulsion are centered on the historic outcome of Marxism. While Marx himself sometimes hedged on the necessity of violent revolution, usually he insisted on it, if only to support his denunciation of the "utopian" socialists, and his followers made it the orthodox doctrine. When Lenin made his revolution in Russia, what came out of it was dictatorship, which under Stalin became a brutal tyranny, still further removed from the supposed "dictatorship of the proletariat," and from the ruling principle of "the full and free development of every individual." Violence is hardly a trustworthy means to brotherhood and perfect justice. Disillusionment with revolutionary Marxism was more bitter because to so many men of good will it had seemed the best hope of realizing the dream of a just society.

Linked with its moral failure was another reason for the revulsion against utopianism, more intimately associated with modern technology. From Saint-Simon to Skinner's *Walden Two*, proposed utopias have seldom been democracies. Even thinkers bred in a democratic tradition, like Edward Bellamy and H. G. Wells, specifically rejected such democratic institutions as a two-party system, representative legislatures, and popularly elected executives. Modern utopias have almost invariably been ruled by an elite, just as Plato's Republic was. In an advanced industrial society this means extensive organization and exclusive rule by technological experts. According to Marx and Lenin the state would "wither away" in the classless society, but an industrial society plainly requires much more organization than agricultural ones. Hence what came out of the Russian Revolution under Stalin was the biggest bureaucracy to date. Workers were its servants and did no dictating at all.

The incessant propaganda by which the rulers of the Soviets induced them to accept a system that denied them any real say brings up still another objection to modern utopias. Their people have to be educated to respect and obey their ruling elite, behave as the ideal society requires. A disagreeable name for such education today is "conditioning," a psychological technique that may be applied alike to animals and people. A worse name for it is "manipulating" people. Now technology has made it possible to do this more deliberately, skillfully, and thoroughly than ever before; and most of us would not entrust such power to any body of rulers.

In general, the basic objections to utopianism spring from ideals of freedom.[1] All the modern utopias are necessarily planned societies, and all are therefore open to the objection Walter Lippmann raised in *The Good Society*: "If a society is to be planned, its population must conform to the plan; if it is to have an official purpose, there must be no private purposes that conflict with it." Likewise the utopians naturally seek uniformity, not necessarily complete, but enough to prevent any serious disagreement over their ideal means and ends, or any open dissent; they do not provide amply enough for the development of personal differences, or the "variety of situations" that John Stuart Mill stressed as an essential condition of freedom and human development. As naturally their ideal society is not open but fixed, essentially closed, since it has in theory attained perfection. True, modern utopias have typically been "dynamic" societies long before Wells claimed this distinction for his version; Saint-Simon's people were forever building, and the motto of Comte's ideal state was Order and Progress (adopted by Brazil as its national motto). Yet the planners do not allow the progress to change the basic order. Few of us welcome the prospect of a closed society so long as we cannot believe that man will ever achieve a perfect one, and as we meanwhile cherish personal freedom. I for one have read of no systematically planned utopia in which I would care to live.

Today the popular visions of progress raise different but no less obvious reasons for complaint, which again are illustrated by the Soviet Union. When Premier Khrushchev told the Russians about all they would be enjoying by 1980, he offered them nothing more inspiring than visions of all kinds of material goods, including 200,000,000 tons of steel a year; in effect he reduced the Marxist utopia to a state-managed America. Such bourgeois materialism helps to explain why Communism no longer stirs

1. I am considering here only serious thinkers—not the unreasoned conservatism of vested interests, such as business interests devoted to private profit. George Katep has conscientiously reviewed the objections at length just because he wishes to preserve the values of utopianism.

much fervor, nor socialism either. In America a shallow pseudo-utopianism produces only as superficial a fervor over the ideal society conceived in terms of affluence, comfort, ease, "happiness" —in short, what Aldous Huxley's Brave New World offers more lavishly. No more inspiring are the "new breed of utopians" studied by Robert Boguslaw, the system designers who offer "computerized utopias." Their primary aims remain efficiency and economy, at best a reduction of human friction, not the attainment of a just society, or any fundamental moral or intellectual improvements. And whereas the old utopians were always dissatisfied with the *status quo*, the new ones are basically satisfied with it.

Much closer to tradition, however, is B. F. Skinner's *Walden Two*, an ultra-modern Utopia based on his "science of human behavior." His stated aims have an agreeable old-fashioned appeal. In defending his book against his many critics, Skinner wrote that he was trying to produce a world "where everyone chooses his own work and works on the average only 4 hours a day, where music and the arts flourish, where personal relationships develop under the most favorable circumstances, where education prepares every child for the social and intellectual life which lies before him, where—in short—people are truly happy, secure, productive, creative, and forward-looking." To assure his ideal state, education in the early formative years is all-important. "All our ethical training is completed by the age of six," his spokesman boasts in *Walden Two*. It assures not only that everybody will be tolerant, cooperative, affectionate, and virtuous, but that they "will be good practically automatically"; virtue becomes simply a habit. Skinner later added that he could produce "a world in which people are wise and good without trying, without 'having to be,' without 'choosing to be.' "

This was precisely what appalled his critics. In general, they were revolted by the idea of psychologists so systematically conditioning people; Skinner could sound as brash in his confidence as John B. Watson in the early days of behaviorism. (His spokesman also boasts, "We can construct groups of artists and scientists who will act as smoothly and efficiently as champion football teams.") In particular critics objected that the guinea pigs in

Walden Two have no real moral consciousness, no power of self-determination, no capacity for growth, no real character. They are denied the moral choices through which people can alone realize both their humanity and their individuality. Skinner is therefore quite consistent in denying them any voice in electing their rulers, the professionals who will condition their children. Similarly he permits no dissent or even discussion. "In no case," says his spokesman, must an inhabitant of Walden Two "argue about the code with the members at large." One might wonder what he would do with people who started talking back, except that as a thoroughgoing behaviorist he denies not only the ideal but the reality of human freedom: "I deny that freedom exists at all. I must deny it—or my program would be absurd." Then one may wonder how he himself managed to devise this very self-conscious, novel program, or how he can be so sure of his own power to turn out people to suit himself; but in any case this ultra-modern utopia is objectionable for the same basic reasons that the classical ones are.[2]

Yet Skinner's outraged critics oversimplified the issues raised by his pretensions to a mastery of the "science" of human behavior. All education may be called a mode of "conditioning" the young; it has always involved some indoctrination, or conditioning in the values of a given society. Within the family too the growing child is "manipulated" by his parents, taught how he should behave; he is not allowed to grow up as he pleases. These terms cannot be as sharply distinguished as we might like. The essential objection is rather to particular modes of educating, conditioning, or manipulating. With Skinner the sound objection is to his conditioning the young without training them to think for themselves or to make their own choices (as presumably he himself learned to do). Some of his critics, however, went further than this. Like Joseph Wood Krutch, they appeared to think that any effort to

2. It differs from them incidentally in being academic or even "reactionary," in line with American utopian tradition, as a small, isolated community or island within the great society. One reason for the popular appeal of Walden Two was that it could be admired from a safe distance, without worry; its "perfection" could never be attained by society at large, in which there are not enough professional behaviorists to condition thoroughly 200,000,000 people.

condition human behavior, any assumption that it could be to some extent controlled, was an affront to the dignity of man. In exalting his power of self-determination they ignored the plain truth that our self is largely determined before we reach maturity. We have learned a great deal about the biological, psychological, social, and cultural determinants of human behavior. And this knowledge is valuable precisely for the sake of our effective freedom, or the intelligent exercise of our powers of choice. Mysterious though these powers may be, it does not help simply to pride ourselves upon the mystery, or upon man's capacity for unpredictable behavior; the reasonable behavior we must aspire to is largely predictable. For that matter, so is much irrational behavior —we can often be pretty sure what fools are going to do. In short, I would not myself trust to the wisdom of B. F. Skinner and his disciples; but considering that much education is notoriously poor, much upbringing in the family as faulty, we might welcome possible light from any source. Since we could do with much more virtue, we might even risk the dread possibility of its becoming a conditioned habit.

At any rate, there is no serious danger of utopians taking charge in America today. As I see it, there is more need of appreciating the values of their mentality than dwelling on its limitations or excesses. Those to whom life in most utopias looks relatively dull would not enjoy either the life that most people have known throughout history. "The bulk of life must be ordinary," Katep reminds us; "the bulk of men must be ordinary; the bulk of the experience of even extraordinary men must be ordinary." Despite their addiction to rule by an elite, the utopians have sought to make humane provision for ordinary people, "just people," and have shown more respect for them than have most of our elitists, who sometimes appear to think that a decent contentment with a conventional life is simply indecent.

In particular the revulsion against the utopians has tended to obscure the valid grounds for their belief in the possibility of a much better life. The too easy scorn of "progress" has minimized the material progress that now makes feasible the ruling principle of "the full and free development of every individual." Thus anti-utopians took a morose or morbid pleasure in Jacques Ellul's lopsided account of the technological society. They have obscured as well the real progress in understanding—of the natural world, of man himself, of his history on earth, and of society. If all the knowledge we have acquired can never assure a better life, it remains at least a possible means, ground for hope. And however impossible their dreams, the utopians kept alive a sense of rich possibility. We could do with more of their "visionary" mentality at a time when we have so many no less fallible practical men. Paul Goodman, for example, has offered many proposals that make better sense than the vastly expensive projects of realists like Robert Moses, and that may seem absurd only because they conflict with a way of life that clearly has not been working well. Something of the utopian spirit is indispensable to any hope of saving the American city, doing justice to the Negro, or making America more like a Great Society.

All this brings me—at last—to my main concern, what we can know and do about our future. As the specialists now say, we are "future-oriented"; our technology has made us so. To be sure, the technological determinants of our society give us nothing like a free choice in futures, but in a real sense have put much of our future behind us. We have to go on living with our science and technology, all the knowledge that we could not possibly forget even if we wanted to. Yet there remain the obvious advantages they give us, beginning with a much wider range of choices than men ever had before, and much more effective means of realizing our choices. Having repeatedly emphasized that computers and systems analysis cannot make the value-judgments necessary in determining our social goals, or cannot tell us what we *ought* to do, I should now emphasize that they are very helpful in estimating costs, which must be considered when making choices, and in assisting us in what we *can* do. Above all, we have the plain advantage of much more conscious effort at planning and control.

Hence the many specialists who are busy studying trends, attempting forecasts. However hazardous, forecasting has become a very practical business, no mere pleasant indulgence in speculation, and may indeed be considered one of our major responsibilities; never has there been greater need of understanding, anticipating, judging, and controlling change. About technological developments forecasting is fairly easy, usually amounting to a simple extrapolation of current trends; one does not have to be a Rand expert to foresee the wide use of thermonuclear power as a source of energy, the farming of the sea for food and raw materials, more reliable long-range weather forecasting and possibly some weather control, "high-IQ" computers, and so forth. Much harder to foresee are the consequences of such developments, the social and cultural changes that must be expected, but much more thought is being given to them too than before the World War—sober thought, neither utopian nor anti-utopian, that assumes we still have some control of our technology, some choice in futures, some hope of a better future.

Thus in France Bertrand de Jouvenel is directing a study of what he calls "futuribles"—possible futures. "There is a continual dying of possible futures," he wrote. "And two mistakes are common: to be unaware of them while they are, so to speak, alive, and to be unaware of their death when they have been killed off by lack of discovery." He echoed the utopian tradition when he raised a simple but fundamental question: Now that every year we are better able to achieve what we want, what do we want? Although an economist, he has suggested such noneconomic goals as pleasant surroundings and not merely more leisure but for every man a man-sized job, in which he could take joy and pride. (He incidentally suggested that a symbol of what people appear to want today is the outboard motorboat: "a toy for grownups for moving about rapidly to nowhere with a great deal of noise.") These were among the goods of the "new life" described by William Morris in *News From Nowhere*, the most attractive utopia I know of. The trouble with Morris, de Jouvenel points out, was his very feeble explanation of how he proposed to institute his earthly paradise. Presumably he himself is studying practical economic and political means of achieving his proposed ends.

I am using as a springboard into what is now being called (alas) "futurology" the report of a similar project in America, the Commission on the Year 2000. Set up by the American Academy of Arts and Sciences, and headed by Daniel Bell, the Commission has issued a long preliminary report on "Work in Progress." For its members there was none of the magic in the millennial number 2000 that has been dazzling editors of popular journals, who can still seem more naive than the utopians of the past.[3] The main theme of the Commission's reports was problems. "The only prediction about the future that one can make with certainty," Bell announced at the outset, "is that public authorities will face more problems than they have at any previous time in history." Accordingly the business of the Commission was merely an effort to *anticipate* likely problems, and then to suggest strategies for dealing with them, possible solutions that would give society a choice in futures. This was a staggering enough task, which reminded Bell of the saying of Yeats, "In dreams begin responsibilities"; and it looked more staggering as the various specialists began offering their different dreams, grappling with their common responsibilities. At that the Commission concentrated largely on problems of public policy, putting off those that might be expected in the still less predictable future of culture.

One apparent reason for its choice, that its members are mostly social or behavioral scientists, itself raises questions about this future. The Commission includes no artists, creative writers, or literary intellectuals—types who no doubt lack the technical knowledge needed for its task, but who might contribute to the dreams we also need. Michel Crozier, a French sociologist, has warned it against the temptations of "the arrogance of rationality," an excessive faith in scientific techniques, methodology, or "social engineering." Ithiel Pool, a political scientist on the Commission, may illustrate the need of such a warning. While granting that philosophy and literature "have their value," he has written

---

3. The hoopla in these journals may recall a once popular prediction about the city of the future. Among other things it would have "abundant functions: fresh air, fine green parkways, recreational centers, all results of plausible planning and design. No building's shadow will touch another. Parks will occupy one third of the city area." This, noted Bell, was the vision of the American city of 1960 presented in 1939 in Futurama, the General Motors exposition at the World's Fair in New York. To enthrall its own customers, who thirty years later would drive to work bumper to bumper, General Motors added that soaring among quarter-mile-high towers would be "four-level, seven-lane directional highways on which you can surely choose your speed—100, 200 miles an hour."

that the knowledge by which "men of power are humanized and civilized" comes from psychology, sociology, political science, and systems analysis—a faith that is hardly warranted by the recent political record of America. Pool is also typical of the technical intelligentsia who dominate the Commission in that he is critical of public policy, but neither anticipates nor proposes any basic change in the economic and political establishments. I would have welcomed a few radical voices among them, more basic criticism of the establishments, and of the rules of the current professional game. As it is, the strategies they have so far suggested are seldom bold and may recall Jouvenel's observation that professional planners make no choices for society, only suggesting better means for achieving its current ends.

Still, we have to count on specialists for the purposes of forecasting or anticipating problems. I assume that political neutrality may be warranted by their effort to be properly "objective," since I do not anticipate either any revolutionary political change in America. It also seems safe to forecast that the technical intelligentsia will continue to ride high in the world of thought, and to have much more influence on government than philosophers or literary people. The rest of us may therefore rejoice that those on the Commission are often concerned with values other than technical rationality, beyond what were sometimes called "feedback processes."[4] In their wide-ranging discussions and reports they made no effort at consensus beyond an agreement that their "work in progress" is exploratory and tentative. Laymen may accordingly feel freer to be critical of their efforts and to raise more questions, as I shall. But even so I think it may be helpful to consider what some modest professionals have to say about our prospects, the future we have been so busy making. And we might even be heartened by what appalled Bell and other sociologists, that the Kennedy and Johnson administrations "discovered" the problems

4. Although the reports of the Commission are civilized, readers should be warned that they involve considerable technical language, including occasional jargon. One who tries to keep up with "futurology" has to put up with much reference to problem areas, the planning process, resultant feedbacks, motivations, attitudinal reactions, value orientations, etc.

of poverty, education, urban renewal, and air pollution as if they were quite new—problems that could easily have been anticipated years earlier. It is something that many more people are growing alert to our problems, trying to look ahead, and facilitating the public discovery of the obvious.

20

The Specialists on the Future

S ince I have been dismissing cavalierly the optimistic visions of scientists, while also stressing that we have the technological means of doing almost anything we want, let us consider what a symposium of eminent scientists had to say some years ago about "The Next Hundred Years." In introducing it, William L. Lawrence of the *New York Times* said: "We stand on the threshold of what is, without doubt, the greatest era of the history of man on earth in all his existence"—an era in which "he could literally turn this earth of ours into a veritable Garden of Eden" and "fulfill all his great potentials," not only physiological but intellectual and spiritual. The scientists went on to dwell especially on the possibilities opened up by the extraordinary advances in biochemistry. In time man will be able to develop processes of artificial photosynthesis much more efficient than the natural process. Having stolen "the secrets of the plants," he will learn how to make food out of sunshine, most of which is now wasted. He will also learn how to create the simplest forms of life. With people he will be able to control metabolism, curb disease, and regulate the reproductive cycle, or direct his heredity. Other great potentials will be fulfilled by triumphs in other branches of science and technology. Education will be speeded up by electronic memory banks that might even transmit coded electrical information directly into the nervous system. Voyages to the moon will become a commonplace. The earth will be surrounded by a family of artificial satellites, "all of them accepted as members in good standing of our solar system." A few of them will suffice to handle all the private and official mail between points more than five hundred miles apart, delivering it in no more than an hour after it was sent.

In a similar vein Glenn T. Seaborg, chairman of the U.S. Atomic Energy Commission, recently spelled out some possibilities of nuclear power. He foresees the development of "breeder reactor systems" that "could make available to mankind, in the form of abundant very low-cost energy, a force for progress unparalleled in history." Among the possibilities he finds especially interesting is the maintenance of a manned colony on the moon; nuclear reactors could supply power to help provide all the immediate necessities of life. More important meanwhile is the boon that breeder reactor plants could be to the poor countries on earth. Seaborg says they could desalt billions of gallons of sea water at a very low cost and so make possible the irrigation of large areas, while extracting chemicals from the brine; they could then become "agro-industrial" centers, serving as "food factories," producing fertilizers, supplying electricity, and processing minerals; and so they might soon become "the basis for a self-sustaining, growing economy" for the whole region.

Now, I take it that all these visions of the future represent real possibilities, or even strong probabilities as far as science goes. As a layman who has no idea what a "breeder reactor system" is, I cannot question the authority of leading scientists. I can say, however, that I am not always dazzled by their visions. When Seaborg added that a colony on the moon would probably have to live underground and subsist on the products of some kind of food factory, I could not imagine a less attractive habitat for man. The food of the future seems as unattractive, no less when a scientist said that children brought up only on synthetic steaks, made chewy by a plastic matrix, will think they are wonderful. But chiefly I remain troubled by the obvious reasons for doubting that we stand on the threshold of the greatest of all eras, or a veritable Garden of Eden. Seaborg took for his motto a saying of Mme. Curie that epitomizes the optimistic faith of science: "Nothing in life is to be feared—it is only to be understood." My experience has been that a deeper understanding of a technological society gives better reasons for fear of life in it. I agree with Gunnar

Myrdal, who in a paper on the necessity and the difficulty of planning the future concluded: "We need not the courage of illusory optimism but the courage of almost desperation."

At the symposium on "The Next Hundred Years" most of the scientists added conscientiously, even somberly, that their visions could be realized only if mankind did not destroy itself, but learned to live together. Likewise Glenn Seaborg concluded that all the "progress" he envisaged, and much more, could be achieved "if we in science could effect one 'human breakthrough'—if we could somehow convince our fellowmen that we now live in an age when fear, mistrust, and blind passions based on and regenerated by past ignorance and error must give way to a new level of understanding and reason among men." Once more it is a very big *if*; and though he added that "now the light of truth glows brighter than ever," I still think that science as such can provide no more than a hope of "somehow."[1] At the symposium one of the scientists confessed that as a breed they did not understand "the emotional structure of other people." But the message of another one, a psychologist who might be expected to have some such understanding, could make a layman hope they were only as naive as they sometimes sounded.

According to him there was no real *if*—a wondrous "human breakthrough" can be effected by science. "The important principles concerning the thinking processes, as they relate to creative imagination, will be worked out, and the procedure will be so systematized that man should be able to generate creative ideas at will." Similarly a mastery of the principles of "group dynamics" will tell us "how to form groups, how to develop group goals, how to select group leaders, how to reach effective group decisions." Perhaps more plausible, but if so more disturbing, were the psychologist's hopes of "the biochemistry of the central nervous system," means of increasing the efficiency of nerve cells and

---

1. His article appeared in the January 1968 issue of the *Bulletin of the Atomic Scientists*, which readers should know is not given to optimism. The clock on its cover, which in 1945 was set at seven minutes to midnight, was twice moved back in response to encouraging developments, such as the limited test-ban treaty; but in this issue it was again set close to midnight.

networks. "We would then be able to change man's emotions, desires, and thoughts by biochemical means, as we are now doing in a rather gross way with tranquilizers." To him this meant that by such techniques we could eliminate psychological insecurity, mental illness, maladjustment, crime, and psychopathy, and learn how to bring up our children to be secure, self-confident, happy adults. To me, a layman unwilling to let psychologists mess around with his "emotions, desires, and thoughts," it looks as if "we" might be led into Brave New World.

So I turn to the Commission on the Year 2000, which is much more concerned about the problems raised by such scientific possibilities. Because I think efforts at long-range forecasting or anticipating are so necessary I would again emphasize—as various members did in their reports—that it is necessarily a highly uncertain business. Although we can make out basic trends readily enough, these may change; with peace, for instance, depression might come again. The hardest thing to forecast, the social changes to be expected, is the most important.[2] We have no adequate theory to guide research into social change, and for the many who want more statistics first, no clear idea of just what we ought to measure. Always there remain such variables and immeasurables as the political response to problems, and such unpredictable unique events as the decisions of leaders, the possible appearance of another Lenin or Hitler. The "objective" truth is that the vital decisions about policy in America are made by government and big business, that a decision made tomorrow may drastically restrict the possibilities of the future (as the unplanned war in Vietnam did), and that it is impossible to predict the future if only because we cannot assume confidently that either government or big business will act rationally and responsibly in the public interest. No more can we be confident that the prosperous majority

---

2. In a recent forum on "Systems Analysis and Social Change," the experts in such analysis began to realize that their elegant methods didn't work when they got closer to human beings and their goals. The incoming president of the Operations Research Society confessed: "We're very good at hardware and tactical problems and starting well-defined research and development programs. We're lousy at strategic and philosophical problems."

of Americans will as taxpayers become willing to make the substantial sacrifices necessary if we are to deal effectively with such problems as poverty and the injustice to the black ghetto dwellers. I think the Commission's reports might have included more notes of "almost desperation."

But then we may repeat that modern technology, which created a need for forecasting not felt by leaders in the past, has also made it possible for men to anticipate change better than they ever could before. A vast deal of *present* activity—economic, political, scientific—is based on forecasting or calculated plans for the future. In government the difficulties seem worse because planning has so often been tardy, hasty, and short-sighted, as in the "unconditional war on poverty" announced by President Johnson that has got nowhere because no preparations were made for it.

In the Commission's initial set of reports Herman Kahn and Anthony J. Wiener led off with "A Framework for Speculation," an effort at purely objective anticipation without regard for goals. They expect all the basic trends of our technological society to continue, at an increased tempo and universality. Specifically, they list a hundred important technological innovations that they believe "will almost certainly occur" by the year 2000. Many of these are improved techniques that most people would regard simply as "progress," and that at least raise no apparent social or political problems. Some, such as a "widespread use of cryogenics," may make laymen aware of how little they know about much that is going on. (I had to look up the word in my dictionary.) A few startled me—"programmed dreams," for example, and "human hibernation" for periods of months to years. Almost all the forecasts seem plausible, however, and so may give a more vivid idea of how extraordinary an age we live in; the authors comment on how easy it is to draw up a list of significant innovations that are almost certain. Only the trouble is, as they also realize, that many of the innovations raise controversial issues, and some are positively alarming.

Most people will welcome the new kinds of very cheap, reliable birth control techniques, and probably also the ability to choose the sex of unborn children. Most will welcome too the prospects of further improvements in medicine, including the postponement of aging and "limited rejuvenation," even though they may aggravate the population problem. Other prospects, such as more synthetic foods and underground buildings, are less engaging for those of us who still enjoy the natural world. Still others raise misgivings. About the "new techniques" to be expected in both adult and child education much will depend on what they are, but a teacher may not relish anyway still more dependence on techniques, or even the prediction of "chemical methods for improved memory and learning." More disturbing, since Americans are already the most drugged people on earth, is the forecast of new and more reliable drugs "for control of fatigue, relaxation, alertness, mood, personality, perceptions, and fantasies." Most of these might be welcomed by a race that for ages has liked to relax or enliven its moods by drinking, but with others—especially the control of "personality"—we may again have fears of a Brave New World. Similarly we are told to anticipate new and more reliable propaganda techniques for influencing behavior, and new techniques as well for spying on people and controlling them. With these will come "new biological and chemical methods to identify, trace, incapacitate, or annoy people for police and military uses." As for war, Kahn is still resolutely thinking the unthinkable. We must expect "widely available or excessively destructive" weapon systems, and new, "possibly very simple methods" for lethal biological and chemical warfare.

One who reads between the lines may make out other serious problems in the authors' report. The many new techniques they predict include none for control of the pollution of water and air. About the problems of the cities the only helpful innovation they indicate is low-cost buildings for home and business use. They indicate none at all that might facilitate weapons control or disarmament. In general, this "framework" provides ample room for gloomy speculation.

As for the problems I have considered, the Commission has not sought to be comprehensive. Among those it has not looked into yet is the power of the military-industrial complex, which may continue to make adequate domestic programs seem utopian. The Pentagon might be dismayed, however, by the predictions of

Samuel P. Huntington about American foreign policy. He points out that because Americans fear so much the expansion of Communism, which has actually expanded very little since 1949, they have not realized that we have been the most successful expansionist power, as we moved into the vacuums left by the decline of European influence all over the world. He predicts that American power will begin to wane in the last quarter of the century, and that by the year 2000 our world system will be in a state of disintegration and decay; as a result domestic politics will be disrupted. I should say that this process is already well under way, and then stress another serious complication suggested by the leader of a group working on world affairs. He reported that they had not adequately discussed the possible ideological movements in the year 2000, in particular the new ones that might arise in the developing nations; but they do expect "large-scale turmoil" in the interim. To me it seems impossible to forecast what will come out of this turmoil. Meanwhile the trouble at home is still that American foreign policy has not been planned with an eye to long-range consequences, but chiefly has "reacted" to changes neither anticipated nor wanted by our diplomats. They still seem unaware that our technology has made archaic the traditional power politics of nationalism.

In government Kahn and Wiener expect a continuation of the trend to bourgeois, bureaucratic, meritocratic elites, or technocracy—a forecast that few would question. The Commission also appears to agree that since the United States has become a genuinely national society, in which the states are something of an anachronism, the federal government will take on more national problems. Nevertheless Daniel Moynihan believes that beneath the multiplication of new agencies the basic structure of American government will remain much the same. This would mean, I take it, a continuation of the problems of centralization and decentralization, or of the relations between federal, state, and local governments, with the usual bureaucratic confusion and conflict, and no doubt the usual complaints about government interference, coupled with demands for government aid—all further complicated because state and local government have grown much more even than the national government in the last twenty years, likewise without premeditated, coherent plan. Moynihan also believes, however, that just as the "most powerful development" in government in the last generation was the development of a political economy able to direct economic events, so the most powerful development in the next generation will be "the emergence of a social science coupled with and based upon a system of social accounting that will give government an enlarged capacity to comprehend, predict, and direct social events." He has fairly high hopes of this even though he grants that the social sciences are still very rudimentary and a major breakthrough is unlikely, if not impossible.

I would have higher hopes of it if government leaders or their advisers heed the proposal of Lawrence K. Frank, who originally suggested the idea of a Commission on the Year 2000 as a first step toward meeting the "urgent need for a new social philosophy." Now he translates this into the urgent need of a new political theory to replace an eighteenth-century theory that has grown ever more inadequate and frustrating. Thus people hang on to the traditional beliefs about the limited powers and responsibilities of government in a Service State that provides all kinds of assistance to business, but still meets opposition when it provides assistance to individuals and families, lacks a political theory to justify and rationalize its increasing activities, and so nibbles away at its vast problems.[3] Frank adds that we might be guided by the example of the Founding Fathers, who freely formulated a political theory to suit their new order without benefit of science or quantitative research: similarly "we should not look hopefully to

---

3. When Congress was demanding that spending be cut to the bone, the Senate voted to give the Boeing Aircraft Company $142 million for work on the supersonic plane, despite the objection that this would probably cost the government $2 or $3 billion before it was finished. Russell Baker summed up the prevailing philosophy: "Building S. S. T.'s is what the United States does well. Technology is the name of the game. We may be a bust at all that business about the good life for everybody and liberty and justice for all, but, Buster, one thing we can do is build machines, and as long as we can build machines we're still on top because machines are what the world is about nowadays, and save all that about the wretched of the earth and the free enterprise system for my Fourth of July speech. We don't do long sentences."

the bewildering array of contemporary findings and research techniques nor try to invoke science as the source of our procedure." He did not predict that by the year 2000 we would be operating on an adequate political theory, but concluded by merely insisting again on our need of one: "A social order that cannot reaffirm its aspirations, goals, values and also revise and reconstruct its institutions must succumb to increasing disorder and conflict or decline as the torch of human advance is taken over by the new nations." This could make gloomy reading in the light of the 1968 presidential campaign, in which the winner rehearsed the vague generalities of a musty political philosophy, coupled with the old slogans about the preeminent virtues of free private enterprise.

Other proposals of the Commission were likewise not so much predictions of political developments as recommendations. One popular theme was the need of "participatory democracy," just because the government has to do so much more in a society growing more communal. Leonard Duhl (a psychiatrist) maintained that change should be "the product of a wide consensus, not the fiat of an expert"; the initial mistake in the poverty program was that the "patient" was not involved in it. Daniel Moynihan also stressed the lesson of this program: "it is *good* for people to participate in government—not just a right but a remedy." All this was in line with Bertrand de Jouvenel's suggestion that the essential condition of a better future is a better understanding of human beings, and that whatever progress may have been made in this subject by scientists, "no trace thereof has come through upon the public scene." Daniel Bell, however, suggested that the need of solving more and more problems politically will increase community conflict. Presumably the conflict would not be eased if people do participate more in government, or try directly to influence it as many demonstrators have been doing.

Daniel Moynihan also introduced a theme congenial to President Nixon, whose staff he has since joined: "As government tries to do more, it will find it accomplishes less." He therefore anticipates the "rediscovery of the market" as a means of achieving social objectives: the planners and bureaucrats may learn to pay more heed to the wishes of the people, including the poor, as expressed in the market. No doubt it can be helpful in public as well as private enterprise, as the Soviet leaders have been learning; few believers in democracy would trust all decisions to the wisdom of bureaucrats. But mainly I would emphasize again the inadequacies of the market for public purposes, above all for the problems of the cities.

Kahn and Wiener anticipate that by the year 2000 half of the American people will be living in three huge megalopolises: "Boswash," with up to 80 million people; "Chipitts," extending from Chicago to Pittsburgh; and "Sansan," extending from San Diego to Santa Barbara or maybe even San Francisco. Other forecasters expect that between 80 and 90 per cent of a minimum of 280 million people will be living in cities, or about twice as many as now do. In a report to the Commission Harvey S. Perloff of Resources for the Future, Inc. begins by saying that "we can imagine a very exciting urban future," but only if some difficult problems are solved. He proceeds to paint a gruesome picture of what the cities will be like in the year 2000 if present trends continue: pollution grown so dangerous that relatively pure air and water would be among the scarcest and costliest of natural resources; horrendous traffic congestion in the air above; open space so scarce that the use of parks and other recreational areas would probably have to be rationed; central cities more segregated than ever, with more poor people living in slums and a greater gap between them and the privileged suburbs; some of the larger cities "ungovernable"; and so forth. At that he barely mentions the automobile. To reverse present trends, he sums up, "or even to make a dent in them," will take "tremendous resources, imaginative new solutions, and effective political infighting." The future of the city seems exciting to him because he is confident that "the grossest failures will be recognized as such, and changes will be made"; and also, perhaps, because he does not venture to estimate how many billions of dollars the changes would cost.

As for positive suggestions, Perloff begins by calling for a new look at natural resources. They have now come to include such "new" resources in the cities as relatively pure air, the space above the city, and the open space around it. We must not only conserve nearby rural lands in the traditional sense, leaving them untouched, but make positive use of them, for example by maintaining farms that city children could visit for educational and recreational purposes. We could also use the outlying regions of metropolitan areas to build New Towns, with federal assistance to assure housing for Negroes and lower-income people. We could build as well "New Towns—in town," with better schools and other public services, "lighted centers" for fun and shopping, a great variety of housing, and other means of making city life attractive that would bring middle-class whites back into the city.

Since the realization of such possibilities will require considerable government enterprise, Perloff calls for "creative federalism." City governments urgently need to be modernized, thoroughly reorganized; state governments, whose creatures they are, will have to redefine the sadly outmoded rules of the game in their relations; and the federal government, which has been mired in neighborhood projects, will have to consider the metropolitan region as a whole. But he retains an optimistic air. Thus when he writes that a new breed of mayor is clearly needed, he adds "and probably will be forthcoming." More hopefully he writes that "concepts of private property will undoubtedly undergo some significant changes" because of the growing demands for space. The need of "enterprising developmental organizations, both private and public," leads him to suggest various things that "might" or "could" be done, usually with the implication that at least something of the sort will be done. He concludes: "While new technologies can be enormously helpful, and probably will be, the really great problems for urban development in the last third of the century will be to resolve certain basic value conflicts, to overcome some outmoded ideas, and to experiment with new institutional arrangements to achieve agreed-upon social goals."

I have reviewed his essay in some detail because I largely agree with his proposals about what could and should be done, feel even more strongly that the problems of the city present at once an urgent challenge and perhaps our most exciting opportunity, but still very much doubt that enough will be done about them, at least for some years to come. As I write, the basic value conflicts are far from being resolved, the outmoded ideas are still prevailing in Washington and the business world, and Americans are not at all agreed on social goals. Technology is not being enormously helpful, but is still adding to the congestion and the pollution.

There is no prospect of any basically new institutional arrangements. I contemplate the sad state of John Lindsay, the most promising of the "new breed of mayors." It disposes me to agree with Edward C. Banfield, who elsewhere developed the thesis "Why Government Cannot Solve the Urban Problem." First it would have to solve the problems of poverty, chronic unemployment, and racial injustice. Except for the few representatives elected from the city ghettos, there is no hay for politicians in a war on these social evils. The vested interests defending the status quo (now including teachers' unions) have far more political power than the ghetto-dwellers. By the year 2000 American politics may be more enlightened and responsible; but first most suburban Americans would have to become much more willing to make some sacrifice of their privilege.

Although the problems of the mass media are among those that the Commission has not gone into, John R. Pierce of the Bell Telephone Laboratories had something to say about television in a report on "Communication." He believes that it has had tremendous effects as "the greatest unifying force ever to act upon man." It may therefore have still greater impact on the emerging nations in the world, which have yet to achieve national unity; the communications satellites may make it more available to them. At home Pierce sees the wave of the future as the spread of cable television, which can offer cities large or small not only a better signal but as many channels as are available in the biggest cities. It would therefore seem that in the year 2000 America will be more unified than ever. But like Marshall McLuhan, the archtribalist, Pierce does not go into the content of television programs, the kind of message that unifies America. He says only that mass communication is one-way, "aimed from the few to the many"; so I remain concerned about the purposes of the few.[4]

While Kahn and Wiener forecast such innovations as pocket

4. In a working session Fred Iklé incidentally noted both the kind of harm and the good that the mass media can do. Twenty years ago *Time* magazine ridiculed the pioneering demographers who were worrying over the population problem, foreseeing the "explosion." In 1961, however, James Reston of the *New York Times* joined the Population Council and began writing editorials; whereupon "the whole news media started on the problem."

phones and video communication in both home and office, Pierce dwells on the immense influence of electrical communication, beginning with the telegraph and telephone, and now on the new wonders in store through micro-electronics. He does not pretend to know what men will do with these tools. His main point is that as electrical communication altered the human environment, so the latest advances will create "an entirely new environment in which life will again be different," though again in ways about which he can only guess. I see no serious problems raised by these innovations except perhaps a threat to privacy, or to the *desire* for privacy, but I see no promise either that they will make language crisper or communication more civilized. (Pierce is perhaps unfair to modern linguists when he remarks in passing that computers being taught to translate revealed how little we understand language, since linguists were unable to supply them with a satisfactory grammar for any language; my guess is that this revealed rather the limitations of the machines.) The Commission did not take up the deeper problems of communication in a society with two or twenty-two cultures.

On intellectual institutions, which have become so all-important in a post-industrial society, a working party anticipates that if the universities try to meet the many demands on them—technical, economic, cultural, intellectual—they will be transformed as much in the next generation as they have been in the last one; but while recognizing some weakness in cultural affairs, it appears inclined to believe that they will rise to the challenge. Stephen R. Graubard foresees specifically "a new kind of university in a new kind of urban environment"—university cities whose main business will be education, to a greater extent than it now is in Cambridge or Berkeley. They will shelter a number of strong institutions that will cooperate in fulfilling their many different functions, in a growing realization of their common interests and a growing willingness to experiment. Teachers and students will migrate more freely among them. Graubard cannot be sure that either the federal or the state governments will assist in the construction of such cities, especially because we will no doubt retain all our present institutions, but in any case the main need now is "new bridges" between these institutions, not new buildings.

Otherwise the Commission's reports so far anticipate chiefly technical developments in education. The new techniques predicted by Kahn and Wiener include "home education via video and computerized and programmed learning." George A. Miller, a psychologist worried over the problem of how to "motivate" youngsters to study, thinks it can be solved by the new computer systems, with machines that enable students to go at their own pace. Daniel Bell anticipates a "national information-computer-utility system, with tens of thousands of terminals in homes and offices 'hooked' into giant central computers," which would provide library as well as business information services—what would appear to be something like a scholar's paradise (if not one I would care to enter). After the obvious forecast of continued specialization, Harold Orlans adds that "by the year 2000, the schools will have to impart a common culture strong enough to withstand the specialization," but he then drops this problem, which seems to me urgent right now inasmuch as there is no agreement on what universities should teach—except everything under the sun. He dwells on such expectations as the greatly enlarged role of independent research institutions, "new kinds of research-cum-action organizations," and "advanced techniques of storing, retrieving, reproducing, and transmitting information," in "machine-recoverable form." Cheerfully he foresees all kinds of closer cooperation. Altogether, the Commission's reports on education might have been less optimistic had they addressed themselves to John Gardner's complaint that the universities are not training leaders, only aides and advisers to leaders. This would call for more attention to the problem of a common culture, which can hardly be solved by new techniques, or by the mass of "information" stored and retrieved by machines.

About traditional culture the Commission had something to say even though it has put aside the problems to be anticipated in it. In their expectation that the basic, long-term trends would continue, Kahn and Wiener led off with a trend to increasingly "Sensate" cultures, using Sorokin's term: that is, "empirical, this-worldly, secular, humanistic, pragmatic, utilitarian, contractual, epicurean, or hedonistic." Sensate art, they go on, is not idealistic, heroic, noble, or in general uplifting. They think we may be entering a "Late Sensate" stage, which they characterize by a long string of mostly disagreeable adjectives: "underworldly, expressing protest or revolt, over-ripe, extreme, sensation-seeking, titillating, depraved, faddish, violently novel, exhibitionistic, debased, vulgar, ugly, debunking, nihilistic, pornographic, sarcastic, or sadistic." Since these tendencies are conspicuous, I should add that there is also considerable "protest or revolt" against them, even some effort to be uplifting, if chiefly in "midcult," and that the culture explosion is keeping traditional art alive. I would not hazard a guess about what art as a whole will be like in the year 2000, except that I would expect as much variety, confusion, and conflict as we now enjoy, or don't enjoy.

On science as an Establishment a working party is considering the changing attitude of the scientific community. Long inclined to regard itself as the major agent of progress, it is now growing more critical of technological innovations; so it "may become a much more conservative force than it has been, or it may adopt an ambivalent attitude toward change." But the Commission has ventured no specific predictions on this score.[5] Meanwhile its scientists contributed a few reports on the largest, most difficult subject—the changes to be expected in people by the year 2000.

Ernst Mayr, a biologist, sensibly discounted both the enthusiasm and the alarm over the possibilities of controlling human heredity. Biological man will be the same in another generation, or for that matter several centuries hence. It would take generations to make any significant change in his hereditary constitution even if we knew and could do much more about it than we now can. As it is, we will have to learn a great deal about the genetic contribution to human traits before we can propose "any mean-

---

5. Since its report, there has been some alarm in the scientific community because Congress in its economy drive has made deep cuts in appropriations for research and development. "American science is in financial trouble" began a report in the *Bulletin of the Atomic Scientists*; for the first time since World War II it appeared that federal support for R & D may well decline. Still, the financial trouble is only relative, since scientists still get billions to work with. I assume that except perhaps for basic research they can count on ample support in the years to come, and that they might benefit from some reduction by giving more heed to priorities in research.

ingful program for a genetic improvement of mankind." Mayr adds, however, that inasmuch as geneticists are learning much more about the possibilities of eugenics we should begin thinking seriously about them now, reexamining our notions about valuable traits and individual rights. Over the centuries genetic changes are inevitable, so "man must ask himself whether he wants to adopt a *laissez-faire* attitude toward these changes or to be the master of his own fate."

Gardner C. Quarton, a psychiatrist, took up the much more immediate possibilities for controlling behavior and radically changing personality. Although present techniques are crude, we must expect increasingly effective ones. Quarton asks the fundamental questions: "What will these techniques be? Who will use them? And for what purposes?" Besides drugs, the techniques include the use of sexual hormones, brain surgery and stimulation, brainwashing, and conditioning by behaviorists. They are used by various kinds of experimenters and specialists, including doctors and psychiatrists. The purposes may or may not be benevolent; thus tranquilizers are used to relieve patients of anxiety, but also to relieve doctors of inconvenience. In the drug industry the purposes are primarily commercial; the value and the need of the many new drugs it will develop will be exaggerated by its admen. Yet its millions of customers suggest to me the probability that a great many people will gladly take to the drugs, which may become more literally the opium of the masses. As to whether people will submit to extensive, systematic use of the new techniques by specialists, Quarton thinks society is likely to resist the more radical, efficient ones. He concludes: "An understanding of behavior-control technology is to be encouraged because even though extensive knowledge may bring some undesirable applications, it is also necessary to develop an alert, well-informed public that will watch for abuses." I think developing such a public will be a considerable job.

In James Q. Wilson's report on "Violence" Americans may be comforted by the news that the rate of murder has been steadily declining over the last generation. The evident reason is the growth of affluence, since murder is primarily a lower-class crime; so with continued prosperity we may expect the decline to continue. Yet this is small comfort. Given the long American tradition of violence, the homicide rate remains much higher than in the European democracies. (In Manhattan alone there are more murders than in all of Britain.) The far more common crime of assault has been increasing in the central cities, making streets notoriously unsafe at night. Above all, the nation is faced with the recent growth of collective violence, which has been only most conspicuous in the Negro riots. Wilson stresses the violence among whites too, ranging from the growth of racist organizations to students storming the Pentagon and seizing university buildings. Young people, both black and white, have become self-conscious about "the social functions and therapeutic value of violence." Although Wilson ventures no confident prediction, he is inclined to believe that disorder and violence will become more common in the years ahead with the growing habit of marches, protests, demonstrations, and other forms of direct collective action, variants of the "participatory democracy" that has all along been the political style of the middle class. He concludes that the threat of lower-class violence that dominated politics in the last century may be replaced by the threat of middle-class violence. Since he wrote, the increasing violence of the blacks on campus suggests that we are faced by both threats.

Harry Kalven, Jr., a professor of law, is no more reassuring on the problems of privacy. He believes that the technology of eavesdropping by remote control will unquestionably continue to improve at a rapid rate, so much so that "by the year 2000 it will be possible to place a man under constant surveillance without his ever becoming aware of it." The use of these snooping devices may become common with private parties for various business or personal purposes. (Popular journals now advertise them.) With the help of computers, the government's increasing demand for special information for its income tax, social security, welfare programs, and other purposes raises the prospect of "a formidable dossier on every member of society." The increasing emphasis on consumer credit also deprives people of their once valued privacy about their financial status. Behavioral scientists invade privacy by questionnaires, personality tests, manipulative studies, and experiments that lead people to betray themselves (as did the

subjects of the Yale psychologist I mentioned who thought they were cooperating in an experiment on learning). Because of constant exposure to unwanted communication, such as billboards, sound trucks, junk mail, and unsolicited telephone calls, society may become largely a captive audience. And dramatic changes in privacy may result from cultural changes already marked—the decline of the family, of religion, and of the habit of reading. These involve what to me is the worst possibility, that people will no longer value privacy.

The most obvious protection remains the law. Kalven notes that traditionally it has protected privacy under varying conditions, and lately has grown more interested in the right to it. Otherwise he only raises a few questions for discussion, such as the possibility of a program of education for privacy. Least heartening to me is what seems to him the "most interesting" possibility, the development of new institutions designed to insure "some private moments in unprivate lives." One might be secular retreats somewhat like the old religious retreats. "It may be a final ironic commentary on how bad things have become by 2000," he concludes, "when someone will make a fortune merely by providing, on a monthly, weekly, daily, or even hourly basis, a room of one's own." I would prefer to bank more on the ambivalent attitude toward privacy, since in spite of all the togetherness most people still want and enjoy some. As for all the super-snooping devices to come, I am less confident that the FBI and business will refrain from using them (even if J. Edgar Hoover proves to be less immortal than he now seems).

Erik Erikson writes more hopefully in a "Memorandum on Youth," my own chief concern. On the assumption that adolescents are struggling to define relevant new modes of conduct, he qualifies the common report that they are doing so in a context historically unprecedented because of a "skepticism of all authority" and an "extraordinary hedonism," meaning broadly "a desacralization of life and an attitude that all experience is permissible and even desirable." Actually the youth are not skeptical of *all* authority, only distrustful of people "who act authoritatively without authentic authority or refuse to assume the authority that is theirs by right and necessity"; their elders are more to blame.

Similarly Erikson sees in their "hedonism" little relaxed joy, but often a "compulsive and addictive search for *relevant* experience," again because of the kind of world we have bequeathed them. We desacralize their lives, among other things by a naive scientism and irresponsible technological expansion, and he sees more of a search for *re*sacralization in them than in their elders. ("All you have to do is to see some of these nihilists with babies.") The youth also have to bear the greatest strains because of the unprecedented divorce between *our* traditional culture and the tasks of *their* society. In the past rebels and romantics might come back to society, but even if our youth felt like doing so, "there may soon be no predictable society to 'come back to.'"

Looking ahead, Erikson assumes that masses of young people will adapt themselves to a technological society, find their identity in it, and run the risk of feeling too much at home in it as it is, too little concerned with ethical values. But a radical enlightenment "forces some intelligent young people into a seemingly cynical pride, demanding that they be human without illusion, naked without narcissism, loving without idealization, ethical without moral passion, restless without being classifiably neurotic, and political without lying: truly a utopia to end all utopias." These "humanist" types too are beset by dangers because they lack as yet "the leadership that would replace the loss of revolutionary tradition, or any other tradition of discipline." Nevertheless Erikson thinks that "technological" and "humanist" youth may yet come together. In a revolutionary world the one is likely to be forced to entertain radically new modes of thought, the other to find some accommodation with the machine age, and both to realize that they share a common fate. We can help the older ones in particular by recognizing and cultivating their ethical capacity. Erikson risks the prediction that in the years ahead "youth will force us to help them to develop ethical, affirmative, resacralizing rules of conduct." In all this he is perhaps being over-generous to the youth, but I think he is describing fairly the many thoughtful ones, and stating as fairly adult responsibility for their problems and their education.

For the rest, the Commission's discussions included many scattered suggestions about the state of people in the coming

generation, on the whole not cheering. Since no one anticipated the resurgence of either religious faith or a high sense of national purpose, many people will presumably continue to be troubled by the problem of how to make life more meaningful. David Riesman wrote that the morale even of the meritocratic elite may be undermined because its scientific and rationalist temper has no religious basis and the system no transcendent aims, no goal beyond its own further advance. Others were troubled by a possible growth of irrationality, as a reaction against the spread of functional rationality, or specifically the widening gap between planners and people. (One example is the campaigns against fluoridation.) In general, the discussions gave few intimations of a richer life, becoming to the abundance and leisure to be expected. Their tenor, it seemed to me, was at best that life might not be so bad as many writers think. I am inclined to have no higher hopes, though with a reservation. Life will still hold exciting possibilities for the spirit in man that seeks truth, beauty, and goodness, and many people will realize these possibilities, as many now do. This remains the excuse for dwelling chiefly on what they—especially the young—will be up against, or need to look out for.

21

Coda:
The Problem of Human Nature

Paul Goodman, who writes on an exceptionally wide variety of subjects, has said that he has only one subject, "the human beings I know in their man-made scene." In ranging all over our technological society, dealing with very large issues, I have tried to keep an eye on these human beings, but have also emphasized how hard it is to generalize about them. In lieu of a resounding finale, which is impossible for this book, I now propose to consider more closely the ultimate issue they raise—the nature of their "human nature." This is implicit in all the talk about the "alienation" of modern man, the "dehumanizing" tendencies of his technology, the "unnatural" life he is leading, and so on, as well as in the growing insistence that planners need a better understanding of human beings.

Now, at the outset I likewise emphasized how hard it is to say what is the "natural" life for man. On the record human nature is extraordinarily plastic, adaptable to innumerable different ways of life. We have seen man move from the cave to the village, then take to the city with its civilized life, then develop an industrial society, a very new kind, and now a more revolutionary post-industrial society; the "man-made scene" today is radically different from all the scenes that shaped his nature throughout his history. In the cave he also began evolving different cultures and thereafter developed a fantastic variety of them, all of which seemed natural to the human beings brought up in them; they make plain that human nature is largely a second nature, man-

made. Over the last century we have learned a vast deal about him, through the study of history and the sciences of man and society, but as a result there has been much more radical disagreement over his nature than ever before. A creature of Original Sin who became perfectible, fit for utopias like Marx's classless society, also emerged as economic man, Darwinian man, Neitzschean man, Freudian man, existentialist man, technological or one-dimensional man, or what have you.

Looking ahead—as I shall continue to do in this chapter—we have as wide a choice in futures suited to the creature. In the Communist world it is still taught that man will fully realize his humanity in an ideal classless society. In our world more thinkers believe he will end in something like Huxley's Brave New World, toward which America and the Soviets are converging. Jacques Ellul, once more, thinks man will become thoroughly conditioned to a way of life he insists is thoroughly unnatural. Lewis Mumford takes seriously Roderick Seidenberg's prediction that the triumph of organized intelligence will eventually lead to the disappearance of individuality and then consciousness, the most human thing about him. Nevertheless Mumford retains utopian hopes: in *The Transformations of Man*—a review of the history of his nature—he declares that man's principal task today is "to create a new self," or new nature, and looks forward to "One World man," who will be just the opposite of the bureaucrat and technocrat. In *The Five Ages of Man*, a psychological history, Gerald Heard agrees that man can change his nature, but has a different idea of means and ends: he has man entering the fifth stage of post-individual or "Leoptoid man," who is growing conscious of the vast resources of "nonself-consciousness." Norman O. Brown, who sees repression as the source of all evil, has visions of the emancipation of genital or hipster man. And all these thinkers—and many others—are pointing to real human potentialities.

Given such diverse potentialities, prediction becomes still more difficult. One who has no hopes of utopia might like to say at least that "human nature" just wouldn't stand for Brave New World; only it looks like a real possibility, considering the nature of technological man and affluent man in America. As it is, we still hear too much provincial talk, echoing the traditional conservative wisdom, that you can't change human nature—human nature is always and everywhere the same. Usually the stress is on the disagreeable potentialities of man, including the selfishness that grew more pronounced with the rise of civilization and privileged classes, the aggressiveness or lust for power that was encouraged by our highly competitive society, and the "acquisitive instinct" that Americans take for granted; this makes it easier to blame on human nature evils that may be more plainly due to the organized power of a few. Pierre Bertaux has observed too that by "man" we usually mean a "Western white civilized adult," whereas by a statistical index the largest typical group is young Chinese females.

Yet we of course can and must talk about human nature, as we do about "man," "mankind," the "human condition," and our "common humanity." Biologically man is a distinct species, with the same structure and needs that he had fifty thousand years ago. "It is absolutely certain," writes René Dubos, "that all of his physiological needs and drives, his potentialities and limitations, his responses to environmental stimuli, are still determined by the same twenty thousand pairs of genes that controlled his life when he was a palaeolithic hunter or a neolithic farmer." Dubos explains that we know very little about the pathological effects on man of a polluted, crowded environment and the stresses of modern life because the problems they create appear chiefly late in life, past the reproductive age that could call into play the mechanisms of natural selection; but he is certain that the new environment and ways of life have had deleterious effects. Likewise biologists and sociologists know hardly anything about man's adaptive potentialities, which would require studies of long-range consequences; but again it is certain that man cannot adapt himself successfully to any kind of life in any kind of environment. Meanwhile technology keeps plunging on in ignorance, in ways largely determined by economic expediency, with little regard for the basic nature and needs of man. So the accelerating pace of change may make one wonder: Just how much change can man stand? Can he adapt himself successfully to a continuously revolutionary world?

As a humanistic scientist who believes that "quiet, privacy, independence, initiative, and open space" are "real biological necessities," Dubos recalls me to my own premises. At the outset I stated my belief that despite the conspicuous diversity of cultures and relativity of values we *can* talk of absolute values, permanent as long as man exists; or in other words, these derive from "human nature." The psychologist William Maslow maintains that even the "higher" human needs, as for friendship, love, dignity, self-respect, and self-fulfillment, are biological in origin, and human nature has been sold short because they have not been so considered by most scientists; but in any case they have long been vital needs. As for our own society, he attributes the growing discontent and rebelliousness of young people to the frustration of the idealism natural to them. Academic psychology, economics, and most social science offer them a limited view of human nature, with emphasis on the lower or material needs. The dominant positivistic and behavioristic theories reduce the highest values and virtues to mere appearances or illusions. The whole society operates on a low conception of human nature. "How could young people not be disappointed and disillusioned?" Maslow concludes. "What else could be the result of *getting* all the material and animal gratifications and then *not being happy*, as they were led to expect, not only by the theorists, but also by the conventional wisdom of parents and teachers, and the insistent grey lies of the advertisers?"

The youth also best exemplify the peculiar complication of human nature today. A person's genes determine not primarily his traits but his potentialities; man is distinguished by the remarkable range of his potentialities, most of which the individual never realizes or develops. Our technological society has carried much further the ambiguous tendencies that arose with civilization, which at once widened the range of opportunities for self-realization and for most people narrowed it by a division of labor as well as inequality of opportunity. Young people now have an extraordinary range of choice in profession, habitat, and way of life, while most of them also feel impelled to become specialists and conform to the requirements of a technological society, in an urban environment becoming increasingly uniform. And collective man, always at once creator and creature of his culture, has become both to a more marked extent. While developing immense power to change the environment that shapes his nature, and now more power to change his nature directly, technological man—the most thoroughly conditioned creature of his culture—feels impelled to go on acquiring still more power, however dangerous, and to use it for purposes that many young people find revolting. They have been growing more aware of rich potentialities unrealized in the American way of life. Change is now so rapid that it has taken only a few years to create a wide gap between the generations.

In this view of human nature "alienation" can become more meaningful as a neglect or frustration of higher needs. First I would repeat that it is not a plight peculiar to modern man. By virtue of his plastic nature and his powers of choice, man has always been capable of stupid, perverted, self-defeating, possibly fatal behavior, as his whole history abundantly proves. Yet alienation in modern America is in some respects historically new. It may spring from mechanical routines to which man is not clearly adapted, biologically or psychologically. The many symptoms of it in ordinary people, including the feelings of emptiness and impotence, are quite different from the fatalistic resignation to which the peasant masses were always prone. And the very achievements of America, in creating opportunities and incentives for self-fulfillment, have alienated many people because they have higher expectations, make higher demands on their society than people did in the past.

Hence the question now is, are Americans likely to get more alienated in the years ahead? Much will depend on what the nation does about all its serious problems, including the man-made scene. Whether in particular the many millions of poor, above all the blacks, remain alienated or get more so depends wholly on whether it recognizes its debt to them. Their prospects are dimmer because the problem in my present terms is to improve not merely their material but their spiritual condition, by building up their self-respect, enabling them to have more feeling of dignity, giving them more opportunities for self-fulfillment, bringing them into the society as full-fledged members; and this calls for more patience, humility, tact, and concern for the higher needs than are displayed by either most technologists or most prosperous Americans. While not at all confident that the nation will commit itself to a serious effort to eliminate poverty, discrimination, and social injustice, I take it that at best the problem will not be solved by the year 2000. When dealing with such problems one may realize, as William James did in his later years, that for the purposes of a lifetime human nature isn't so damn plastic after all.

As for the subtler "alienation" of prosperous Americans, I would not venture a prediction. Increasing wealth might fortify their complacence and their hedonistic tendencies, or it might intensify their underlying discontents and anxieties. If the latter, it might produce more violent outbursts of irrationality and inhumanity, or more concern for basic human values and higher human needs. Most likely the affluence will have all these effects on people, and then some, but in proportions impossible to predict with assurance in a society changing so rapidly. All I should say is that our hope lies in the growing awareness of these conflicting tendencies, the persistent criticism of our society, and the characteristic theme of "challenges." Having announced that alienation is "almost total," Erich Fromm nevertheless writes *The Revolution of Hope: Toward a Humanized Technology.* Maslow believes that a "Humanist Revolution" is under way in all fields of thought, with a centering of concern on human needs. Ralph Siu, a fervent technologist, likewise believes that we are on the threshold of an "Age of Holistic Humanism." If so, people may in time grow better adapted to affluence and leisure, as over the ages they have to toil.

At least the symptoms of alienation might lessen the fear that we are headed toward Brave New World. They belie the assumption of Huxley and other critics that people are growing ever more content with their affluence, their mindless role as consumers, or the satisfaction of merely material needs. In this respect "human nature" does appear to be resisting the trend toward Brave New World. But here we run into all the other technological threats to humanistic ideals. In particular they threaten the ideals of individuality and personal freedom, the most obvious basis for resistance to the whole trend.

As I have said, these ideals cannot on the historic record be considered an absolute need of man, or permanent values, but are nevertheless clearly basic potentialities of human nature. All societies have enabled the individual to achieve some measure of self-fulfillment through the development of skills. Since our own civilization has developed the fullest consciousness of selfhood, the value of self-realization, and then of personal freedom as a necessary means to this end, these possibilities of human life can for us be regarded as among the higher needs of man. "The excellent becomes the permanent," Aristotle observed, "—once seen it is never completely lost." Or in historical terms, human nature is a product of a development that has included some values the human race has hung on to. Hence the widespread alarm over the trend to uniformity and the pressures to conformity is quite "natural" even though it is historically novel. No less natural are the fears of increasing organization. As Roderick Seidenberg sees it, organization breeds organization, it accomplishes what used to be the work of creative individuals, it steadily shrinks the autonomy of the individual, and it moves inexorably toward uniformity and universality, the ideal of frictionless mechanism or perfect efficiency. Lewis Mumford agrees that already the foundations have been laid for a society that could produce Seidenberg's "post-historic man," lacking all individuality:

> If the goal of human history is a uniform type of man, reproducing at a uniform rate, in a uniform environment, kept at constant temperature, pressure, and humidity, living a uniformly lifeless existence, with his uniform physical needs satisfied by uniform goods, all inner waywardness brought into conformity by hypnotics and sedatives, or by surgical extirpation, a creature under constant mechanical pressure from incubator to incinerator, most of the problems of human development would disappear. Only one problem would remain: Why should anyone, even a machine, bother to keep this kind of creature alive?

There is no immediate prospect, however, of any such uniformity in society or human nature. Granted all the plain pressures against the individual, the signs of resistance to them have become plainer too, especially among the many critical or rebellious young people. If this may be a passing fashion, the individual remains very much alive in ordinary capacities. Even in systems analysis he is still considered real and important as a means of "feedback." Administrators of public welfare programs soon learn (as August Heckscher did in the Parks Department of New York) that they are not dealing with atomized mass-men, for they run into plenty of positive opinions from spokesmen of local groups or communities. And they make plainest that the technological threats to individuality are not all simply threats.

A planned society need not be totalitarian, and if the experience of the Soviets means anything, it cannot be totally or permanently planned. Karl Mannheim's idea of "planning for freedom" may be questionable, but at least planning is not necessarily fatal to freedom and individuality, any more than would be a program for eliminating poverty; some planners want to get rid of much unnecessary interference with people. Neither, to repeat, is organization necessarily fatal. This is obviously necessary not only for accomplishing the massive action called for by some problems, but for combatting the organizations of vested interests, as the young radicals have discovered. While feeling most strongly that the "Establishment" is hostile to the ideal of self-determination, these radicals are also learning that the old-time uncompromising individualist, or the loner like Thoreau, cannot hope to be effective today, that in organized movements they have to provide for cooperativeness and tolerance of diverse opinion, and that in these movements they can nevertheless enjoy more sense of personal freedom and self-fulfillment. (No class, incidentally, is more addicted to "groupism" than the hippies.) On a national scale the extraordinary number of diverse organizations—public, private, professional—likewise afford the individual opportunities for self-fulfillment, and suggest that the real problem is to develop a different kind of individualist. Nor is rationality or intelligence so blind or ultimately stupid as Seidenberg pictures it. Intelligence can of course be sensitive to its own abuses, just as what men used to call "reason" can be aware of its limitations and its excesses. It is reason or intelligence that warns us against the dangers of organization, the common excesses of technical rationality, the trends toward Brave New World. Human nature as it has developed has some capacity for reasonableness, which is the necessary basis for what hope we can muster.

In any case Seidenberg is looking ahead thousands of years, to a future we need not worry about at the moment. Thus he sees society becoming ever more unified, moving toward a state of complete and permanent stability; but we have to deal with the realities of a divided people and dynamic change, a highly uncertain future. He likewise argues that freedom must disappear for the sake of perfect justice, since injustice would make impossible the frictionless mechanism to which organization aspires; but we still have plenty of social injustice to cope with, for millions of Americans it remains the chief threat to their effective freedom, and we can count indefinitely on ample friction. For such reasons too there is little or no prospect of Brave New World in the near future. Meanwhile it is too easy to blame everything on technology. Our most serious problems are political and moral, strictly up to us. With these problems we could do with more utopian thinking, appealing to the best in the American people, whose tradition has it that human nature can be improved by democratic institutions, or appealing to the many who still are concerned with the higher human needs.

But so far I have been writing with an eye chiefly to the nature of Americans in their man-made scene. For the sake of perspective I wish finally to take a brief look at the rest of mankind, specifically the non-Western world, which remains the most obvious reason why we will not have to worry about Brave New World for a long time to come. It also throws some light on the nature of man, who has grown capable of much more rapid adaptation to change, but at enough biological and psychological cost to make one wonder how much change he can stand. Today the disorders all over the world, including the student revolts, suggest that the whole human race may be facing a crisis.

Herman Kahn and Anthony Wiener forecast that by the year 2000 Japan will have joined the United States and the Western democracies as post-industrial societies; so presumably it will know all the blessings and the problems of affluence. A few countries, including the Soviet Union and several of its satellites, will have reached the early post-industrial stage, and more the mass-consumption stage. But such large countries as China, India, Pakistan, Indonesia, Brazil, and Nigeria, with an estimated total population then of over 3 billion, will still be only partially industrialized and will have a per capita income ranging from only $200 to $600. Behind them will still be most of Africa, two-thirds of the Arab world, and some scattered small countries, with a total population of 750 million, greater than that of the post-industrial countries. Hence an island of wealth will be surrounded by "misery." Although the poor countries will have a higher standard of living, they will be much farther behind the affluent ones than they now are.

This prospect, which will give Americans more trouble than they realize even though they have begun to learn that we have fewer friends in the world than ever, is still not bringing out the best in them. The $30 billion we have spent on aid to pre-industrial countries represent much less than one per cent of the national income, and of late the foreign aid program has been steadily whittled as nominally Christian politicians and voters complain of "give-aways." Apparently they do not know that American business has profited from the program, inasmuch as about 80 per cent of the aid is spent here. Now more of it is being tied to exports, while seldom giving the poor countries better opportunities to export their products. Although the disappointing results of the aid program have been largely due to the common inefficiency and corruption of government in the poor nations, and the selfishness of their wealthy class, they have also been due to politics at home. Backed by Congress, our aid missions have usually favored industrialization projects more than projects for improving agriculture; these might lead to competition with U.S. farm products. In short, Uncle Sam has been neither the Santa Claus nor the sucker portrayed in cartoons.[1] Not to mention that a people very proud of their own Revolution remain about the least capable of understanding or tolerating revolutions in the rest of the world.

---

1. In fairness, the Soviet Union has been no more disinterested. With increasing affluence both it and the European democracies could easily afford more generous aid, but my guess is that the poor countries had better not count on it. Meanwhile the U.S. Arms Control and Disarmament Agency has reported that the soaring cost of the global arms race has reached more than $182 billion a year, averaging $53 for every man, woman, and child on earth—more than many of them have to live on; and it estimates that if the increasing cost continues, the world will spend $4 trillion in the next ten years. Such is the cost of a "security" that no nation enjoys, including America, the leading spender.

I shall not go into the staggering economic and political problems of the new nations, except to say that they cannot be solved by any magic of free private enterprise, or merely economic rationality. My concern here is some questions raised by their recognition of the goods of modern technology, their confirmation of its apparently irresistible drive, and the consequences of their efforts to modernize, in the name of "progress." They have already made enough progress to produce a growing uniformity, immediately recognizable in their airports and in the new buildings and boulevards of their capitals. Industrial societies are bound to be much more alike than agricultural, handicraft societies. Railroads, automobiles, cement factories, steel mills, power plants, and all kinds of machinery and machine products are everywhere basically the same, alike under Communism or capitalism. People grown used to them might be expected to be more alike.

Yet industrialism by no means requires the total elimination of traditional cultures. When Japan, for example, developed a factory system, it was quite different from the American system. A worker employed by a company assumed he would work for it the rest of his life and would not quit to take a higher paid job in another factory, while the company felt as obligated to him, only under extreme circumstances discharging him or even laying him off temporarily. Still by far the most successful of the non-Western countries in modernizing, Japan accordingly refutes the assumption of Jacques Ellul that technology is wholly autonomous and automatically imposes its requirements. In India, where Nehru gave up Gandhi's opposition to industrialism, leaders are still trying to adapt it to their traditional culture; some resent the term "underdeveloped" because it implies that economic growth and technological advance are the supreme goals for a nation. Although leaders are typically eager to modernize, few of the non-Western peoples set such store by economic values as Americans do, or assume that economic motives are the most powerful.[2] They are

---

2. Americans might be bewildered by the attitude of peasants in Java. When the price of coconuts goes up, they are likely to bring fewer instead of more coconuts to market, for the sensible reason that

not simply awed either by American know-how or modes of efficiency, which are usually not suited to their culture and environment.

For such reasons they are likely to be handicapped in industrializing, especially because their traditional ways often conflict with the regular habits and technical rationality it requires. Nevertheless their leaders also have a potential advantage in being aware of the evils that can come with it. Although most of them now tend to be arrogant and short-sighted in the exercise of their new powers, the "developing" nations (to use the politer term) may hope to escape the inhumanity of early or "palaeolithic" industrialism, and to retain more respect for esthetic and other human values than most Americans appear to have. Kahn and Wiener, who remarked that the "Sensate" culture of the West has been spreading over the whole world, consider it an open question whether this trend will continue in the next generation. The tendencies to "Late Sensate" are certainly much less conspicuous in the non-Western world.

In the very long run, if or when all countries reach the post-industrial stage, I suppose it is conceivable that there might be an almost complete uniformity, and "human nature" will everywhere be more literally the same. Even by the year 2000 there may be a considerable degree of drab uniformity, especially if television "unifies" peoples. In my own travels abroad I have been mostly depressed by the changes that have taken place in just thirty years, because of both technology and tourism. Yet it seems to me unlikely that the world will ever have but one culture, any more than that all people will speak the same language and worship the same God. Countries with advanced technology can still retain much of their traditional culture and have different styles of living, as European countries do; my generalizations about the nature of Americans and their man-made scene would not all apply in Europe. In the non-Western world an intense nationalism —a sentiment that Americans think of as just human nature, but that was actually acquired under Western domination—creates a

---

the higher price makes it unnecessary for them to work as hard to satisfy their needs.

more conscious pride in traditional values or native variations, and strengthens a resolve to be not just like Americans.

Unfortunately it also heightens the dangers of war—the all too real possibility that there may be no long run for modern man. The non-Western world throws into relief the most tragic aspect of the failures of America in its professed role as leader of the "free world." As an Italian historian remarked to Walter Lippmann, we have come down off the pedestal and entered history, joined all the other nations that have insoluble problems and suffer defeats; yet the tragedy remains that there is no nation clearly fit to take our place or carry the torch for the non-Western world. Despite de Gaulle's dreams of *gloire*, France is no more ready or able than Great Britain, West Germany, or Japan. The Soviet Union still adheres to a narrow conception of its own national interests; it antagonized most of the world by its ruthless suppression of the Czechs. The future may belong to teeming China, but who would now look to it for world leadership? Or to India, Pakistan, Indonesia, Egypt, Nigeria? By the year 2000 there will no doubt be surprising changes in the status and the alignment of nations, but the critical period is the coming generation, during which the danger of a nuclear explosion will be most acute; and as we face it, the fact is that no nation on earth has both the material means and the will to provide the enlightened aid needed by the poor countries.

Otherwise the plainest lesson for Americans in the turbulent drama of the non-Western world is the need of not only more sympathetic understanding of the different ways of other peoples, but more self-understanding, a more critical attitude toward their own accepted values and notions of the good life, or simply more awareness that their own nature is not just human nature, their ways are too often a perversion or defiance of the developed higher needs of man. Else there is also a real possibility that America—the pace-setter in technology, and the country in which above all it has run wild—may go down as the greatest failure in history, or that it will be spared this ignominy only because history will have come to an end.

Selected Bibliography

The following bibliography is of course far from comprehensive. Made up chiefly of contemporary nontechnical books that I found useful for my broad purposes, it is designed as a guide for the general reader, not for specialists. There is also a vast periodical literature, even apart from the many professional journals, but space does not permit an adequate selection from all the pertinent articles. I have made an exception of the journal *Daedalus* because its issues are usually focused on a single topic, with a wide spectrum of opinion.

GENERAL AND HISTORICAL

Boulding, Kenneth E. *The Organizational Revolution*, New York, 1949.

Burke, John G., ed. *The New Technology and Human Values*, Belmont, 1966.

Childe, V. Gordon. *Man Makes Himself*, London, 1941.

Daedalus, Spring 1962. *Science and Technology in Contemporary Society*.

Ellul, Jacques. *Propaganda*, New York, 1964.

———. *The Technological Society*, New York, 1964.

Hutchings, Edward and Elizabeth, eds. *Scientific Progress and Human Values*, New York, 1967.

Juenger, Friedrich Georg. *The Failure of Technology*, Chicago, 1949.

Kennedy, Gail, ed. *Democracy and the Gospel of Wealth*, Boston, 1949.

Krantzberg, Melvin and Pursell, Carroll W., Jr., eds. *Technology in Western Civilization*, 2 vols., New York, 1967.

Muller, Herbert J. *Freedom in the Modern World*, New York, 1966.

Mumford, Lewis. *Technics and Civilization*, New York, 1934.

Rostow, W. W. *The Stages of Economic Growth*, Cambridge, Mass., 1960.

Schon, Donald A. *Technology and Change*, New York, 1967.

Theobald, Robert. *The Challenge of Abundance*, New York, 1961.

Walker, Charles R., ed. *Modern Technology and Civilization*, New York, 1962.

WAR (Chapter 7)

Clarke, Robin. *The Silent Weapons*, New York, 1968.

Daedalus, Fall 1960. *Arms Control*.

Kahn, Herman. *On Thermonuclear War*, Princeton, 1960.

————. *Thinking about the Unthinkable*, New York, 1964.

Lapp, Ralph. *The Weapons Culture*, New York, 1968.

Lewin, Leonard C. *Report from Iron Mountain on the Possibility and Desirability of Peace*, New York, 1967.

McNamara, Robert. *The Essence of Security*, New York, 1968.

Schelling, Thomas C. *Arms and Influence*, New Haven, 1966.

Stone, I. F. *In a Time of Torment*, New York, 1967.

SCIENCE (Chapter 8)

Barzun, Jacques. *Science: The Glorious Entertainment*, New York, 1964.

Bronowski, Jacob. *Science and Human Values*, New York, 1956.

Daedalus, Winter 1965. *Science and Culture*.

Lerner, Daniel, ed. *The Human Meaning of the Social Sciences*, New York, 1959.

Matson, Floyd W. *The Broken Image: Man, Science, and Society*, New York, 1964.

Natanson, Maurice, ed. *Philosophy of the Social Sciences: A Reader*, New York, 1963.

Price, Derek J. de Solla. *Science since Babylon*, New Haven, 1961.

————. *Little Science, Big Science*, New York, 1963.

Snow, C. P. *The Two Cultures: and A Second Look*, New York, 1964.

Whitehead, Alfred North. *Science and the Modern World*, New York, 1925.

GOVERNMENT (Chapter 9)

Dahl, Robert A. and Lindblom, Charles E. *Politics, Economics, and Welfare*, New York, 1953.

Frankel, Charles. *The Democratic Prospect*, New York, 1962.

Gerth, H. H. and Mills, C. Wright, eds. *From Max Weber*, New York, 1958.

Jouvenel, Bertrand de. *Power, The Natural History of Its Growth*, London, 1945.

Lichtheim, George. *The Concept of Ideology and Other Essays*, New York, 1968.

Mills, C. Wright. *The Power Elite*, New York, 1956.

Nelson, William R., ed. *The Politics of Science*, New York, 1968.

Price, Don K. *Government and Science*, New York, 1954.

Schumpeter, Joseph A. *Capitalism, Socialism, and Democracy*, New York, 1942.

BUSINESS (Chapter 10)

Bazelon, David T. *The Paper Economy*, New York, 1963.

Beard, Miriam. *A History of Business*, vol. 2, Ann Arbor, Mich., 1963.

Berle, Adolf. *The Twentieth Century Capitalist Revolution*, New York, 1954.

Boguslaw, Robert. *The New Utopians: A Study of System Design and Social Change*, Englewood Cliffs, N.J., 1965.

Burck, Gilbert and the editors of *Fortune: The Computer Age*, New York, 1965.
Daedalus, Winter 1969. *Perspectives on Business.*
Dichter, Ernest. *The Strategy of Desire,* New York, 1960.
Drucker, Peter. *The Future of Industrial Man,* New York, 1942.
Galbraith, John Kenneth. *The Affluent Society,* Boston, 1958.
———. *The New Industrial State,* Boston, 1967.
Heilbroner, Robert. *The Worldly Philosophers,* New York, 1953.
Hook, Sydney, ed. *Human Values and Economic Policy,* New York, 1967.
Masters, Dexter. *The Intelligent Buyer and the Telltale Seller,* New York, 1966.
Mayer, Martin. *Madison Avenue U.S.A.,* New York, 1958.
Rae, John Bell. *The American Automobile,* Chicago, 1965.
Silberman, Charles E. and the editors of *Fortune: The Myths of Automation,* New York, 1958.

LANGUAGE (Chapter 11)
McLuhan, Marshall. *The Gutenberg Galaxy,* Toronto, 1962.
———. *Understanding Media,* New York, 1964.
Muller, Herbert J. *The Uses of English,* New York, 1967.
Ong, Walter J. *The Presence of the Word,* New Haven, 1967.
Steiner, George. *Language and Silence: Essays on Language, Literature, and the Inhuman,* New York, 1967.

HIGHER EDUCATION (Chapter 12)
Ashby, Eric. *Technology and the Academics,* London, 1958.
Bell, Daniel. *The Reforming of General Education,* New York, 1966.
Daedalus, Fall 1964. *The Contemporary University: U.S.A.*
Goodman, Paul. *Compulsory Mis-education and the Community of Scholars,* New York, 1964.
Jencks, Christopher and Riesman, David. *The Academic Revolution,* New York, 1968.
Kerr, Clark. *The Uses of the University,* Cambridge, Mass., 1963.
Plumb, J. H., ed. *Crisis in the Humanities,* Middlesex, 1964.

ENVIRONMENT, NATURAL AND SOCIAL (Chapters 13 and 14)
Blake, Peter. *God's Own Junkyard,* New York, 1964.
Briggs, Asa. *Victorian Cities,* New York, 1965.
Carson, Rachel. *Silent Spring,* Boston, 1962.
Daedalus, Fall 1967. *America's Changing Environment.*
Dubos, René. *Man, Medicine and Environment,* New York, 1968.
Ewald, William R., Jr., ed. *Environment for Man,* Bloomington, Ind., 1967.
———. *Environment and Change,* Bloomington, Ind., 1968.
———. *Environment and Policy,* Bloomington, Ind., 1968.
Gruen, Victor. *The Heart of Our Cities: The Urban Crisis: Diagnosis and Cure,* New York, 1964.

Handlin, Oscar and Burchard, John, eds. *The Historian and the City*, Cambridge, Mass., 1963.
Jacobs, Jane. *The Death and Life of Great American Cities*, New York, 1961.
Mumford, Lewis. *The City in History*, New York, 1961.
———. *From the Ground Up*, New York, 1956.
Whyte, William H. *The Last Landscape*, New York, 1968.
Willbern, York. *The Withering Away of the City*, Bloomington, Ind., 1966.

POPULAR AND TRADITIONAL CULTURE (Chapters 15 and 16)
Casty, Alan, ed. *Mass Media and Mass Man*, New York, 1968.
Daedalus, Winter 1960. *The Visual Arts Today*.
Friendly, Fred W. *Due to Circumstances beyond Our Control*, New York, 1967.
Gellner, Ernest. *Thought and Change*, New York, 1965.
Haftmann, Werner. *Painting in the Twentieth Century*, 2 vols., New York, 1960.
Hauser, Arnold. *The Social History of Art*, vols. 3 and 4, New York, 1958.
Hoggart, Richard. *The Uses of Literacy: Changing Patterns in English Mass Culture*, Fair Lawn, N.J., 1957.
Jacobs, Norman, ed. *Culture for the Millions?*, Boston, 1964.
Lowenthal, Leo. *Literature, Popular Culture, and Society*, Englewood Cliffs, N.J., 1961.
MacDonald, Dwight. *Against the American Grain*, New York, 1962.
Mumford, Lewis. *Art and Technics*, New York, 1952.
Ortega y Gasset, José. *The Revolt of the Masses*, New York, 1932.
Toffler, Alvin: *The Culture Consumers*, Baltimore, 1965.
Williams, Raymond. *Culture and Society 1780-1950*, New York, 1958.

RELIGION (Chapter 17)
Cogley, John, ed. *Religion in America*, New York, 1958.
Cox, Harvey. *The Secular City*, New York, 1966.
Herberg, Will. *Protestant, Catholic, Jew*, Garden City., N.Y., 1960.
Muller, Herbert J. *Religion and Freedom in the Modern World*, Chicago, 1963.
Niebuhr, Reinhold. *The Godly and the Ungodly*, London, 1958.
Ong, Walter J. *The Presence of the Word*, New Haven, 1967.
Stone, Ronald H., ed. *Faith and Politics*, New York, 1968.
Tillich, Paul. *The Protestant Era*, Chicago, 1957.

"PEOPLE, JUST PEOPLE" (Chapter 18)
Daedalus, Fall 1963. *The Professions*.
———, Spring 1964. *The Woman in America*.
———, Winter 1962. *Youth: Change and Challenge*.
Erikson, Erik H. *Identity, Youth, and Crisis*, New York, 1968.
Goodman, Paul. *Growing Up Absurd*, New York, 1960.
Grazia, Sebastian de. *Of Time, Work and Leisure*, New York, 1962.

Harrington, Michael. *The Other America*, New York, 1962.

Hoffer, Eric. *The Ordeal of Change*, New York, 1964.

Josephson, Eric and Mary, eds. *Man Alone: Alienation in Modern Society*, New York, 1962.

Keniston, Kenneth. *The Uncommitted: Alienated Youth in American Sociey*, New York, 1965.

Marcuse, Herbert. *One-dimensional Man*, Boston, 1964.

Riesman, David. *The Lonely Crowd*, New York, 1953.

———. *Individualism Reconsidered*, Glencoe, Ill., 1954.

Whyte, William H. *The Organization Man*, New York, 1956.

TOWARD THE YEAR 2000 (Chapters 19, 20, 21)

Amis, Kingsley. *New Maps of Hell: A Survey of Science Fiction*, New York, 1960.

Bell, Daniel, ed. *Toward the Year 2000: Work in Progress*, New York, 1968.

Daedalus, Spring 1965. *Utopia*.

Drucker, Peter. *The Age of Discontinuity: Guidelines to Our Changing Society*, New York, 1969.

Fromm, Erich. *The Revolution of Hope: Toward a Humanized Technology*, New York, 1968.

Huxley, Aldous. *Brave New World*, Garden City, N.Y., 1932.

———. *Brave New World Revisited*, London, 1959.

Kahn, Herman and Wiener, Anthony J. *The Year 2000*, New York, 1967.

Katep, George. *Utopia and Its Enemies*, Glencoe, Ill., 1963.

Mumford, Lewis. *The Transformations of Man*, New York, 1956.

Orwell, George. *Nineteen Eighty-four*, New York, 1949.

Seidenberg, Roderick. *Post-historic Man*, Chapel Hill, N.C., 1950.

Skinner, B. F. *Walden II*, New York, 1948.

Young, Michael. *Rise of the Meritocracy 1870-2035*, Baltimore, 1967.

Index

DLAND BOOKS